Interaction of the Chemical Senses with Nutrition

THE NUTRITION FOUNDATION
A Monograph Series

HORACE L. SIPPLE AND KRISTEN W. MCNUTT, EDS.: *Sugars in Nutrition*, 1974

ROBERT E. OLSON, ED.: *Protein-Calorie Malnutrition*, 1975

ANANDA S. PRASAD, ED.: *Trace Elements in Human Health and Disease, Volume I, Zinc and Copper*, 1976; *Volume II, Essential and Toxic Elements*, 1976

MORLEY R. KARE AND OWEN MALLER, EDS.: *The Chemical Senses and Nutrition*, 1977

JOHN N. HATHCOCK AND JULIUS COON, EDS.: *Nutrition and Drug Interrelations*, 1978

CLIFFORD F. GASTINEAU, WILLIAM J. DARBY, AND THOMAS B. TURNER, EDS.: *Fermented Food Beverages in Nutrition*, 1979

MORLEY R. KARE, MELVIN J. FREGLY, AND RUDY A. BERNARD, EDS.: *Biological and Behavioral Aspects of Salt Intake*, 1980

JENNY T. BOND, L. J. FILER, JR., GILBERT A. LEVEILLE, ANGUS THOMSON, AND WILLIAM B. WEIL, EDS.: *Infant and Child Feeding*, 1981

ROBERT H. CAGAN AND MORLEY R. KARE, EDS.: *Biochemistry of Taste and Olfaction*, 1981

MELVIN J. FREGLY AND MORLEY R. KARE, EDS.: *The Role of Salt in Cardiovascular Hypertension*, 1982

DONALD C. BEITZ AND R. GAURTH HANSEN, EDS.: *Animal Products in Human Nutrition*, 1982

BARBARA A. UNDERWOOD, ED.: *Nutrition Intervention Strategies in National Development*, 1983

MORLEY R. KARE AND JOSEPH G. BRAND, EDS.: *Interaction of the Chemical Senses with Nutrition*, 1986

Interaction of the Chemical Senses with Nutrition

Edited by

MORLEY R. KARE
JOSEPH G. BRAND

*Monell Chemical Senses Center
and University of Pennsylvania
Philadelphia, Pennsylvania*

1986

ACADEMIC PRESS, INC.
Harcourt Brace Jovanovich, Publishers
Orlando San Diego New York Austin
London Montreal Sydney Tokyo Toronto

COPYRIGHT © 1986 BY ACADEMIC PRESS, INC.
ALL RIGHTS RESERVED.
NO PART OF THIS PUBLICATION MAY BE REPRODUCED OR
TRANSMITTED IN ANY FORM OR BY ANY MEANS, ELECTRONIC
OR MECHANICAL, INCLUDING PHOTOCOPY, RECORDING, OR
ANY INFORMATION STORAGE AND RETRIEVAL SYSTEM, WITHOUT
PERMISSION IN WRITING FROM THE PUBLISHER.

ACADEMIC PRESS, INC.
Orlando, Florida 32887

United Kingdom Edition published by
ACADEMIC PRESS INC. (LONDON) LTD.
24–28 Oval Road, London NW1 7DX

Library of Congress Cataloging in Publication Data
Main entry under title:

Interaction of the chemical senses with nutrition.

Includes index.
1. Chemical senses—Congresses.   2. Nutrition—
Congresses.   3. Flavor—Congresses.   I. Kare,
Morley Richard, Date .   II. Brand, Joseph G.
QP455.I525   1986   599'.01826   85-23003
ISBN 0–12–397855–6 (alk. paper)

PRINTED IN THE UNITED STATES OF AMERICA

86 87 88 89   9 8 7 6 5 4 3 2 1

Contents

Contributors	xiii
Participants	xvii
Preface	xix
Samuel Lepkovsky	xxi

Part I Effects of Nutritive State on Chemical Senses

1 Importance of Saliva in Diet–Taste Relationships
CAROL M. CHRISTENSEN

I.	Introduction	3
II.	The Salivary Glands	4
III.	Dietary Influences on Saliva	6
IV.	Salivary Influences on Taste Perception	11
V.	Research Needs	20
	References	21

2 Factors Affecting Acceptance of Salt by Human Infants and Children
BEVERLY J. COWART AND GARY K. BEAUCHAMP

I.	Introduction	25
II.	Responses to Salt and Sugar in Early Development: A Review	26

III.	Developmental Shifts in Salt Acceptance	35
IV.	Research Needs	42
	References	42

3 Effects of Dietary Protein on the Taste Preference for Amino Acids in Rats

KUNIO TORII, TORU MIMURA, AND YASUMI YUGARI

I.	Introduction	45
II.	Taste Preference and Protein Intake in Rats during Growth	47
III.	Changes of Taste Preference in Rats Fed a Diet with or without L-Lysine Deficiency	52
IV.	Relation among Protein Intake, Taste Preference, and Genetic Predispositions	61
V.	Research Needs	63
VI.	Conclusion	65
	References	65

4 Preference Threshold for Maltose Solutions in Rats Treated Chronically with the Components of an Oral Contraceptive

MELVIN J. FREGLY

I.	Introduction	72
II.	Methods	72
III.	Results	76
IV.	Discussion	82
V.	Summary	85
	References	86

5 The Chemical Senses and Nutrition in the Elderly

CLAIRE MURPHY

I.	Introduction: Nutritional Status in the Elderly	87
II.	Chemosensory Loss in the Elderly: Review of the Literature	89
III.	Chemosensory Preference and Biochemical Indexes in the Elderly	93
IV.	Discussion and Conclusions	101

V.	Research Needs	102
	References	103

6 Micronutrients and Taste Stimulus Intake

SHARON GREELEY, CHARLES N. STEWART, AND MARY BERTINO

I.	Introduction	108
II.	Effects of Deficiencies on Taste Preferences	110
III.	Experimental Data	115
IV.	Future Research	126
	References	126

7 Effect of Non-Insulin-Dependent Diabetes Mellitus on Gustation and Olfaction

LAWRENCE C. PERLMUTER, DAVID M. NATHAN, MALEKEH K. HAKAMI, AND HOWARD H. CHAUNCEY

I.	Introduction	129
II.	Methods	131
III.	Results	133
IV.	Discussion	138
V.	Summary	141
	References	141

Part I Discussion   143

Part II Effects of the Cephalic Phase on Digestion and Absorption

8 Intragastric Feeding of Fats

ISRAEL RAMIREZ

I.	Introduction	151
II.	Review of Intragastric Feeding	152

III.	Research Needs	162
	References	163

9 The Stomach and Satiety
PAUL R. McHUGH AND TIMOTHY H. MORAN

I.	Introduction	167
II.	Gastric Emptying of Liquids	168
III.	Intestinal Control of Gastric Emptying	170
IV.	The Two Phases of Gastric Emptying	171
V.	The Stomach and Glucose Consumption	173
VI.	The Stomach and Chow Intake	175
VII.	Cholecystokinin and Gastric Distention	176
VIII.	Conclusions	178
	References	179

10 The Cephalic Phase of Gastric Secretion
MARK FELDMAN AND CHARLES T. RICHARDSON

I.	Introduction	181
II.	The Cephalic Phase of Gastric Secretion	181
III.	Research Needs	191
	References	191

11 The Gut Brain and the Gut–Brain Axis
WENDY R. EWART AND DAVID L. WINGATE

I.	Introduction	193
II.	The Gut Brain	195
III.	The Gut–Brain Axis	200
IV.	Peptides	205
V.	Research Goals	206
VI.	Summary	207
	References	208

12 Cephalic Phase of Digestion: The Effect of Meal Frequency
KATHERINE A. HOUPT AND T. RICHARD HOUPT

I.	Introduction	211
II.	Critical Review and Discussion of Subject Matter	212

III.	Research Needs	236
	References	237

Part II Discussion 241

Part III Consequences of Food Palatability to Nutrition

13 Changing Hedonic Responses to Foods during and after a Meal

BARBARA J. ROLLS, MARION HETHERINGTON, VICTORIA J. BURLEY, AND P. M. van DUIJVENVOORDE

I.	Introduction	247
II.	Sensory-Specific Satiety: Basic Studies	248
III.	Nutrient-Specific Satiety	253
IV.	The Role of Sensory Properties of Food in Satiety	255
V.	Changes in the Palatability of Uneaten Foods	259
VI.	Effects of Variation in the Sensory Properties of Foods on Food Intake	263
VII.	Conclusions	267
	References	267

14 Role of Variety of Food Flavor in Fat Deposition Produced by a "Cafeteria" Feeding of Nutritionally Controlled Diets

MICHAEL NAIM, JOSEPH G. BRAND, AND MORLEY R. KARE

I.	Introduction	269
II.	"Cafeteria" Feeding as a Model for Dietary Obesity	271
III.	Preference Tests for Food Flavors and Texture	273
IV.	"Cafeteria" Feeding Experiments with Nutritionally Controlled Diets	281

V. Conclusions 289
References 290

15 Quantitative Relationship between Palatability and Food Intake in Man
HARRY R. KISSILEFF

I. Introduction 293
II. A Brief-Exposure Taste Test for Measuring Intrinsic Palatability 298
III. Relationship of Intrinsic Palatability to Food Consumption 307
IV. Needs for Future Research 313
V. Summary and Conclusions 315
References 315

Part III   Discussion 319

Part IV   Interplay of Chemical Senses with Nutrient Metabolism

16 Taste and the Autonomic Nervous System
RALPH NORGREN

I. Introduction 323
II. The Afferent Limb 325
III. The Efferent Limb 329
IV. The Central Projections 334
References 340

17 Caudal Brainstem Integration of Taste and Internal State Factors in Behavioral and Autonomic Responses
HARVEY J. GRILL

I. Introduction 348
II. The CBS Receives Input from Oral Exteroceptors That Evaluate the Sensory Characteristics of Food 348

III.	The CBS Is a Site of Metabolic Interoceptors	349
IV.	The CBS Contains Simple Reflex Connections between Oral Exteroceptor Input and Autonomic and Behavioral Effector Output	354
V.	The CBS Contains Connections between Exteroceptive Input and Behavioral Effector Output for the Production of Discriminative Responses to Taste	358
VI.	Interoceptive Input from Food Deprivation and Insulin-Induced Hypoglycemia Is Integrated with Oral Afferent Information within the CBS to Control the Ingestive Consummatory Behavior of Chronically Decerebrate Rats	365
VII.	Conclusion	369
	References	369

18 Possible Participation of Oro-, Gastro-, and Enterohepatic Reflexes in Preabsorptive Satiation

MAURICIO RUSSEK AND RADU RACOTTA

I.	Introduction	374
II.	Regulation of Glycemia	375
III.	Hepatic Receptors and Control of Food Intake	377
IV.	The Hepatic Hypothesis of Feeding	379
V.	Preabsorptive Satiation	381
VI.	Oropharyngeal Receptors	383
VII.	Gastric Distention Receptors	384
VIII.	Gastrointestinal Chemoreceptors	385
IX.	Gastrointestinal Hormones	386
X.	Possible Relation between Satiation and Lipostasis	387
XI.	Conclusions	387
	References	388

19 Effects of Protein and Carbohydrate Ingestion on Brain Tryptophan Levels and Serotonin Synthesis: Putative Relationship to Appetite for Specific Nutrients

JOHN D. FERNSTROM

I.	Introduction	395
II.	Diet, Brain Tryptophan Uptake, and Serotonin Synthesis	396

III.	Diet, Brain Tryptophan and Serotonin, and Appetite	404
IV.	Summary and Conclusions	412
	References	413

20 Time Course of Food Intake and Plasma and Brain Amino Acid Concentrations in Rats Fed Amino Acid-Imbalanced or -Deficient Diets

D. W. GIETZEN, P. M. B. LEUNG,
T. W. CASTONGUAY, W. J. HARTMAN, AND
Q. R. ROGERS

I.	Introduction	415
II.	Time Course of the Food-Intake Response	420
III.	Feeding Patterns	423
IV.	Dietary Choice	425
V.	Amino Acid Concentrations in Plasma, Brain, and Cerebrospinal Fluid	433
VI.	Operant Response	437
VII.	Brain Areas Implicated	442
VIII.	Amino Acid and Neurotransmitter Concentrations in Brain	444
IX.	Monoamines in the Prepyriform Cortex	449
	References	452

Part IV Discussion   457

Part V Conclusion

21 Concluding Remarks

D. MARK HEGSTED

Text   463

Index   469

Contributors

Numbers in parentheses indicate the pages on which the authors' contributions begin.

Gary K. Beauchamp (25), Monell Chemical Senses Center, and School of Veterinary Medicine, University of Pennsylvania, Philadelphia, Pennsylvania 19104

Mary Bertino (107), Monell Chemical Senses Center, Philadelphia, Pennsylvania 19104

Joseph G. Brand (269), Monell Chemical Senses Center, and University of Pennsylvania, Philadelphia, Pennsylvania 19104

Victoria J. Burley (247), Department of Experimental Psychology, University of Oxford, Oxford OX1 3PS, England

T. W. Castonguay (415), Department of Nutrition and Food Intake Laboratory, University of California at Davis, Davis, California 95616

Howard H. Chauncey (129), Veterans Administration Outpatient Clinic, and Harvard School of Dental Medicine, Boston, Massachusetts 02108

Carol M. Christensen (3), Monell Chemical Senses Center, and School of Dental Medicine, University of Pennsylvania, Philadelphia, Pennsylvania 19104

Beverly J. Cowart (25), Monell Chemical Senses Center, and the Thomas Jefferson University, Philadelphia, Pennsylvania 19104

Wendy R. Ewart (193), Gastrointestinal Science Research Unit, The London Hospital Medical College, London E1 2AJ, England

Mark Feldman (181), Veterans Administration Medical Center, and Department of Internal Medicine, University of Texas Health Science Center, Dallas, Texas 75216

John D. Fernstrom (395), Department of Psychiatry, Western Psychiatric Institute and Clinic, and The Center for Neuroscience, University of Pittsburgh School of Medicine, Pittsburgh, Pennsylvania 15213

Melvin J. Fregly (71), Department of Physiology, University of Florida, College of Medicine, Gainesville, Florida 32610

D. W. Gietzen (415), Department of Physiological Sciences and Food Intake Laboratory, University of California at Davis, Davis, California 95616

Sharon Greeley (107), Monell Chemical Senses Center, Philadelphia, Pennsylvania 19104

Harvey J. Grill (347), Department of Psychology, University of Pennsylvania, Philadelphia, Pennsylvania 19104

Malekeh K. Hakami (129), Harvard School of Dental Medicine, Boston, Massachusetts 02108

W. J. Hartman (415), Department of Physiological Sciences and Food Intake Laboratory, University of California at Davis, Davis, California 95616

D. Mark Hegsted (463), New England Regional Primate Center, Southborough, Massachusetts 01772

Marion Hetherington (247), Department of Experimental Psychology, University of Oxford, Oxford OX1 3PS, England

Katherine A. Houpt (211), Department of Physiology, New York State College of Veterinary Medicine, Cornell University, Ithaca, New York 14853

T. Richard Houpt (211), Department of Physiology, New York State College of Veterinary Medicine, Cornell University, Ithaca, New York 14853

Morley R. Kare (269), Monell Chemical Senses Center, and University of Pennsylvania, Philadelphia, Pennsylvania 19104

Harry R. Kissileff (293), Obesity Research Center, St. Luke's-Roosevelt Hospital, and Department of Psychiatry, Columbia University College of Physicians and Surgeons, New York, New York 10025

P. M. B. Leung (415), Department of Physiological Sciences and Food Intake Laboratory, University of California at Davis, Davis, California 95616

Paul R. McHugh (167), Department of Psychiatry and Behavioral Sciences, The Johns Hopkins University School of Medicine, Baltimore, Maryland 21205

Toru Mimura (45), Life Science Laboratories, Central Research Laboratories, Ajinomoto Company, Inc., Yokohama, 244 Japan

Timothy H. Moran (167), Department of Psychiatry and Behavioral Sciences, The Johns Hopkins University School of Medicine, Baltimore, Maryland 21205

Claire Murphy (87), Department of Psychology, San Diego State University, San Diego, California 92182

Michael Naim (269), Faculty of Agriculture, The Hebrew University of Jerusalem, Rehovot, Israel, 76-100

David M. Nathan (129), Harvard Medical School, and Massachusetts General Hospital, Boston, Massachusetts 02108

Ralph Norgren (323), Department of Behavioral Science, and The Milton S. Hershey Medical Center, Pennsylvania State University, Hershey, Pennsylvania 17033

Lawrence C. Perlmuter (129), Veterans Administration Outpatient Clinic, and Harvard School of Dental Medicine, Boston, Massachusetts 02108

Radu Racotta (373), Department of Physiology, National School of Biological Sciences, National Polytechnic Institute, Mexico City 17, Mexico

Israel Ramirez (151), Monell Chemical Senses Center, Philadelphia, Pennsylvania 19104

Charles T. Richardson (181), Veterans Administration Medical Center, and Department of Internal Medicine, University of Texas Health Science Center, Dallas, Texas 75216

Q. R. Rogers (415), Department of Physiological Sciences and Food Intake Laboratory, University of California at Davis, Davis, California 95616

Barbara J. Rolls[1] (247), Department of Experimental Psychology, University of Oxford, Oxford OX1 3PS, England

Mauricio Russek (373), Department of Physiology, National School of Biological Sciences, National Polytechnic Institute, Mexico City 17, Mexico

Charles N. Stewart (107), Department of Psychology, Franklin and Marshall College, Lancaster, Pennsylvania 17604

Kunio Torii (45), Life Science Laboratories, Central Research Laboratories, Ajinomoto Company, Inc., Yokohama, 244 Japan

---

[1]Present address: Department of Psychiatry and Behavioral Sciences, The Johns Hopkins University School of Medicine, 600 N. Wolfe Street, Meyer 4-119, Baltimore, Maryland 21205.

P. M. van Duijvenvoorde (247), Department of Experimental Psychology, University of Oxford, Oxford OX1 3PS, England

David L. Wingate (193), Gastrointestinal Science Research Unit, The London Hospital Medical College, London E1 2AJ, England

Yasumi Yugari (45), Life Science Laboratories, Central Research Laboratories, Ajinomoto Company, Inc., Yokohama, 244 Japan

Participants

*The Third International Conference
on the Chemical Senses and Nutrition
Monell Chemical Senses Center
October 10–12, 1984*

Gary K. Beauchamp
Mary Bertino
Peter Boyle
Joseph G. Brand
Robert H. Cagan
Sidney M. Cantor
Mabel Chan
Howard H. Chauncey
Carol M. Christensen
Beverly J. Cowart
William Darby
Jeannette Desor
Wendy R. Ewart
Mark Feldman
John D. Fernstrom
Melvin J. Fregly
Mark Friedman
Victor Fulgoni
Sharon Giduck
Stanley Gershoff
Walter Glinsmann
Sharon Greeley
Harvey J. Grill
Arthur Hecker
David Heckert
D. Mark Hegsted
Marion Hetherington
Sally Hite
Katherine A. Houpt
Taufiqul Huque

Peter Huth
Morley R. Kare
Chor San Khoo
Harry R. Kissileff
Albert Kolbye
Paul R. McHugh
Owen Maller
Katherine Massey
Richard Mattes
David Mela
Joe Mullen
Claire Murphy
Michael Naim
Ralph Norgren
George Preti
Israel Ramirez
Q. R. Rogers
Barbara J. Rolls
Mauricio Russek
Gary Schwartz
Gerald Soffen
Alan Spector
Charles N. Stewart
Rose Threatte
Kunio Torii
Theodore Van Itallie
Judith Wellington
David L. Wingate
June Yantis
Yasumi Yugari

Preface

Most human food is obtained from a relatively few plant and animal species. Through breeding programs and genetic engineering, nutrient value and productivity of these plants and animals can be optimized. However, the physiological implications of the sensory properties of these foods and the role that these properties play in nutrition have traditionally not been areas of intense research. Confirming evidence for this observation is the absence of a specialized forum for such information.

This book, which is the result of the third symposium on the Chemical Senses and Nutrition, provides such a forum. The progress in research reported by the participants in each chapter and the accounts of the discussion following each session serve to further focus many remaining problems in our understanding of the relationship of taste and smell to nutrition.

In the United States alone, processed food and beverages constitute an annual market of over 350 billion dollars. The primary concerns of industry with the sensory quality of food and beverages usually relate to marketing, toxicity, or behavior. The importance of the hedonic aspects of eating and drinking is obvious. Yet evidence has accumulated that the flavor of food can have significant physiological effects influencing not only ingestion, but also digestion and metabolism. In turn, the nutritional state can affect perception of taste and smell.

Taste and smell, particularly as they relate to food, are of concern to nearly everyone at least several times a day. However, relative to other senses, our knowledge of the biochemical and physiological mechanisms of the chemical senses is incomplete. Research in these areas is expanding, and significant gains have recently been made. In addition, the physiological and nutritional functions of the chemical senses are only now beginning to be understood in greater depth. A keener appreciation for the role of the chemical senses in, for example, the cephalic phase of digestion has recently emerged, and the influence of flavor and variety on food selection and intake is being more closely and rigorously investigated.

The paucity of information in the field of the chemical senses as it impacts on nutrition can be explained, in part, by the difficulties of performing research on these senses. One of the first stumbling blocks is the enormous perceptual diversity of taste and smell in man. Further, by ecological design, animals do not share man's sensory world, even if occasionally it overlaps with that of one or another species. Animal models need to be chosen, therefore, with an assurance that the stimuli have appropriate meaning for the species being studied. Since stimuli are almost unlimited in number and their qualities are chemically specific and concentration dependent, research on the chemical senses defies simple experimentation. Parameters as basic as stimulus delivery are difficult to control. To overcome these and other problems, a multidisciplinary approach to investigating the chemical senses and their relation to nutrition is almost essential.

There is evidence that these symposia are alerting nutritionists to consider sensory quality as a variable in their research. Similarly, the nutritive state of the experimental subject is now more frequently a controlled variable, if not a focus in itself, of research on taste and smell. In the coming decade, we anticipate that there will be a growing interest in the physiological impact of the sensory qualities of food. For example, the concern for sensory acceptance of the array of sweetners will be shared with questions of their effects on digestion. Many of the studies published in this volume will form the basis for these continuing investigations.

The conference and the publication were supported by the International Life Sciences Institute/Nutrition Foundation, the National Science Foundation (grant no. BNS 83-13966), and the Food and Drug Administration. The assistance of the staff of the Monell Center in running the conference and in the preparation of this book is gratefully acknowledged.

In the preparation of a collaborative publication such as this, it is often impossible to resubmit the manuscripts for review to the many contributors. The editors acknowledge responsibility for any resulting errors.

Morley R. Kare
Joseph G. Brand

Samuel Lepkovsky
1899–1984

To those who knew Sam Lepkovsky, a summary of his life and achievements is superfluous. To those who were not so fortunate, I can only say that you missed knowing one of our century's great scientists in the disciplines of biochemistry, physiology, poultry husbandry, nutrition, and the brain sciences. He had the intellect and vision to integrate all these areas into an understanding of the emerging new frontier of nutrition: the role that the central nervous system and the hypothalamus play with the sensory properties of foods in nutrition and behavior.

Sam was special to each of us who knew him well. His home was the laboratory where he sought truths and understanding of scientific phenomena. He had keen powers of observation; that is how he discovered xanthurenic acid. He believed in a certain axiom of nutrition, namely, "In nutrition, it is equally as important to know what comes out of an animal as what goes into it." He had keen investigative powers undergirded by a thorough knowledge of organic chemistry, sharp intuitive powers, persistence, and willingness to work hard; that is how he won the race for the crystallization of vitamin B_6 against almost insurmountable odds.

He worked in all phases of nutritional research—with macronutrients (the sparing action of fats on the B vitamins) as well as with micronutrients (riboflavin, pyridoxine, etc.). He worked with all types of animals, from turtle to man.

Sam was a patriot, carrying a "musket" (as he called a rifle) as a doughboy in World War I, and leading the U.S. Army nutritional ration research effort in World War II. Here, he worked with man as the experimental animal. In this case, the laboratory was a large army unit training under the simulated hardship conditions of battle—in the mountains. He lived with the troops, ate with them, talked with them, and shared their experiences by being with them day and night. In addition to talking with the troops about the rations they were being fed, he also made measurements. One of his most important measurements was the amount of garbage after each meal! Out of this latter experience grew the basis for that new area of nutrition that he pioneered and in which he was to remain an active force from about 1947 until 1983, a year before he passed away.

For Sam Lepkovsky, "the era of the essential nutrients" was over. He would leave the refinements of quantities of nutrients needed by different populations for others to "tidy up." He intended to explore and exploit a new and uncharted field. The ideas for this new field were based on his work with the troops in the field, which is beautifully chronicled in the acceptance paper which he gave on receiving the Babcock–Hart Award of the Institute of Food Technologists in 1959. It was this hard work, plus the work of his associates guided by and working with him in the laboratory, that opened new pathways and began the emerging new fields of the role of the brain and the hypothalamus in nutrition. It was his pioneering thinking and research that served as the basis for programs such as the one now in existence at the Monell Chemical Senses Center.

Sam was a giant in nutrition when the field was full of giant intellects—Funk, Drummond, Boyd–Orr, Lord Todd Hart (his teacher whom he revered), Babcock, Woolley, Steenbock, Elvehjem (who at one time was Sam's laboratory assistant), McCollum (whom Sam knew well and to whom he paid his respects whenever he could be visiting McCollum in Baltimore; in fact

Lep introduced the writer to him). Lep "was always at the head of the pack" to quote Wayne Woolley. Yet, with all his scientific achievements, Sam Lepkovsky was a very modest individual—quick to give credit to others; self-deprecating about his own achievements. He was a writer, par excellence—just read his classic "Bread in War and Peace," published in *Physiological Reviews*. He had legions of friends and admirers. Several of us once thought of forming an organization of "Les Amis de Sam Lepkovsky."

He developed into a great oenophile (studying wines in Edinburgh in the late 1950s while on a sabbatical in the Poultry Research Center at Edinburgh). His "laboratory" was the main dining room of the George Hotel in Edinburgh which had a fine wine cellar at reasonable prices, and Sam did a good laboratory study each evening at dinner. How his liver withstood that oenophillic assault, Lord only knows. But, he learned his oenology well, using what Pasteur called "la methode experimentale."

He was a superb dancer—I don't know when and how he became an expert in that field. He was a raconteur of the first order. He was a true Zionist and loved Israel and traveled there fairly frequently. He was honored by the Hebrew University by being made a Fellow. He was generous with his time and with his money. Sam was recognized by his adopted Alma Mater, the University of California at Davis, by being awarded an honorary L.L.D. on the occasion of his so-called "retirement."

He was well known as an habitue of "Sam's Grill" on Bush Street in San Francisco, where three generations of the family who owned and ran Sam's Grill greeted him by "two B_6's equal B_{12}." There was always a table in the corner ready for The Professor.

He was known at Claridges in London where he had a very well known friend who occupied a prized table in an alcove in the main dining room (on a permanent basis). This individual married one of Sam's former students and this is how Sam met him. I saw this table once again just this past June when Diana and I had Sunday dinner at Claridges with three of my M.I.T. colleagues. I made a much greater impression on these colleagues by the discussion I had with the Wine Steward (who has been at Claridges for twenty-five years) about the occupants of that simple little table in that prized location, in the alcove, the place of honor, than I have ever impressed them by either my professional work or any other of the events in my life.

The saga of Samuel Lepkovsky's scientific achievements is in the literature. The latter tell us one side of his multifaceted life. But like a diamond of great value, Sam had many facets to his life, each with its own particular brilliance. I have attempted to give you some of my thoughts about a few gems that are not recorded in the scientific literature. I miss him greatly. I cannot pick up the phone to call him and to hear the familiar high-pitched voice reply "Junior, it's good to hear your voice. When are you coming out here?" But I do not mourn for him. I just think of how blessed I was to have known him so well; to have learned so much from him about so many different subjects; to have met him in so many different places over a period of more than two decades: Edinburgh, Tel Aviv, Geneva, Zurich, New York, London, and many other cities all over the world. He enriched my life and that of my whole family just as he enriched the lives of many others. "He left his camping ground a better place than when he found it," and we, his friends, scientific colleagues, and, in fact, all mankind are the better for it.

Samuel A. Goldblith
Professor and Vice President
Massachusetts Institute of Technology
Cambridge, Massachusetts

Interaction of the Chemical Senses with Nutrition

PART I

Effects of Nutritive State on Chemical Senses

1

Importance of Saliva in Diet–Taste Relationships

CAROL M. CHRISTENSEN
Monell Chemical Senses Center and
School of Dental Medicine
University of Pennsylvania
Philadelphia, Pennsylvania

I.	Introduction	3
II.	The Salivary Glands	4
	A. Flow Rate	4
	B. Composition	5
III.	Dietary Influences on Saliva	6
	A. Effects on Salivary Flow Rate	6
	B. Effects on Salivary Composition	9
IV.	Salivary Influences on Taste Perception	11
	A. Sensory Adaptation to Salivary Components	12
	B. Modification of Tastants by Saliva	15
V.	Research Needs	20
	References	21

I. INTRODUCTION

Saliva is likely to be an intermediate link in the relationship between the chemical senses and nutrition, or more specifically, between taste and diet. Indirect evidence for this link comes from two separate lines of research demonstrating that (1) diet can alter salivary characteristics, and (2) saliva affects taste sensation. In neither area is there an extensive number of studies, and there are, as yet, no studies directly investigating the possible tripartite relationship between diet, saliva, and taste in which taste changes would occur as a result of dietary-induced changes in saliva. However, current evidence is supportive of the view that future research will uncover the existence of such a relationship.

This article reviews the evidence that saliva is involved in diet–taste relationships through dietary effects on saliva and salivary effects on taste perception. Salivary flow rate and composition will be treated separately in the following discussion, although they are not totally independent characteristics of saliva. It is useful to consider salivary flow and composition separately because dietary factors can influence one component more than another, and because predictions of the effects of saliva on taste function can be different depending upon whether flow or compositional parameters are being considered. This article begins with a brief description of the anatomical and physiological features of the salivary glands relevant to the discussion of diet–taste relationships.

II. THE SALIVARY GLANDS

In humans, saliva is secreted into the oral cavity principally by three pairs of major glands: parotid, submandibular, and sublingual. Major glands are defined by the presence of a ductal network which empties the saliva produced by a large group of secretory cells, the acinar cells. Minor salivary glands are much smaller clusters of secretory cells; they are numerous and are identified by their location: minor sublingual, lingual, labial, buccal, palatine, and glossopalatine.

A. Flow Rate

Salivary secretion is primarily controlled by the autonomic nervous system and, in humans, flow rate is most closely related to activity in the parasympathetic branch. Individual differences in salivary flow rate are large, as evident from Fig. 1, and it is common to find more than eightfold differences in whole mouth flow rates (saliva of mixed origin collected from the oral cavity) among groups of healthy subjects (Jenkins, 1966; Navazesh and Christensen, 1982). The relative contribution of each pair of major glands to total salivary volume depends upon whether the glands are in a resting or stimulated state. The resting state is defined as the absence of any obvious source of stimulation, and is so defined because isolated salivary glands from experimental animals rarely secrete spontaneously (Burger and Emmelin, 1961). The submandibular glands contribute approximately 69%, the parotid glands 26%, and the sublingual glands 5% of the total major gland volume in the resting state (Schneyer and Levin, 1955a). Minor gland secretions account for approximately 10% of total salivary volume (Dawes

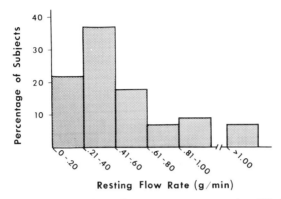

Fig. 1. Whole mouth resting salivary flow rates among a group ($n = 85$) of healthy, young adults. (From Christensen, 1985.)

and Wood, 1973). In response to stimulation, salivary flow may increase two- to fourfold from resting levels (Navazesh and Christensen, 1982; Kerr, 1961). The parotid fraction of total salivary volume increases disproportionately during stimulation (Kerr, 1961; Schneyer and Levin, 1955b; Shannon, 1962) and can even contribute more saliva than the other major glands. Salivary flow also displays a circadian rhythm that is most evident in resting measures of saliva; whole mouth flow rate is lowest in the morning and reaches a peak in late afternoon (Dawes and Ong, 1973).

B. Composition

Saliva is more than 99% water but contains a large number of organic and inorganic constituents (Ellison, 1979; Schneyer et al., 1972). The proteins are the largest class of organic constituents and include a number of enzymes, most notably the digestive enzyme amylase. The main electrolytes comprising the inorganic fraction of saliva are sodium, potassium, calcium, chloride, bicarbonate, and inorganic phosphate. The concentrations of salivary sodium and chloride are less than the same ion concentrations in blood, whereas the salivary concentrations of potassium and bicarbonate are higher (Schneyer et al., 1972).

The concentrations of salivary constituents are variable, and a major source of this variation is the degree of salivary stimulation. Chiefly as a result of acinar cell stimulation and changes in ionic exchange in the salivary ducts, protein and major ion concentrations reliably fluctuate with changes in salivary flow rate (Schneyer et al., 1972; Dawes, 1969; Dawes, 1974). For example, the concentration of salivary sodium increases when salivary flow rate increases, and the resulting salivary sodium levels can be double or triple the concentration of resting saliva. Salivary bicarbonate, pH, calcium, and protein also increase with salivary flow rate. Potassium concentration is largely unaffected by salivary flow rate, whereas urea levels in saliva decrease with increasing flow rates.

The composition of whole mouth saliva is affected both by the relative contribution of the major glands to the total salivary pool and by the source of neural stimulation. The parotid glands are composed of serous cells which produce a watery saliva. The submandibular and sublingual glands contain both serous and mucous cells, the latter producing a more viscous saliva that is rich in muco- and glycoproteins. As described above, the parotid gland contribution to total saliva increases with oral stimulation, at least stimulation with acids, and thus total salivary composition may be expected to change to reflect the greater input from the parotid glands. Another source of varying salivary composition is the relative contribution of different autonomic fibers in salivary gland stimulation. Parasympathetic stimulation produces a more watery secretion than sympathetic stimulation, and adrenergic stimulation, notably β-adrenergic stimulation, produces a saliva richer in proteins (Young and Schneyer, 1981).

III. DIETARY INFLUENCES ON SALIVA

A. Effects on Salivary Flow Rate

Dietary sources of salivary stimulation are varied. Chewing is a very effective stimulant of salivary flow, and increasing the size or hardness of the bolus being chewed will further increase flow rate (Kerr, 1961). Food powders produce greater salivary flow than food pieces (Pangborn and Lundgren, 1977), presumably because powders absorb more saliva and lead to a drying of the oral mucosa which is, itself, a stimulus for salivary secretion.

Certain classes of chemicals, especially those producing oral sensations of irritation, are also potent stimulants of salivary flow. For example the pungent food flavorings, capsicum, vanillyl nonamide, and piperine produce salivary flow increases following oral rinses with these compounds (Lawless, 1984). Interestingly, the observed increases in salivary flow parallel the perceived intensity of oral irritation produced by the different chemicals (see Fig. 2). Chemicals that are inhaled may also produce salivary increases, although at much reduced rates. Salivary increases appear to be mediated by trigeminal rather than olfactory stimulation because irritant chemicals, and not pure olfactory chemicals, appear to stimulate salivary flow (Lashley, 1916; Pangborn and Berggren, 1973). Further research is needed to better

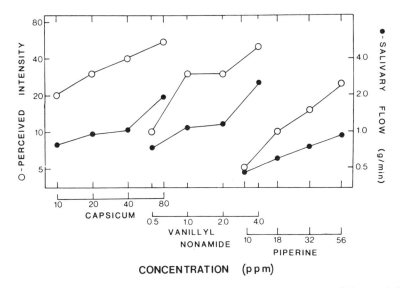

Fig. 2. Median perceived intensity of irritation (open circles) and mean whole mouth flow rate (filled circles) as a function of concentration for three compounds. All scales are logarithmic. (From Lawless, 1984.)

characterize the nature of olfactory stimulation of salivary glands; it may be that salivary stimulation following inhalation occurs not because of stimulation of olfactory or trigeminal receptors in the nasal cavity but because oral cavity receptors are stimulated by chemicals entering through the nasal cavity.

Among chemicals that are taste stimuli, acids are the most effective stimulants of salivary flow (Chauncey et al., 1963; Makhlouf and Blum, 1972); they are also more potent oral irritants than other standard taste stimuli. Chauncey et al. (1963) examined the effects of a wide range of acids on unilateral parotid flow rate and found that, for a given acid, there was a positive linear relationship between H^+ concentration and parotid secretion rate. However, hydrogen ion concentration was not the sole predictor of an acid's salivary stimulatory effectiveness; the anionic species were apparently also important. Interestingly, a significant correlation was found between an acid's effectiveness as a salivary stimulant and its reported taste threshold value; generally, acids that were more effective sialogogues were also perceived at lower concentrations in taste threshold tests.

Other taste stimuli can produce increases in salivary flow rate; however, the flow increases are much lower than those obtained with acids whether the taste compounds are matched with acids on perceived intensity (Funakoshi and Kawamura, 1965) or concentration (Chauncey and Shannon, 1960).

Salivary flow in response to single tastants and taste mixtures has also been measured (Feller *et al.*, 1965; Speirs, 1971). By and large, the results demonstrate that salivary flow increases to taste mixtures fall short of a simple addition of flow rate increases for the separate taste components, even when the suppression of perceived intensity commonly observed in taste mixtures is taken into account (Speirs, 1971).

Salivary flow to the sight of food may also occur. This response is presumably a conditioned (learned) response, unlike the reflexive salivary responses to oral stimulation described above. Salivation does not occur in response to the sight of all foods, rather it is most frequently observed in response to the sight of foods with a high acid content (Jenkins and Dawes, 1966; Christensen and Navazesh, 1984). The results are mixed as to whether hunger and satiation affect salivary flow rate; some investigators find increased salivary flow in the hunger state, whereas others report no changes (Sahakian, 1981).

In addition to the immediate effects on salivary flow induced by stimulation, dietary variables can also have longer term effects on salivary flow rate. Animal studies (e.g., Ekström and Templeton, 1977) demonstrate that both salivary gland size and physiological reactivity to stimulation are affected by the history of oral stimulation (use and disuse of salivary glands). In these studies, liquid diets were found to promote salivary gland atrophy, whereas high bulk solid diets promoted hypertrophy of the glands, presumably as a result of decreases and increases, respectively, in the amount of masticatory stimulation of salivary flow. In humans, significant salivary flow reductions (34% in resting parotid flow rates) were found in a group of subjects after only 7 days of maintenance on a liquid diet, and normal flow was restored within a week after resumption of a normal diet (Hall *et al.*, 1967). In another study, increases in salivary flow rate occurred in a group of subjects who increased masticatory activity by chewing four pieces of gum per day for an 8-week period (Edgar and Jenkins, 1981). The increase in resting whole mouth flow rates was evident not only during the test period but even 2 months after cessation of the experiment.

As early as Pavlov, researchers knew that the state of hydration affected salivary flow rate; flow rates are reduced by dehydration and increased by excessive hydration. Cellular dehydration can be induced by a variety of methods, and some of these methods result in reduced salivary flow rates. For example, dehydration due to excessive perspiration will reduce salivary flow, and the amount of salivary flow reduction roughly corresponds to the amount of weight loss accompanying fluid loss (Holmes, 1964). Salivary flow decreases also occur when intracellular fluid depletion is induced by intravenous injections of hypertonic saline (Holmes, 1964). Excessive hydration produces increases in resting (Shannon and Chauncey, 1967) but not stimulated salivary flow rates (DeWardener and Herxheimer, 1957), and, as illus-

1. Importance of Saliva in Diet–Taste Relationships 9

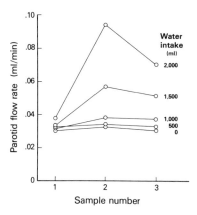

Fig. 3. The effect on resting parotid flow rate of the ingestion of different volumes of water. Sample 1 was obtained prior to consumption, whereas Samples 2 and 3 were obtained 45 and 90 min after consumption. (From Shannon and Chauncey, 1967.)

trated in Fig. 3, the consumption of at least 1.5 liters of water produces a relatively long-lasting increase in resting salivary flow rate.

B. Effects on Salivary Composition

Dietary constituents can also alter salivary composition. Preabsorptive routes for these changes are through oral stimulation that results in flow-linked changes in salivary composition and stimulation-induced changes in the relative contribution of each of the major glands to the total salivary pool (see Section II,A and B). Another route for changes in salivary composition is through absorption of dietary components with subsequent expression in saliva. Relatedly, postabsorptive events may also trigger the release of hormones that affect salivary gland function; as an example, aldosterone increases the uptake of sodium both in the kidney and in salivary ducts (Wotman *et al.*, 1973; Thrasher and Fregly, 1980).

Preabsorptive influences of dietary constituents on salivary composition appear to be chiefly a consequence of changes in salivary flow rate (Dawes and Jenkins, 1964). Thus, after oral stimulation by acids, the saliva contains more bicarbonate than saliva stimulated by other taste stimuli. This is because acids are the most potent gustatory salivary stimulants, and increased bicarbonate production is one consequence of increased salivary gland stimulation. However, not all changes in salivary composition induced by oral stimulation are entirely flow linked; as illustrated in Fig. 4, oral stimulation with fructose produces a saliva higher in total protein, including the enzyme amylase, than the saliva produced by citric acid stimulation, even though the

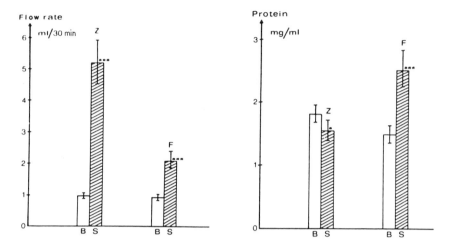

Fig. 4. Parotid gland flow rate before (B) and upon (S) stimulation with citric acid (Z) or fructose (F) (left figure). Salivary protein concentration is indicated in the right figure. (From Kemmer and Malfertheiner, 1983.)

latter stimulates greater salivary flow (Kemmer and Malfertheiner, 1983). The pattern of results suggests that the two tastants may stimulate somewhat different neural pathways. As described above, stimulation of the sympathetic system, particularly β-adrenergic fibers, produces a saliva rich in protein, and it may be that fructose differentially stimulates these adrenergic fibers. There is also evidence that oral stimulation with NaCl produces a saliva higher in protein than that produced by acids (Dawes, 1984).

The postabsorptive effects of diet on salivary composition have not received much study (Dawes, 1970). However, there is evidence that diet can influence the concentrations of several salivary constituents that are also recognized as tastants. The relationship between dietary and salivary levels of sodium has received the most attention, with several studies demonstrating that severe restriction of dietary sodium results in lowered concentrations of salivary sodium (Wotman et al., 1973; Thorn et al., 1956; McCance, 1938). In one study (Thorn et al., 1956), stimulated parotid sodium levels decreased approximately 25%, whereas salivary potassium levels rose by nearly the same amount in three hypertensive individuals maintained on a severely restricted sodium diet of 4 mEq Na^+/day; the salivary sodium levels declined to an asymptote after 2 weeks on the diet. Recently, we reported that even moderate reductions in dietary NaCl (maintenance on approximately 75 mEq Na^+/day) significantly lowered resting and stimulated whole mouth salivary sodium levels (Christensen et al., 1986b). In that study, healthy adult subjects were maintained on reduced sodium diets for approximately 2 months, and during that period salivary sodium concentrations fell 25% for resting and 17% for stimulated saliva. Decreases in salivary sodium levels following moderate or severe dietary reductions in NaCl probably reflect the action of hormones that conserve sodium, chief among them aldosterone. In contrast to salivary changes following dietary reductions in NaCl, increases in salivary sodium following increased sodium ingestion have not been found (Wotman et al., 1973; Prader et al., 1955; Powan, 1955). For example, ingestion of 10 g of NaCl did not increase salivary sodium levels when measures were obtained 2, 4, and 9 hr after NaCl consumption (Prader et al., 1955). On occasion in our laboratory, however, we have observed rises in salivary sodium following NaCl ingestion (see Section IV,A). It may be that there is a significant but transient rise in salivary sodium after NaCl ingestion reflecting the same transient rise in blood sodium levels (Yoshimura et al., 1963).

Two bitter-tasting components of saliva, urea and caffeine, are also influenced by diet. Unlike sodium, caffeine is not an endogenous constituent of saliva, but rather appears in saliva only after its consumption. Caffeine is rapidly absorbed by the body and is present in approximately equal concentrations in plasma and saliva (Axlerod and Reichenthal, 1953; Cook et al., 1976). Following ingestion of two cups of coffee at 8:00 AM and two cups at 12 noon (80 mg caffeine/cup), plasma caffeine levels at 1:00 PM were measured at 2.3 mg/liter (Axlerod and Reichenthal, 1953). Assuming salivary levels are comparable dietary additions of caffeine to saliva can be substantial (2.3 mg/liter is equivalent to 1×10^{-5} M), and may affect taste because the salivary levels overlap taste threshold values for many bitter tastants.

Urea is an endogenous component of saliva, and because it passively diffuses between blood and saliva fluid compartments, the levels in saliva are directly proportional and sometimes equivalent to urea concentrations in blood (Shannon and Prigmore, 1961). Urea levels are not as tightly regulated by the body as sodium levels are, and as a consequence, ingestion of some dietary components, such as protein, alters circulating urea levels (Addis et al., 1947).

This review has focused on salivary constituents that are recognized as taste stimuli. However, other salivary organic and inorganic constituents, such as proteins and phosphate, are also affected by diet (see review by Dawes, 1970). Research on these compounds is not described here because their role in the sensory perception of foods is not known.

IV. SALIVARY INFLUENCES ON TASTE PERCEPTION

It is reasonable to expect that saliva influences taste perception because it bathes the taste receptors and mixes with and dissolves food chemicals in the oral cavity. Despite its assumed importance in taste perception, there have been only a handful of studies investigating salivary effects on taste. These studies demonstrate that saliva is capable of modifying taste perception, although currently there is only a sketchy understanding of the mechanisms underlying salivary–taste interactions and little knowledge of the breadth of these interactions among different taste stimuli and concentrations.

There are studies demonstrating profound taste losses associated with long-term, severe reductions in salivary flow occurring as a result of damage to salivary glands (Conger, 1973; Mossman et al., 1982; Henkin et al., 1972). An adequate flow of saliva is required for the integrity of oral tissues, including taste receptors (Cano et al., 1978; Cano and Rodriguez-Echandia, 1980), and this is the apparent explanation for taste losses in these abnormal clinical groups. Under conditions of normal salivary function, there is evidence for two mechanisms controlling salivary–taste interactions: (1) sensory adaptation of the peripheral taste system to salivary components, and (2) modification of tastant properties by saliva. As described below, evidence for sensory adaptation comes from studies examining the relationship between salivary sodium levels and taste responses to NaCl. Evidence for changes in taste perception as a result of salivary modification of taste stimuli comes from experiments demonstrating a relationship between saliva-induced changes in the pH of acid tastants and sour perception.

A. Sensory Adaptation to Salivary Components

Sensory adaptation describes the phenomenon of decreased sensitivity to a stimulus that is continually present. The taste system, like other sensory modalities, will exhibit adaptation (Meiselman, 1968; Gent and McBurney, 1978). However, complete taste adaptation (total loss of sensation) is difficult to demonstrate because the taste system is easily disadapted by modulations in the adapting stimulus such as those produced by tongue movements or the addition of saliva (McBurney, 1969). When these problems are overcome, the rate of adaptation is observed to be exponential, and complete adaptation occurs within 2 min for a wide range of tastants and concentrations (Gent and McBurney, 1978). One important consequence of complete adaptation is that the taste threshold for the adapted compound is shifted to a concentration above that of the adapting stimulus.

In the oral cavity, salivary constituents that are also tastants, such as sodium, potassium, and urea, are thought to provide a source of constant stimulation that adapts taste pathways sensitive to these compounds. This possibility has received some experimental support in studies comparing salivary sodium levels with the perception of sodium chloride, particularly concentrations near threshold. Detection and recognition thresholds for

NaCl can be shifted by the sodium concentration in the oral cavity (McBurney and Pfaffmann, 1963; Morino and Langford, 1978; Bartoshuk, 1978), and the results from one such study (McBurney and Pfaffman, 1963) are illustrated in Fig. 5. In that study, detection thresholds for NaCl were lowest when deionized water was continuously flowed over the tongue; thresholds were higher and generally above the concentration of NaCl solutions flowed over the tongue or above the sodium content of saliva that was allowed to bathe the tongue. Other supportive evidence for the occurrence of salivary adaptation comes from a positive relationship ($r = 0.64$) observed between NaCl recognition thresholds and salivary sodium levels, which is illustrated in Fig. 6 (Morino and Langford, 1978). Evidence for salivary adaptation is also found at suprathreshold levels; in a single subject, the perceived saltiness of weak NaCl solutions (near threshold) decreased in intensity following prolonged stimulation of salivary flow, which has the effect of increasing salivary sodium levels (Bartoshuk, 1978).

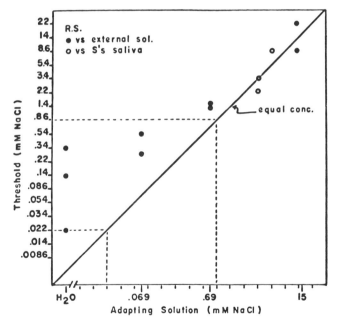

Fig. 5. NaCl recognition thresholds as a function of the adapting solution for a single subject (R. S.). Closed circles represent values obtained with adapting NaCl solutions flowed over the anterior portion of the tongue. Open circles represent values obtained when the tongue was held in the mouth to allow salivary adaptation. [From McBurney, D. H., and Pfaffmann, C. (1963). Gustatory adaptation to saliva and sodium chloride. *J. Exp. Psychol.* **92**, 523–529. Copyright (1963) by the American Psychological Association. Reprinted by permission of the author.]

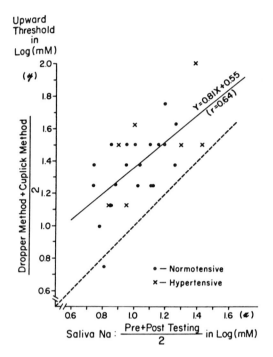

Fig. 6. The relationship between salivary sodium levels (average of two samples) and NaCl recognition thresholds that were obtained using an ascending series of concentrations (average of two taste testing methods). Also indicated is the correlation coefficient and regression equation and line (solid line). The broken line represents the identity function. [From Morino and Langford (1978). Reprinted with permission from *Physiology and Behavior* **21**, Morino, T. and Langford, H. G. Salivary sodium correlates with salt recognition threshold. Copyright (1978), Pergamon Press.]

An adaptation model would predict that thresholds for NaCl should be lower among individuals maintained on low NaCl diets because salivary sodium levels are lower (see Section III,B). In animal studies, such a relationship has been observed when the sensory measure is the preference threshold for NaCl (Contreras and Catalanotto, 1980). Lowered NaCl thresholds during periods of extreme NaCl dietary restriction (to the point of mild salt deficiency) have also been found in humans (Yensen, 1959). Salivary sodium values were not measured in that study, but in another study using the same dietary regimen (McCance, 1938), resting whole mouth sodium values fell 44% during the sodium restriction period. We have not been able to observe decreases in detection thresholds for NaCl when the reductions in dietary NaCl are moderate, even though salivary sodium concentrations declined (Christensen *et al.*, 1986b, and unpublished studies). It may be that salivary sodium levels must decline substantially before effects on NaCl thresholds can be observed, and this would be consistent with our finding in a large group of subjects (50 twin pairs), that no relationship exists between NaCl detection thresholds and salivary sodium levels when a normal range of salivary sodium values is encountered (Christensen et al., 1982). However, in that study, we also encountered a seemingly anomalous finding that individuals with unusually high salivary sodium levels (stimulated Na^+ levels greater than 600 μg/ml) had lower than average NaCl detection thresholds. According to an adaptation model, one would expect that NaCl thresholds should be higher in this group. This finding was corroborated in another experiment in which salivary and taste measures were obtained from a small group of subjects before, during, and following maintenance on a high sodium diet (Christensen, 1985). Two of the subjects tested showed a significant rise in salivary sodium when tested during the high dietary sodium period (12 g additional dietary NaCl) and, coincident with this rise, was a 10-fold decline in NaCl detection thresholds. An explanation for these findings is not immediately apparent; it may be that high concentrations of salivary sodium have effects on the taste system that supercede the effects produced by adaptation.

Adaptation of the taste system to salivary components may not be as functionally important as some of the results designed to observe adaptation suggest. With standard taste testing procedures and during natural taste stimulation as occurs during meals, adaptation to salivary components should be considerably weakened by the continuous presentation of tastants in the form of solutions or foods. On the other hand, although disadapting influences are normally present, most studies will record recognition thresholds for the salty taste of NaCl at concentrations above salivary sodium levels, which is, of course, predicted by an adaptation model. Clearly, more research is needed to understand the influences of salivary adaptation on the taste perception; in particular a systematic examination of the effects of partial adaptation on taste function is required. The possibility that longer term adaptation to salivary components is occurring should also be examined; that is, that the functional capacities of taste receptors may be influenced by the chemical media of saliva and blood in which they develop.

B. Modification of Tastants by Saliva

With sensory adaptation, only the concentration of salivary constituents is important to salivary effects on taste. However, our studies reveal that salivary flow rate also plays an important role and thus adaptation is not the only source of salivary–taste interactions. Our findings suggest that the probable mechanism for flow rate effects on taste are saliva-induced changes in the efficacy of taste stimuli. Significant changes occur in the chemical properties of taste solutions after brief exposure to saliva in the oral cavity, and these changes can have significant consequences for taste perception. The finding that taste stimuli are modified by saliva was unexpected because it has commonly been asssumed that not enough saliva is available to affect tastants held briefly in the mouth. This assumption is not valid because the quantity of saliva added to taste stimuli during taste testing is substantial. We find that individuals with average salivary flow rates add approximately 0.8 ml of saliva to water or weak taste solutions, and this is the same amount of saliva that has been found to remain in the oral cavity after swallowing (Lagerlöf and Dawes, 1984). Thus, the salivary contribution to a 10-ml quantity of tastant can be 10% of the total volume, with an even greater salivary contribution when tastants also stimulate salivary flow.

Dilution is the most likely consequence of salivary mixing with tastants because saliva is more than 99% water and contains relatively low concentrations of substances that are themselves tastants or taste modifiers (e.g., Na^+ or Cl^-). However, dilution is not the only consequence of salivary mixing; salivary buffers can dramatically change the pH of acid tastants, as described below, and very dilute tastants can be concentrated by exposure to saliva. An example of the latter is the addition of salivary sodium to NaCl solutions near detection threshold levels. The detection threshold for NaCl occurs at a concentration below that of salivary sodium, and NaCl solutions at the detection threshold will nearly double in concentration after they are briefly held in the oral cavity (Christensen, 1985). The addition of salivary sodium to NaCl solutions may partially underlie the observation that individuals with high salivary sodium levels have lower than average detection thresholds (see Section IV,A).

The degree to which saliva modifies tastants will, of course, depend on the volume of tastant; small volumes will exaggerate and large volumes will minimize salivary effects. Individual differences in salivary flow rate and composition should also influence the magnitude of salivary effects on tastants. The quantity of a salivary component added to a taste solution will depend both on flow rate and the concentration of that salivary component; but because individual differences in flow rate are very often greater than those for concentration, flow rate is likely to contribute more significantly to individual differences in the salivary modification of tastants.

The initial studies suggesting that salivary modification of tastants influenced taste perception came from measures of threshold and suprathreshold taste function in humans following reduction of salivary flow by pharmacologic means (Christensen et al., 1983; Christensen et al., 1984a). The pharmacologic agents produced a 25–82% reduction in salivary flow during the taste-testing period, and accompanying these flow reductions were reliable changes in recognition thresholds for sour (citric acid). Thresholds for other taste qualities were not affected by the salivary flow reductions. As illustrated in Fig. 7, following administration of atropine (1.0 mg orally), significantly *lower* recognition thresholds occurred for both aqueous and nonaqueous citric acid stimuli. The specificity of the drug's effect to a single taste quality suggested that there may have been drug-related changes in salivary composition as well as in flow rate. The effect of atropine on salivary composition has been examined (Shannon *et al.*, 1969) and, of the salivary constituents measured, bicarbonate levels were most affected by atropine. In that report, the concentration of salivary bicarbonate dropped from 0.77 to 0.24 mEq/liter following an oral dose of 1.3 mg of atropine sulfate. Bicarbonate is the principal buffer in saliva; changes in its concentration could be expected to affect sensitivity to acids more than other tastants if one assumes that the buffering action of saliva significantly alters the pH of an acid taste solution. We hypothesized that the increased sensitivity to sour tastants following salivary flow reductions occurred because less salivary buffer (function of buffer concentration and quantity of saliva) was available to increase the pH of acidic taste solutions. This explanation would be consonant with the greater effect of atropine on sour thresholds when smaller quantities (5 ml) of acid tastant were used (Fig. 7). This hypothesis was tested in the next series of experiments.

In one experiment (Christensen *et al.*, 1984b) we determined whether pH changes in acid tastants occurred after they were held briefly in the oral cavity and, further, whether these changes in taste solution pH were related to individual salivary measures of flow rate, pH, and total buffering capacity. Fourteen subjects participated in a single test session in which saliva was collected under oil and then a series of acid solutions were sipped and expectorated. Both resting and stimulated whole mouth saliva were collected for determination of flow rate, pH, and total buffering capacity. Subjects then sipped and expectorated (after holding the solutions for 3 sec in the oral cavity) 4- and 20-ml quantities of acetic, citric, and HCl solutions prepared in four concentrations: 1×10^{-3}, 10^{-4}, 10^{-5}, and 10^{-6} M. Each solution was sampled three times, and presentation of the total solution series was randomized.

The pH values for the prepared (untasted) and expectorated acetic acid solutions are depicted in Fig. 8 (similar patterns were observed for the other acids). Three findings should be noted. First, over a wide range of concentrations that span threshold and suprathreshold taste perception, briefly holding acid solutions in the oral cavity dramatically changes their pH. Second, at most concentrations, there are differences in the degree of salivary buffering of 4- and 20-ml volumes of acid. These volume-related differences in expectorated solution pH permit us to test whether salivary-

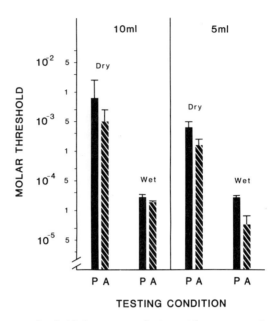

Fig. 7. Recognition thresholds (mean ± standard error) for aqueous and nonaqueous citric acid stimuli during placebo (P) and atropine (A) conditions. The left portion of the figure depicts the results when 10-ml quantities of aqueous tastant were used, and the right portion when 5-ml quantities were used. [From Christensen *et al.* (1984a). Reprinted with permission from *Archives of Oral Biology* **29**, Christensen, C. M., Navazesh, M., and Brightman, V. J. Effects of pharmacologic reductions in salivary flow on taste thresholds in man. Copyright (1984), Pergamon Press.]

induced pH changes in acid tastants affect taste perception by examining whether there are differential taste responses to the two volumes of solution. Third, the pattern of pH changes over the concentration range provides strong indirect evidence that salivary buffers are producing the pH changes rather than a simple dilution of the tastant by saliva. There is a "ceiling" on the pH of expectorated solutions near a value of 6.3 (which is the pK of bicarbonate) for very low concentrations of acid, whereas for the highest concentrations, salivary changes in solution pH become increasingly smaller.

Individual measures of salivary flow rate and total buffering capacity were positively correlated with changes in the pH of expectorated acid solutions. The higher an individual's resting flow rate or total buffering capacity, the greater the changes in the pH of acid solutions held in the oral cavity. The Pearson correlation values relating resting flow rate and total buffering capacity to the average pH change in all acid solutions were 0.58 and 0.57,

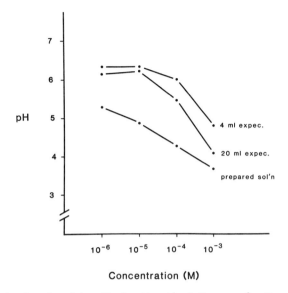

Fig. 8. A log–log plot of the pH of acetic acid solutions as a function of concentration measured before (prepared sol'n) and after 4- or 20-ml quantities were held in the oral cavity for 3 sec and then expectorated.

respectively ($p < 0.05$). A linear regression using both salivary characteristics to predict changes in solution pH produced a significant multiple correlation value of 0.75.

The results of this experiment suggested that there were large changes in acid tastant pH, particularly near recognition threshold levels. Thus, in another experiment, we determined whether salivary changes in tastant pH affected recognition thresholds for sourness (Christensen et al., 1984b). Different volumes of tastant were used because, over the concentration range bracketing acid recognition thresholds, saliva less completely buffers larger volumes of tastant; thus, if salivary changes in acid pH affect taste perception, then larger volumes should be more easily perceived.

Nine subjects participated in two testing sessions. During each session, resting and stimulated saliva were collected for later measurement of flow rate, pH, and total buffering capacity and then taste testing was performed. Recognition thresholds were obtained for 4- and 20-ml quantities of sucrose, acetic acid, and hydrochloric acid. During each session subjects received both acid (a different acid during each session) and sucrose solutions. A modified staircase method was used to determine thresholds for both taste qualities, and subjects were instructed to indicate whether a solution was either sour, sweet, or neither sour nor sweet. The results are summarized in

TABLE 1 Taste Recognition Thresholds (Molar)

Tastant volume	Tastant		
	Acetic acid	HCl	Sucrose
4 ml	2.28×10^{-4}	7.31×10^{-5}	3.92×10^{-3}
20 ml	7.06×10^{-5}	4.66×10^{-5}	3.95×10^{-3}

Table I. As predicted from the expectorated solution pH data in the previous experiment, subjects were significantly more sensitive to 20-ml quantities than 4-ml quantities of acid ($p < 0.05$). Sucrose thresholds were not affected by volume, which indicates that the effect of volume on the perception of acids was not attributable merely to a volume-related response bias.

Additional supporting evidence that sour taste perception is influenced by salivary alteration of tastant pH comes from a comparison of sour taste perception in groups with naturally low or high salivary flow rates. As described above, individuals with high salivary flow rates increase the pH of acid taste solutions more than those with low flow rates and thus should demonstrate less sensitivity to acid taste stimuli. Indeed, Norris et al. (1984) reported that high salivators perceived suprathreshold sour solutions as less sour than a group with lower flow rates. More recently, we demonstrated (Christensen et al., 1986b) that low and high salivators also differ in their ability to perceive sour solutions at threshold levels. A group with with naturally low salivary flow rates had lower detection and recognition thresholds for hydrochloric acid than a group with high flow rates, whereas both groups had similar thresholds for sucrose.

V. RESEARCH NEEDS

As this review demonstrates, there is a small but convincing body of evidence indicating that (1) a broad range of dietary variables affects salivary characteristics and (2) saliva affects taste perception. Neither of these relationships has been explored to their fullest, particularly the effects of salivary flow and composition on taste and other food-related sensations. Investigators have used only simple, aqueous systems in the study of saliva–taste interactions because their purpose was the characterization of mechanisms for these interactions. However, this has left unexplored a description of the effects of saliva for the entire range of sensory experiences associated with foods. Significant effects of saliva on the perception of solid foods are expected because these foods usually remain in the oral cavity longer than liquids, thus, permitting greater interaction with saliva. For dry, solid foods, saliva is also essential for the expression of taste (perhaps also flavor) because saliva dissolves the chemicals from the structural matrix. The perception of food texture should also be affected by saliva because, in concert with mastication, saliva aids in food breakdown.

As described in the Section I, saliva is likely to be one possible source of relationships between the chemical senses and nutrition. Given that interactions between diet–saliva and saliva–taste have been found, it is surprising that research has not been directed to establishing the existence of the tripartite interaction, diet–saliva–taste. The current evidence is supportive that such a relationship will be found in which saliva forms an essential intermediate link in the relationship between dietary variables and taste perception.

REFERENCES

Addis, T., Barrett, E., Poo, L. J., and Yuen, D. W. (1947). The relation between the serum urea concentration and the protein consumption of normal individuals. *J. Clin. Invest.* **26**, 869–874.

Axlerod, J., and Reichenthal, J. (1953). The fate of caffeine in man and a method for its estimation in biological material. *J. Pharmacol. Exp. Ther.* **107**, 519–523.

Bartoshuk, L. M. (1978). The psychophysics of taste. *Am. J. Clin. Nutr.* **31**, 1068–1077.

Burger, A. S. V., and Emmelin, N. G. (1961). "Monographs of the Physiological Society No. 8. Physiology of the Salivary Glands." Edward Arnold, London.

Cano, J., Roza, C., and Rodriguez-Echandia, E. L. (1978). Effects of selective removal of the salivary glands on taste bud cells in the vallate papilla of the rat. *Experientia* **34**, 1290–1291.

Cano, J., and Rodriguez-Echandia, E. L. (1980). Degenerating taste buds in sialectomized rats. *Acta Anat.* **106**, 487–492.

Chauncey, H. H., and Shannon, I. (1960). Parotid gland secretion rate as method for measuring response to gustatory stimuli in humans. *Proc. Soc. Exp. Bio. Med.* **103**, 459–463.

Chauncey, H. H., Feller, R. P., and Shannon I. L. (1963). Effect of acid solutions on human gustatory chemoreceptors as determined by parotid gland secretion rate. *Proc. Soc. Exp. Biol. Med.* **112**, 917–923.

Christensen, C. M. (1985). Role of saliva in human taste perception. *In* "Clinical Measurement of Taste and Smell" (H. L. Meiselman, and R. S. Rivlin, eds.). Macmillan, New York. (In press.)

Christensen, C. M., and Navazesh, M. (1984). Anticipatory salivary flow to the sight of different foods. *Appetite* **5**, 307–315.

Christensen, C. M., Bertino, M., Beauchamp, G. K., and Navazesh, M. (1982). Salivary sodium levels and salt perception. Presented at the Association for Chemoreception Sciences, April, 1982.

Christensen, C. M., Navazesh, M., and Brightman V. J. (1983). Effects of pharmacological reductions in salivary flow on judgments of suprathreshold taste stimuli. *Chem. Senses* **8**, 17–26.

Christensen, C. M., Navazesh, M., and Brightman, V. J. (1984a). Effects of pharmacologic reductions in salivary flow on taste thresholds in man. *Arch. Oral Biol.* **29,** 17–23.
Christensen, C. M., Malamud, D. M., Brand, J. C., and Dweck, E. (1984b). Saliva alters the oral perception of acids. Presented at the Association for Chemoreception Sciences, April, 1984.
Christensen, C. M., Malamud, D. M., and Brand, J. C. (1986a). Modification of the chemical properties of taste stimuli: A new mechanism for salivary effects on taste perception. Submitted for publication.
Christensen, C. M., Bertino, M., Beauchamp, G. K., Navazesh, M., and Engleman, C. A. (1986b). A moderate reduction in dietary sodium lowers salivary sodium concentration. Submitted for publication.
Conger, A. D. (1973). Loss and recovery of taste acuity in patients irradiated to the oral cavity. *Radiat. Res.* **53,** 338–347.
Contreras, R. J., and Catalanotto, F. A. (1980). Sodium deprivation in rats: Salt thresholds are related to salivary sodium concentrations. *Behav. Neural Biol.* **29,** 303–314.
Cook, C. E., Tallent, C. R., Amerson, E. W., Myers, M. W., Kepler, J. A., Taylor, G. F., and Christensen, H. D. (1976). Caffeine in plasma and saliva by a radioimmunoassay procedure. *J. Pharmacol. Exp. Ther.* **199,** 679–686.
Dawes, C. (1969). The effect of flow rate and duration of stimulation on the concentration of protein and the main electrolytes in human parotid saliva. *Arch. Oral Biol.* **14,** 277–294.
Dawes, C. (1970). Effects of diet on salivary secretion and composition. *J. Dent. Res.* **49,** 1263–1273.
Dawes, C. (1974). The effects of flow rate and duration of stimulation on the concentrations of protein and the main electrolytes in human submandibular saliva. *Arch. Oral Biol.* **19,** 887–895.
Dawes, C. (1984). Stimulus effects on protein and electrolyte concentrations in parotid saliva. *J. Physiol.* **346,** 579–588.
Dawes, C., and Jenkins, G. N. (1964). The effects of different stimuli on the composition of saliva in man. *J. Physiol.* **170,** 86–100.
Dawes, C., and Ong, B. Y. (1973). Circadian rhythms in the flow rate and proportional contribution of parotid to whole saliva volume in man. *Arch. Oral Biol.* **18,** 1145–1153.
Dawes, C., and Wood, C. M. (1973). The contribution of the oral minor mucous gland secretions to the volume of whole saliva in man. *Arch. Oral Biol.* **18,** 337–342.
DeWardener, H. E., and Herxheimer, A. (1957). The effect of a high water intake on salt consumption, taste thresholds, and salivary secretion in man. *J. Physiol.* **139,** 53–63.
Edgar, W. M., and Jenkins, G. N. (1981). Can salivary function in man be enhanced by increased mastication? *J. Dent. Res.* **60(B),** 1172 (abstract).
Ekström, J., and Templeton, D. (1977). Difference in sensitivity of parotid glands brought about by disuse and overuse. *Acta Physiol. Scand.* **101,** 329–335.
Ellison, S. A. (1979). The identification of salivary components. In "Proceedings; Saliva and Dental Caries" (I. Kleinberg, S. A. Ellison, and I. D. Mandel, eds.). Special supplement to Microbiology Abstracts.
Feller, R. P., Sharon, I. M., Chauncey, H. H., and Shannon, I. L. (1965). Gustatory perception of sour, sweet, and salt mixtures using parotid gland flow rate. *J. Appl. Physiol.* **20,** 1341–1344.
Funakoshi, M., and Kawamura, Y. (1965). Relations between taste qualities and parotid gland secretion rate. In "Olfaction and Taste II" (T. Hayashi, ed.), pp. 281–287. Pergamon Press, New York.
Gent, J. F., and McBurney D. H. (1978). Time course of gustatory adaptation. *Percept. Psychophys.* **23,** 171–175.

Hall, H. D., Merig, J. J., Jr., and Schneyer, C. A. (1967). Metrecal-induced changes in human salivary glands. *Proc. Soc. Exp. Biol. Med.* **124**, 532–536.

Henkin, R. I., Talal, N., Larson, A. L., and Mattern, C. F. T. (1972). Abnormalities of taste and smell in Sjogren's sydrome. *Ann. Int. Med.* **76**, 375–383.

Holmes, J. H. (1964). Changes in salivary flow produced by changes in fluid and electrolyte balance. *In* "Salivary Glands and their Secretions" (L. M. Sreebny and J. Meyer, eds.), pp. 177–195. Pergamon Press, New York.

Jenkins, G. N. (1966). "The Physiology of the Mouth." F. A. David, Philadelphia.

Jenkins, G. N., and Dawes, C. (1966). The psychic flow of saliva in man. *Arch. Oral Biol.* **11**, 1203–1204.

Kemmer, T., and Malfertheiner, P. (1983). Der differenzierte einfluB der beschmacksqualitaten "suB" and "sauer" auf die parotissekretion. *Res. Exp. Med.* **183**, 35–46.

Kerr, A. C. (1961). "The Physiological Regulation of Salivary Secretions in Man." Pergamon Press, New York.

Lagerlöf, F., and Dawes, C. (1984). The volume of saliva in the mouth before and after swallowing *J. Dent. Res.* **63**, 618–621.

Lashley, K. S. (1916). Reflex secretion of the human parotid gland, *J. Exp. Psychol.* **1**, 461–493.

Lawless, H. (1984). Oral chemical irritation: Psychophysical properties. *Chem. Senses* **9**, 143–155.

McBurney, D. H. (1969). Effects of adaptation on human taste function. *In* "Olfaction and Taste III" (C. Pfaffmann ed.), pp 407–419. Pergamon Press, New York.

McBurney, D. H., and Pfaffmann, C. (1963). Gustatory adaptation to saliva and sodium chloride. *J. Exp. Psychol.* **65**, 523–529.

McCance, R. A. (1938). The effect of salt deficiency in man on the volume of the extracellular fluids and on the composition of sweat, saliva, gastric juice, and cerebrospinal fluid. *J. Physiol.* **92**, 208–218.

Makhlouf, G. M., and Blum, A. L. (1972). Kinetics of the taste response to chemical stimulation: A theory of acid taste in man. *Gastroenterology* **63**, 67–75.

Meiselman, H. L. (1968). Magnitude estimations of the course of gustatory adaptation. *Percept. Psychophys.* **4**, 193–196.

Morino, T., and Langford, H. G. (1978). Salivary sodium correlates with salt recognition threshold. *Physiol. Behav.* **21**, 45–48.

Mossman, K., Shatzman, A., and Chencharick, J. (1982). Long-term effects of radiotherapy on taste and salivary function in man. *Int. J. Radiat. Biol. Relat. Stud. Physics, Chem. Med.* **8**, 991–997.

Navazesh, M., and Christensen, C. M. (1982). A comparison of whole mouth resting and stimulated salivary measurement procedures. *J. Dent. Res.* **61**, 1158–1162.

Norris, M. B., Noble, A. C., and Pangborn, R. M. (1984). Human saliva and taste responses to acids varying in anions, titratable acidity, and pH. *Physiol. Behav.* **32**, 237–244.

Pangborn, R. M., and Berggren, B. (1973). Human parotid secretion in response to pleasant and unpleasant odorants. *Psychophysiology* **10**, 231–237.

Pangborn, R. M., and Lundgren, B. (1977). Salivary secretion in response to mastication of crisp bread. *J. Texture Stud.* **8**, 463–472.

Powan, G. L. S. (1955). Studies on the salivary sodium/potassium ratio in man. *Biochem. J.* **60**, xii.

Prader, A., Gautier, E., Gautier, R., Naf, D., Semer, J. M., and Rothschild, E. J. (1955). The Na and K concentration in mixed saliva: Influence of secretion rate, stimulation, method of collection, age, sex, time of day, and adrenocortical activity. *In* "The Human Adrenal Cortex" (G. E. W. Wolstenholme and M. P. Cameron eds.), pp. 382–395. Little, Brown, Boston.

Sahakian, B. J. (1981). Salivation and appetite: Commentary on the forum. *Appetite* 2, 386–389.
Schneyer, L. H., and Levin, L. K. (1955a). Rate of secretion by individual salivary gland pairs of man under conditions of reduced exogenous stimulation. *J. Appl. Physiol.* 7, 508–512.
Schneyer, L. H., and Levin, L. K. (1955b). Rate of secretion by exogenously stimulated salivary gland pairs of man. *J. Appl. Physiol.* 7, 609–613.
Schneyer, L. H., Young, J. A., and Schneyer, C. A. (1972). Salivary secretion of electrolytes. *Physiol. Rev.* 52, 720–775.
Shannon, I. L. (1962). Parotid fluid flow rate as related to whole saliva volume. *Arch. Oral Biol.* 1, 391–394.
Shannon, I. L., and Chauncey, H. H. (1967). Hyperhydration and parotid flow in man. *J. Dent. Res.* 46, 1028–1031.
Shannon, I. L., and Prigmore, J. R. (1961). Effects of urea dosage on urea correlations in human parotid fluid and blood serum. *Arch. Oral Biol.* 5, 161–167.
Shannon, I. L., Suddick R. P., and Chauncey, H. H. (1969). Effect of atropine-induced flow rate depression on the composition of unstimulated human parotid fluid. *Arch. Oral Biol.* 14, 761–770.
Speirs, R. L. (1971). The effects of interactions between gustatory stimuli on the reflex flow-rate of human parotid saliva. *Arch. Oral Biol.* 16, 351–365.
Thorn, N. A., Schwartz, I. L., and Thaysen, J. H. (1956). Effect of sodium restriction on secretion of sodium and potassium in human parotid saliva. *J. Appl. Physiol.* 9, 477–480.
Thrasher, T. M., and Fregly, M. J. (1980). Factors affecting salivary sodium concentration, NaCl intake, and preference threshold and their interrelationship. *In* "Biological and Behavioral Aspects of Salt Intake" (M. R. Kare, M. J. Fregly, and R. A. Bernard eds.), pp. 145–165. Academic Press, New York.
Wotman, S., Baer, L., Mandel, I. D., and Laragh, J. H. (1973). Salivary electrolytes, renin, and aldosterone during sodium loading and depletion. *J. Appl. Physiol.* 35, 322–324.
Yensen, R. (1959). Some factors affecting taste sensitivity in man. II. Depletion of body salt. *Q. J. Exp. Psychol.* 11, 230–238.
Yoshimura, H., Miyoshi, M., Matsumoto, S., Fujimoto, T., and Yamamoto, Y. (1963). Reflection of salt concentrations of blood upon those of saliva. *Jpn. J. Physiol.* 13, 523–540.
Young, J. A., and Schneyer, C. A. (1981). Composition of saliva in mammalia. *Aust. J. Exp. Biol. Med. Sci.* 59, Part 1, 1–53.

2

Factors Affecting Acceptance of Salt by Human Infants and Children

BEVERLY J. COWART[*,†] AND GARY K. BEAUCHAMP[*,‡]
*Monell Chemical Senses Center and
†Thomas Jefferson University
‡School of Veterinary Medicine
University of Pennsylvania
Philadelphia, Pennsylvania

I.	Introduction	25
II.	Responses to Salt and Sugar in Early Development: A Review	26
	A. Newborns	27
	B. Older Infants and Children	29
	C. Summary	34
III.	Developmental Shifts in Salt Acceptance	35
	A. Salt in Water	35
	B. Salt in Foods	38
	C. Discussion	40
IV.	Research Needs	42
	References	42

I. INTRODUCTION

Adult humans often prefer salted foods to those same foods without salt and consume considerably more sodium, primarily in the form of sodium chloride, than is required to maintain sodium balance (Fregly and Fregly, 1982; Meneely and Batterbee, 1976). Whereas a good deal of information is available on the early development of human sweet preference, which is now widely considered to be innate (see Weiffenbach, 1977), there is relatively little known about the developmental course of the apparent preference for saltiness.

As is the case for sugar, consumption of salt in the absence of overt physiological need is common among mammalian species, raising the possibility that salt acceptance too has a genetic basis (Denton, 1982). On the other hand, studies with human infants, which uniformly demonstrate a strong preference for sugar at birth (e.g., Desor *et al.*, 1977), have failed to provide evidence for neonatal salt preference, and a few suggest that newborns may find salty tastes to be slightly aversive (e.g., Crook, 1978). In addition, sodium chloride is primarily consumed by humans as a food additive, in which role it may not only impart a salty taste but also favorably alter other flavor components of a food (Yamaguchi and Kimizuka, 1979). When presented as relatively pure taste stimuli in water, moderate concentrations of salt are, in fact, considered by most adults to be hedonically neutral or frankly unpleasant (Bertino *et al.*, 1982; Pangborn, 1970). Thus, it might even be argued that adult humans do not prefer saltiness per se but have acquired a preference for the tastes of particular foods as modified by salt.

In the present chapter, we briefly review the literature on early salt acceptance in humans, comparing and contrasting those findings with data on the development of sweet preference. We also present results from recent studies with infants and young children demonstrating heightened acceptance of salt in water as well as food, beginning at about 4 months of age. Finally, we argue that in conjunction with neurophysiological and behavioral data on other mammals, the available evidence is most consistent with a hypothesis of postnatal maturation of salty-taste preference in humans, which, like sweet preference, is largely unlearned but may be modified by dietary experience.

II. RESPONSES TO SALT AND SUGAR IN EARLY DEVELOPMENT: A REVIEW

Studies of early childhood responses to taste substances provide the foundation for understanding the developmental bases, course, and modifiability of specific taste preferences. Although published reports of infants' and children's taste responses first appeared over a century ago, it was not until the early 1970s that research attention began to focus on this area. That attention led to the development of a number of interesting and elegant experimental paradigms. However, most investigations of early taste perception have centered on responses to sweets, especially sucrose, and have been limited to study of the neonatal (≤1 week) and, to a lesser extent, preschool (3–5 years) periods.

A. Newborns

Although a variety of techniques have been employed in studies of neonatal taste responses, almost all have relied on the measurement of behaviors believed to be hedonically motivated. Within that framework, a distinction should be made between relative intake measures and others such as facial expression and sucking patterns. Whereas the former provide a measure of ingestive preference or acceptance, the latter are assumed to be indicators of positive or negative affect. By and large, interpretation of the latter responses depends on comparison with the response to sucrose, which has been shown through intake measures to be highly accepted by newborns and is, therefore, presumably perceived as pleasant.

Intake is typically measured during brief exposure periods (0.5–3.0 min). Bottles containing the solvent alone (usually water) and one or more concentrations of a tastant in solution are presented sequentially, with order of presentation counterbalanced within and/or across subjects. Infants are allowed to consume as much as they like during the specified period, and the amount of tastant solution consumed is then compared to consumption of the solvent. Using this basic technique, it has repeatedly been shown that newborns consume larger quantities of sweetened than of unsweetened solutions (e.g., Desor et al., 1973, 1977; Maller and Desor, 1973). Moreover, they appear to prefer sweeter sugars (sucrose and fructose) to less sweet ones (glucose and lactose) and higher concentrations of a given sugar to lower concentrations (Desor et al., 1973). This striking sensitivity and responsiveness is not, however, observed when the stimulus is salt. Quite simply, newborns do not differentially ingest salt water and plain water (Maller and Desor, 1973).

On the assumption that salt might be somewhat aversive to neonates, it has been suggested that their failure to demonstrate differential intake results from an inability to inhibit sucking below the level associated with water consumption. Desor et al. (1975) attempted to circumvent that potential problem by utilizing 0.07 M sucrose, rather than plain water, as a standard. Consumption of this standard was then compared to consumption of 0.07 M sucrose solutions to which sodium chloride (0.05–0.2 M) had been added. Infants still appeared to be indifferent to the salty stimuli.

A similar lack of responsiveness to sodium chloride stimulation is observed when specific facial reactions to taste stimuli are assessed (Steiner, 1973, 1977, 1979). Whereas solutions of sucrose, citric acid, and quinine sulfate all seem to elicit distinctive patterns of facial expression in neonates, salty solutions are not associated with a typical reaction.

These failures to observe differential responding to sodium chloride seem to indicate that newborns find its taste to be completely hedonically neutral and/or that they are not particularly sensitive, in an absolute sense, to salt. On the other hand, some measures of neonatal response suggest that apparently negative reactions to sodium chloride can occur. For example, Crook (1977, 1978) examined the effects of discrete, intraoral presentations of taste solutions on the normal burst–pause pattern of nonnutritive sucking. When drops of sugar water are presented within this paradigm, the number of sucks per burst (burst length) increases, and the length of the pause between successive bursts decreases. Somewhat paradoxically, the rate of sucking within a burst also decreases. All three changes are dependent on sucrose concentration, becoming more pronounced as concentration increases up to $0.375\ M$. Of these parameters, only burst length has been examined following presentation of salt water drops. In contrast to sugar, salt decreases the length of subsequent sucking bursts. This decrease is apparently not concentration dependent, however, being fairly uniform following drops of 0.1, 0.3, and $0.6\ M$ NaCl.

Crook (1978) suggests that the decrease in burst length associated with saline presentation reflects a negative hedonic reaction. Not only is this behavioral change in the direction opposite of that observed following sucrose, but it would also appear to be a change directed toward decreased consumption. However, as noted above, intake measures fail to demonstrate any reliable decrease in neonatal consumption of fluids following the addition of salt. Thus, the decrease in burst length may either be relatively short lived, or its potentially negative impact on total intake may in some way be compensated for by alterations in other parameters of sucking behavior.

Crook's (1978) finding, together with reports of other nonspecific or less consistent reactions to salt (Jensen, 1932; Nowlis, 1973), does indicate that there is some sensory response to sodium chloride in human newborns. However, the subtlety of the observed behavior is such that it does not point to an unambiguous hedonic reaction. Moreover, in the one study designed to determine the lowest concentration of sodium chloride necessary to elicit measurable changes in newborn sucking behavior, consistent responses were not observed below a concentration of approximately $0.08\ M$ NaCl (Jensen, 1932). The median adult threshold for sodium chloride detection is, by contrast, around $0.003\ M$ NaCl (Bartoshuk, 1974), and recognition of a salty taste generally occurs between 0.008 and $0.01\ M$ NaCl (C. Christensen, personal communication).

Nonetheless, the hypothesis that newborns may not be very sensitive to salt has received little attention in previous discussions of neonatal salt responses, including ones by the present authors (Beauchamp, 1981; Cowart, 1981). In part, this is probably because the apparently adultlike sensitivity of newborns to sweet tastes makes an initial presumption of similar sensitivity to other tastants seem parsimonious. Additionally, in the absence of good measures of absolute sensitivity to tastants in human newborns (i.e., measures that do not rely on hedonically motivated behaviors), direct tests of sensitivity simply cannot be made. Recently, however, postnatal changes in neural responsiveness to several salts have been documented in two mammalian species, sheep and rats (Ferrel et al., 1981; Hill and Almli, 1980; Hill et al., 1982; Mistretta and Bradley, 1983). In particular, responsiveness to sodium chloride has been shown to increase, both absolutely and relative to other salts, throughout early postnatal life. Although not directly transposable to the human case, such findings do suggest that more serious consideration should be given to the possibility that the ability to detect salt is limited in human newborns and that there may be quantitative and/or qualitative changes in the neural response to sodium chloride postnatally. If that were the case, it might be inappropriate to attach much significance to the apparent hedonic value of salty stimuli at birth or to consider neonatal salt responses truly contiguous with responses observed in later life.

B. Older Infants and Children

As is true in work with newborns, most investigations of taste function in older infants and children have been concerned with preferential responding. Results of the few attempts to measure absolute sensitivity in young children are difficult to interpret and will not be reviewed here (see discussion by Cowart, 1981).

The preference for sweet evidenced by newborns has, as might be expected, been shown to persist in older infants and preschool children. For example, Desor et al. (1977) found that, like newborns, infants 5–28 weeks of age will consume significantly more sweetened than unsweetened water. Similarly, Filer (1978) demonstrated that 2–6 year olds will eat significantly more sweetened than unsweetened spaghetti.

The results reported by Desor et al. (1977) also seem to suggest that the degree of ingestive responsiveness to sweetened water remains constant in early life. These authors compared intake data from newborns, 5–11 week olds, and 20–28 week olds. Infants in the middle group had just recently been introduced to complex foods other than milk, whereas those in the oldest group were on highly varied diets. The differences between volume of water ingested and intake of 0.1 and 0.2 M glucose and fructose solutions were analyzed, and no effect for age group was observed. However, these groups were relatively small ($n = 12$), and both concentrations of the two sugars employed yield fairly weak sweet tastes. Thus, the generalizability of the conclusion of Desor et al. that ingestive responses to sweet are unaltered by age or postnatal dietary experience would seem to be limited.

In fact, recently reported results from a longitudinal study indicate that early dietary experiences can exert a profound influence on the ingestive expression of sweet preferences. Beauchamp and Moran (1982, 1985) employed relative intake measures to assess sweet preference in a group of children at birth, 6 months, and 2 years of age. They were able to test 199 newborns initially, 140 infants in the 6-month follow-up and 63 children in the 2-year follow-up. At all ages, intake of water was compared to intake of 0.2 and 0.6 M sucrose solutions. In agreement with earlier work, Beauchamp and Moran (1982) found that newborn infants ingested significantly more of both sucrose solutions than of water and significantly more 0.6 than 0.2 M sucrose. A heightened acceptance of 0.2 and 0.6 M sucrose relative to water was also observed at 6 months of age. At that age, however, infants were divided into the following two groups, determined by 7-day dietary histories: those who were regularly fed sweetened water (water sweetened with table sugar, Karo syrup, or honey) and those who were not. The main effect for feeding history proved to be significant in an analysis of variance of the volumes of water, 0.2 and 0.6 M sucrose consumed, as was the feeding history by stimulus interaction. Infants who were regularly fed sweetened water ingested significantly more of each of the two sugar solutions, but not more water, than did infants who were not given sweetened water by their mothers (Fig. 1). The two groups did not differ on other dietary variables, nor was there a relationship in either group between total number of sweet foods ingested, or history of feeding fruit juices, and responses on the taste tests.

In order to examine the relationship between responses at birth and at 6 months, acceptance ratios (milliliters taste solution/milliliters taste solution + water) for 6 month olds were compared to the ratios obtained by those same infants at birth. Feeding history at 6 months again served as a variable. The two groups of infants (those fed and those not fed sweetened water) did not differ in their relative acceptance of sucrose at birth. At 6 months of age, ratio scores of the infants fed sweetened water were still equivalent to those observed at birth. In contrast, acceptance ratios of the 6-month-old group that had not been fed sweetened water had declined almost to the point of indifference (Fig. 2).

When a subset of these same infants were retested at 2 years of age (Beauchamp and Moran, 1985), they were divided into the following three groups based on feeding history: those never fed sweetened water (NF),

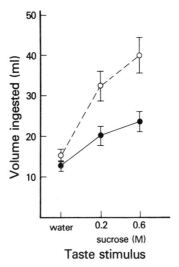

Fig. 1. Volume consumed (± SEM) of water and 0.2 and 0.6 M sucrose solutions in 6-month-old infants fed sweetened water (○, $n = 36$) and not fed sweetened water (●, $n = 95$), as determined from diet histories obtained at 6 months of age. Points are connected for visual purposes only. [From Beauchamp and Moran (1982). With permission from *Appetite* **3**, 139–152. Copyright (1982), by Academic Press Inc. (London) Ltd.]

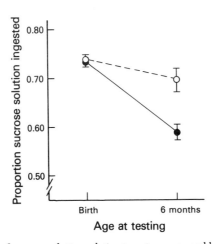

Fig. 2. Proportion of sucrose solution relative to water consumed by infants at birth and at 6 months of age: (○) infants fed sweetened water; (●) infants not fed sweetened water. Feeding experience was determined from diet histories obtained when infants were 6 months of age. [From Beauchamp and Moran (1982). With permission from *Appetite* **3**, 139–152. Copyright (1982), by Academic Press Inc. (London) Ltd.]

those fed sweetened water for 6 months or less (Fed ≤ 6 Mo.), and those fed sweetened water for more than 6 months (Fed > 6 Mo.). A significant ingestive preference for the sucrose solutions was observed only in the two groups who had had postnatal dietary experience with sweetened water. Interestingly, there were no differences between the ingestive responses of the group Fed ≤ 6 Mo. and the group Fed > 6 Mo., suggesting that early exposure to sweetened water, even though it had been discontinued, had lasting effects on the acceptance of sucrose in water. Finally, the 2 year olds in this study were also tested for relative acceptance of 0.6 M sucrose in Kool-Aid. When sucrose was presented in this context, there were no differences between the NF, Fed ≤ 6 Mo., and Fed > 6 Mo. groups. All children ingested significantly more of the sweetened than of the unsweetened Kool-Aid. However, children who had never been given Kool-Aid by their mothers did ingest significantly less Kool-Aid, whether or not it was sweetened, than did children who had previous experience with Kool-Aid.

Beauchamp and Moran's studies (1982, 1985) indicate that ingestive expression of the innate preference for sweet tastes may be subject to modification quite early in life by relatively minor variations in dietary experience. This is not to say that human sweet preference is exceptionally malleable or easily overturned. Rather, the effects of experience on this response seem to be specific to the particular context in which sweet is experienced. At least limited experience with sweetened water seems necessary to maintain the newborn's level of ingestive preference for sugar water. In the absence of such dietary exposure, relative acceptance of sugar water diminishes. At the same time, however, ingestive responses to sweet tastes in the other contexts may be unaffected. In other words, what appears to be shaped through dietary experience is a sense of what should and should not be sweet, rather than generalized hedonic responsiveness to sweetness per se. Although one can only speculate on this point, it seems reasonable to hypothesize that novelty (or, conversely, familiarity) may underlie the context effect. Given a familiar food or beverage product that has been experienced only without added sweetness, the addition of even this highly preferred taste may not enhance the substance's acceptability, and may even render it less acceptable, because the familiarity of its taste is thereby reduced. On the other hand, given a novel food or beverage with and without added sugar, the inherent hedonic value of sweet should lead to an ingestive preference for the sweetened version, assuming neophobia does not suppress ingestion altogether.

Early childhood preferences for salty tastes have also been assessed in a number of studies, although most have been small and somewhat piecemeal in their selection of age groups and stimuli. Reports of responses to salt between birth and 2 years of age are not terribly consistent. In a study of the acceptance of salted foods by infants, Foman et al. (1970) found that 4 and 6 month olds seemed indifferent to the addition of salt to strained baby foods. However, Filer (1978) later suggested that the amounts of salted and unsalted foods consumed in that study might have more accurately reflected the mother's "mechanical skill . . . and determination to feed her infant" (p. 7) than the preference of the infant. Davis (1928) included a bowl of salt crystals among the items regularly offered to three newly-weaned infants in her classic study of self-selection of diet. These infants (7.5 to 9 months of age at the beginning of the study) chose to eat the salt only occasionally and often choked, spluttered, and cried after putting it into their mouths. Nonetheless, they never spit it out and frequently went back for more.

Other studies of salt preference during the first 2 years have employed more traditional intake measures, comparing ingestion of saline solutions to water during brief test periods. In a chapter on the development of flavor preferences, Beauchamp and Maller (1977) cite unpublished work by J. Desor which suggested that by 6 months of age, infants may exhibit a preference for some concentrations of salt water. In the same chapter, however, they report a study by Beauchamp which also included 6 month olds ($n = 12$). Salt concentrations of 0.05, 0.10, and 0.20 M were presented. On average, approximately equivalent amounts of water and each of the salt solutions were consumed by these 6 month olds, suggesting a relative indifference similar to that of neonates. However, inspection of the individual data from this study indicates that for 8 of the 12 infants, the ratio of total saline consumption (0.05 + 0.10 + 0.20 M) to 3 times the infants' water consumption exceeded 1.00. Finally, Vazquez et al. (1982) reported that both malnourished ($n = 55$) and healthy ($n = 41$) Mexican infants between 2 and 12 months of age consumed more 0.10 and 0.20 M NaCl than water. In contrast, groups of malnourished ($n = 41$) and healthy ($n = 12$) infants between 13 and 24 months of age appeared indifferent to these same salt solutions.

Somewhat older children exhibit a fairly consistent preference for salt, at least when it is presented in foods. When Filer (1978) gave 2 to 6 year olds identical jars of salted and unsalted beef stew, they chose to eat the former with significantly greater frequency than the latter. Similarly, Beauchamp (1981) found that when given a choice, 4- to 8-year-old children would select salted pretzels significantly more often than unsalted ones, and Beauchamp and Moran (1985) reported that 2 year olds chose to consume salted carrots significantly more often than unsalted ones. Beauchamp and Moran were, however, unable to demonstrate a preference for salted over unsalted soup broth in these same children. Their subjects underwent multiple taste tests, and mean intakes of both plain and salted soups were low relative to intake of all other taste solutions (i.e., sweetened and unsweetened Kool-Aid and water). Thus, the novelty or unpleasantness of the soup base may have been such that intake was generally suppressed, making differential consumption difficult to demonstrate.

There is only one report of ingestive responses to salted water in toddlers. Beauchamp (cited in Beauchamp and Maller, 1977) included a group of 2 to 3 year olds* ($n = 8$) in the study of 6 month olds described above. Solutions of 0.05, 0.10, and 0.20 M NaCl were again presented. Relative to water, all of the salt solutions were rejected by these children, although there was no effect for sodium chloride concentration.

C. Summary

The evidence concerning sweet acceptance in early development is quite consistent. Human newborns, infants, and children generally demonstrate a preference for sweetened over unsweetened substances, and that preference appears to be unlearned. Nonetheless, dietary experience can significantly alter the acceptance of sweet tastes in particular contexts. Specifically, in the absence of exposure to sugar water, relative acceptance of sweetened over plain water declines such that by 2 years of age, nonexposed children are indifferent to sugar when it is presented in water.

In contrast, the developmental course of salt acceptance appears to be less straightforward. At birth, infants do not discriminate ingestively between salt water and plain water, although subtle behavioral changes, which may be indicative of a negative hedonic reaction, occur to moderate saline concentrations. By 2 to 3 years of age, children clearly prefer many salted foods to their unsalted counterparts, and they may also show an adultlike rejection of salt water. When these groups of observations from the neonatal and preschool periods are juxtaposed, the conclusion that salt acceptance is gradually acquired and quite specific to salted foods appears reasonable. On the other hand, there are some studies which suggest that older infants may exhibit a seemingly anomalous ingestive preference for salty water. Only one of these studies has been fully reported (Vazquez *et al.*, 1982), and, because it involved a somewhat unusual population of children, its results may not represent a general phenomenon. Nonetheless, this is an intriguing finding.

If, as suggested earlier, taste structures underlying sodium chloride responsiveness develop postnatally, one might expect to observe a shift in preferential responding to salt during infancy. Such a shift has been documented in preweanling (Moe, 1986) and just postweanling (Midkiff and

---

*This older group was originally identified as 1.5–3 year olds. However, with the exception of one 22 month old, all subjects were over 2 years of age.

Bernstein, 1983) rats. Further, if that shift were in the direction of preferential ingestion of salt solutions to which newborns are indifferent, and occurred prior to extensive dietary exposure to sodium chloride, a possible implication is that it is the specific rejection of salt in water by older children and adults that is "learned," not a preference for sodium chloride that depends on its interactions with the tastes of foods. In other words, the developmental course of salt preference (and the factors affecting its expression) may actually be quite similar to that of sweet preference, except that the initial emergence of salt preference is delayed by the postnatal development of underlying central and/or peripheral structures.

In order to further investigate this issue, we have initiated a series of studies on the acceptance of salt in water and foods by infants and children between 2 and 60 months of age. The results of some of that work are presented below.

III. DEVELOPMENTAL SHIFTS IN SALT ACCEPTANCE

A. Salt in Water

In our first experiment, 54 healthy human infants between 2.5 and 6.7 months of age were recruited from the Well Baby Clinic of Pennsylvania Hospital (Beauchamp et al., 1986). These infants were tested for acceptance of aqueous solutions of both sodium chloride and sucrose, relative to plain water, using a standard intake paradigm. Concentrations of 0.10 and 0.20 M NaCl and of 0.20 and 0.40 M sucrose were employed. Half of the infants received the sucrose test first, and half received the salt test first.

The results are presented in Fig. 3A and B. The infants were divided into three groups based on age (I = 2.5–3.9 months, n = 18; II = 4.0–5.3 months, n = 17; III = 5.4–6.7 months, n = 19). At each of the two salt and sucrose concentrations, the ratio of tastant consumption to plain water consumption was determined for each infant. Thus, acceptance of salt (or sugar) by an infant was assessed relative to that infant's water intake during the sodium chloride (or sucrose) test. The distributions of these ratio values were normalized by log transformation prior to parametric analysis (Winer, 1971).

Analysis of variance of the resulting sucrose acceptance scores revealed no significant effect for age, although there was a marked trend toward decline in the relative acceptance of sugar water with age (Fig. 3A). In addition, a significant effect for sucrose concentration was observed. Infants in all three age groups consumed more of the sweetened than of the unsweetened solutions, and the ratio of 0.40 M sucrose/water tended to be greater than that of

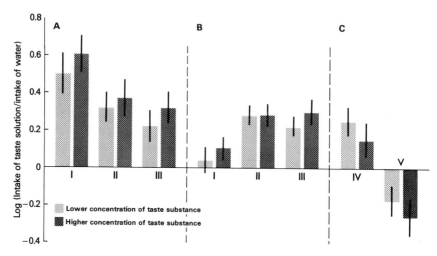

Fig. 3. Log geometric mean ratios (± SEM) of taste solution to water intake. Values greater than 0 indicate consumption of more taste solution than of water; values less than 0 indicate the reverse. (A) Intake of 0.20 and 0.40 M sucrose solution relative to intake of water in three groups of infants (I = 2.5–3.9 months old, n = 18; II = 4.0–5.3 months old, n = 17; III = 5.4–6.7 months old, n = 19). (B) Intake of 0.10 and 0.20 M NaCl relative to intake of water by the same groups listed above. (C) Intake of 0.17 and 0.34 M NaCl relative to intake of water by two groups of older infants and children (IV = 7–23 months old, n = 16; V = 31–60 months old, n = 18). [From Beauchamp et al. (1986). Dev. Psychobiol. **17**, 75–83. Copyright by John Wiley & Sons, Inc.]

0.20 M sucrose/water. These findings are all consistent with the literature on the development of human sweet preference reviewed above.

In contrast, there was a significant age-related change in sodium chloride acceptance in the opposite direction of the age trend observed in sucrose acceptance. There was no effect for saline concentration or the interaction of age × concentration. Relative consumption of both saline solutions was greater in the two older infant groups than in the youngest group (Fig. 3B). Tests to determine if log (NaCl/water) intake ratios differed significantly from log (1/1) suggested that the infants 2.5 to 4 months of age were indifferent to sodium chloride solutions relative to water, consuming both in equal amounts. On the other hand, both groups of older infants ingested significantly more of the saline solutions than plain water. The lack of correspondence between age-related changes in responses to sucrose and to sodium chloride indicates that the observed changes in salt acceptance do not reflect a generalized developmental change in response to the test situation.

These data are generally consistent with those reported by Vazquez et al. (1982) for Mexican infants less than 1 year of age. Indeed, a subsequent reanalysis of the data of Vazquez et al. revealed that, in their study as well, only infants over 4 months of age ingested substantially more saline solution than water. Thus, in three different populations (healthy American infants, healthy Mexican infants, and malnourished Mexican infants), a developmental shift from indifference to heightened acceptance of sodium chloride solutions, relative to water, has been observed at 4 months of age.

Since most adults do not find salt solutions to be particularly pleasant and do not commonly consume them, a further developmental change in the acceptability of sodium chloride in water must occur. The study by Vazquez et al. (1982) suggests that among the infants they studied, this shift may begin in the second year of life, since 13–24 month olds were indifferent to the salt solutions apparently preferred by 4–12 month olds. Beauchamp's (cited in Beauchamp and Maller, 1977) small study further suggests that a real rejection of salty water may be evident by 2–3 years of age. We investigated this question in a second experiment, which included groups of 7–23 month olds ($n = 16$) and 31–60 month olds ($n = 18$) recruited from private day-care centers (Beauchamp et al., 1986).

In this experiment, acceptance of salt in water was again assessed using an intake paradigm. Cups of water, 0.17 M NaCl, and 0.34 M NaCl were presented to each child in counterbalanced orders. The results are depicted as log ratios of tastant intake to water intake in Fig. 3C. Analysis of variance of these log ratios yielded a significant effect for age only. Infants and children between 7 and 23 months of age exhibited a heightened acceptance of salt solutions similar to that observed in the 4–7 month olds in our first experiment. In contrast, children between 31 and 60 months of age tended to reject the salt solutions relative to water. Tests to determine if the log intake ratios differed significantly from log (1/1) suggested that the younger group consumed greater quantities of 0.17 M NaCl, but not of 0.34 M NaCl, than of water. In the older group, 0.34 M NaCl was rejected relative to water, and rejection of 0.17 M NaCl approached significance.

These data support the contention that a second developmental shift in salt-water acceptance occurs such that older children tend to reject aqueous salt solutions which infants over 4 months of age find highly acceptable. There are, however, some inconsistencies between our results and those of the two previous studies which speak to this issue. Whereas Vazquez et al. (1982) reported an indifference to salt solutions in 13–24 month olds, we observed a continued, heightened acceptance of salt solutions throughout the second year of life. In addition, Beauchamp (cited in Beauchamp and Maller, 1977) reported significant rejection, relative to water, of quite low salt concentrations in his sample of 2–3 year olds, but we were only able to demonstrate significant rejection of a fairly concentrated saline solution in somewhat older subjects. Thus, although the shift from acceptance to indifference or rejection does seem to occur during the first few years of life, the exact age at which it is observed may vary.

B. Salt in Foods

Given the heightened acceptance of salt water by older infants, one might expect that a similar, heightened acceptance of salted foods occurs earlier than has previously been documented. We tested this hypothesis in the same groups of children who participated in our second salt-water experiment. A low-sodium chicken broth served as the food base. The plain broth and broths to which concentrations of 0.17 and 0.34 M NaCl had been added were presented in training cups for 30 sec each. Orders of presentation were counterbalanced, and children were simply allowed to consume as much or as little as they wanted from each cup.

Figure 4 depicts the results of this experiment as log ratios of the intake of salted over unsalted broth. In accordance with the above prediction, 7- to 23-month-old children ingested more of the salted than of the unsalted broth. Log intake ratios were compared to log (1/1) at each level of salt concentration, and the results suggest that both 0.17 and 0.34 M NaCl in soup were consumed in significantly greater amounts than the soup alone (one-sample sign rank tests, $p < 0.05$ at each concentration level).

Somewhat disconcertingly, the older group of children (30–60 months of age) did not exhibit a similar heightened intake of the salted relative to the unsalted soup, although they did not reject the salted soups as they had salted water. As noted earlier, Beauchamp and Moran (1985) also observed an apparent indifference to salt in soup in their study of 2 year olds, even though those same children demonstrated an ingestive preference for salted

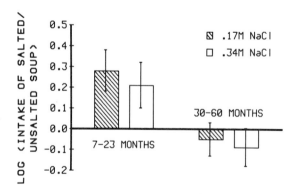

Fig. 4. Log geometric mean ratios (± SEM) of salted to unsalted soup consumption by infants and children 7–23 months of age ($n = 16$) and 31–60 months of age ($n = 18$). Values greater than 0 indicate consumption of more salted than unsalted soup; values less than 0 indicate the reverse.

carrots over unsalted ones. Our subjects, like Beauchamp and Moran's, tended to ingest relatively little of all of the soup samples they received. It may be that it is difficult to find a commercial, unsalted soup base that is sufficiently familiar and pleasant to toddlers and preschoolers to elicit substantial intake. Or, perhaps nonsnack foods in general are not ingested in substantial quantities by older children when presented outside of a meal. This may be a particular problem in settings such as day-care centers, where children routinely receive morning and afternoon snacks in addition to meals and, thus, are unlikely to be hungry when tested. Preliminary data obtained from preschoolers in a clinic setting, most of whom had not eaten for at least 2 hr, indicate that preferential intake of salted (0.17 and 0.34 M NaCl) versus unsalted soup may indeed be observed in this age group under some circumstances. No matter how much they actually consume, however, preschoolers often do volunteer opinions as to which of several samples they think tastes best. Consequently, we have recently begun to systematically examine the preferences of preschool children in paired-comparison tests.

Thus far, subjects tested in this manner have been recruited from the Well Baby Clinic of Pennsylvania Hospital and ranged in age from 36 to 64 months ($n = 15$). Preferences for 0.17 and 0.34 M NaCl in both water and a low-sodium vegetable broth were assessed. Children were presented cups containing small amounts of the unsalted and each of the salted versions of each solvent. All three of the possible pairs in each solvent series were presented. Children were asked to taste the contents of each cup in a pair and to indicate which they liked better. Half of the children received the lower salt sample in each pair first, and half received the higher salt sample first. The children then received each pair again in the reverse order. Half tasted the salt and water samples first, and half tasted the salt and soup first.

The sample from each series which a child chose with the greatest frequency was designated as his or her "most preferred." In all cases, children chose one sample in each series at least 3 of the 4 times it was presented, while choosing no other sample more than twice.

Figure 5 depicts the percentage of subjects who chose each of the three samples in the water and soup series as their most preferred. The results from the water series corroborate intake data from other children in this age range. That is, the vast majority preferred water to both of the aqueous salt solutions, and indeed, none preferred the higher (0.34 M) concentration of sodium chloride in water. In contrast, a majority of these same children most preferred one of the salted forms of the vegetable soup, and two-thirds chose the 0.34 M NaCl in soup as their most preferred. It is perhaps noteworthy that a 0.34 M NaCl concentration is approximately twice that typically found in canned soups, and adults in our laboratory find this soup to have a very salty taste. It therefore seems likely that the preschoolers were responding

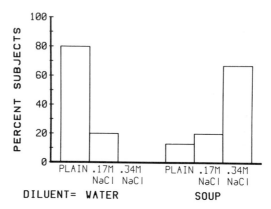

Fig. 5. Percentage of 36- to 64-month-old children ($N = 15$) choosing plain (unsalted), 0.17 M NaCl, and 0.34 M NaCl water and soup as their most preferred.

favorably to saltiness per se as long as it occurred in a reasonably familiar context.

C. Discussion

There is increasing evidence that two distinct changes in the acceptance of the salty taste of sodium chloride occur during early human development. The first is a shift from indifference to relative preference, which seems to appear around 4 months of age and is independent of the interaction of the taste of salt with other tastes in foods. The second is a shift from acceptance to rejection, which may be variable in its time of appearance and depends on the context in which saltiness is experienced.

The initial shift from indifference to heightened acceptance of salt during infancy is, as was argued earlier, consistent with a hypothesis of postnatal maturation of the central and/or peripheral mechanisms underlying salt taste perception. According to the strictest form of such a hypothesis, maturation allows for the emergence of a largely unlearned preference for saltiness. There are, of course, other possible explanations for the change in salt acceptance. For example, changes in the sodium concentration of the saliva bathing the receptors, or in the need for salt, might alter ingestive responses to moderate concentrations of sodium chloride. Although both of these explanations seem unlikely, they cannot be conclusively eliminated on the basis of available data (for further discussion see Beauchamp et al., 1986).

Perhaps the most obvious alternative hypothesis, however, is that this first developmental change in salt acceptance is a function of dietary exposure to sodium chloride. Since we have no accurate figures for exposure to salt among the infants and children who have been studied, the possible influence of differential dietary consumption on salt water acceptance cannot be assessed. However, it has been our general observation that, in our study population, infants up to 6 months of age are primarily fed a diet of breast milk, commercial formula, and/or packaged baby food, all of which are relatively low in sodium. Moreover, as has been noted, a shift toward salt acceptance at 4 months of age has now been observed in several groups of infants who might reasonably be assumed to have had quite different early dietary histories. Thus, experience may not play a major role in the initial emergence of salt acceptance during infancy.

Experience is, however, the probable explanation for the later shift from apparently generalized acceptance of salt to rejection of some salty tasting substances, specifically salt water. There are several lines of evidence which indicate that the strength of neophobia in humans increases with age during early life (see Birch and Marlin, 1982). Thus, children may become increasingly less willing to accept novel or unusual foods and beverages. Preschool children do demonstrate a preference for salt in foods that are commonly experienced with added salt (e.g., Beauchamp, 1981; Filer, 1978). As suggested by the preliminary data presented here, they apparently even prefer very salty soups to less salty ones. Thus, it seems likely that their rejection of aqueous salt solutions reflects the relative novelty of salty water rather than a general change in their acceptance of salty tastes.

Two additional points also seem to support the contention that experiential factors are critical to this second change in response to salted water. First is the obvious parallel between that change and age-related declines in the acceptance of sugar water. In the latter case, declining acceptance has been shown to be linked specifically to a lack of early exposure to sweetened water (Beauchamp and Moran, 1985). Second is the greater variability in the age at which this change is observed than in the age at which heightened salt-water acceptance is first observed.

Certainly the arguments that have been advanced here cannot be considered definitive; a number of alternative hypotheses could account for the observed developmental shifts in salt acceptance. In addition, it is likely that experiential (see, e.g., Yeung *et al.*, 1984) as well as possible physiological and maturational factors beyond those discussed here play a role in determining the ultimate level of sodium chloride acceptance by an individual at any given time. Nonetheless, we believe the current weight of evidence supports the hypothesis of a largely unlearned preference for salty tastes, whose emergence depends upon the postnatal maturation of central and/or peripheral salt-taste receptor mechanisms. The expression of this unlearned preference is then modified and channeled by dietary experience.

IV. RESEARCH NEEDS

A number of lines of research are suggested by the foregoing presentation. Ideally, studies of gustatory sensitivity in infancy and early childhood, as well as of preference, would be desirable. In preference tests, use of a wider range of concentrations than are typically employed in studies with children might enable generation of preference/aversion curves that could be compared at various ages. In addition, studies that directly compare the responses of infants and children with very different histories of salt consumption are needed. Finally, and most generally, greater attention should be paid to the methodologies employed in developmental studies so that we may begin to understand the relationships among various measures and to determine the most appropriate ways of asking the questions we want to answer at different stages of development.

ACKNOWLEDGMENTS

The research reported here was supported by NIH Grant RO1 HL31736 (G.B.) and NIH Training Grant 5 T32 NS07176-05 (B.C.). We thank Dr. A. Freedman and his associates and clients at the Well Baby Clinic of Pennsylvania Hospital and the teachers and children of Penn Children's Center and the ISI Caring Center for Parents and Children for their cooperation. We also thank T. Wiser, S. Hou, E. Chapman, C. Sternberg, and S. Detweiler for assistance in data collection, and Dr. B. Green for his constructive comments on the manuscript.

REFERENCES

Bartoshuk, L. M. (1974). NaCl thresholds in man: Thresholds for water taste or NaCl taste? *J. Comp Physiol. Psychol.* **87,** 310–325.

Beauchamp, G. K. (1981). Ontogenesis of taste preference. In "Food, Nutrition and Evolution: Food as an Environmental Factor in the Genesis of Human Variability" (D. N. Walcher and N. Kretchmer, eds.), pp. 49–57. Masson, New York.

Beauchamp, G. K., and Maller, O. (1977). The development of flavor preferences in humans: A review. In "The Chemical Senses and Nutrition" (M. R. Kare and O. Maller, eds.), pp. 292–315. Academic Press, New York.

Beauchamp, G. K., and Moran, M. (1982). Dietary experience and sweet taste preferences in human infants. *Appetite* **3,** 139–152.

Beauchamp, G. K., and Moran, M. (1985). Acceptance of sweet and salty tastes in 2-year-old children. *Appetite* **5,** 291–305.

Beauchamp, G. K., Cowart, B. J., and Moran, M. (1986). Developmental changes in salt acceptability in human infants. *Dev. Psychobiol.* **19,** 75–83.

Bertino, M., Beauchamp, G. K., and Engelman, K. (1982). Long-term reduction in dietary sodium alters the taste of salt. *Am. J. Clin. Nutr.* **36,** 1134–1144.

Birch, L. L., and Marlin, D. W. (1982). I don't like it; I never tried it: Effects of exposure on two-year-old children's food preferences. *Appetite* **3,** 353–360.

Cowart, B. J. (1981). Development of taste perception in humans: Sensitivity and preference throughout the life span. *Psych. Bull.* **90**, 43–73.
Crook, C. K. (1977). Taste and the temporal organization of neonatal sucking. In "Taste and Development: The Genesis of Sweet Preference" (J. M. Weiffenbach, ed.), pp. 146–160. U.S. Government Printing Office. Washington, D.C.
Crook, C. K. (1978). Taste perception in the newborn infant. *Infant Behav. Dev.* **1**, 52–69.
Davis, C. M. (1928). Self selection of diet by newly weaned infants. *Am. J. Dis. Child.* **36**, 651–679.
Denton, D. A. (1982). "The Hunger for Salt." Springer-Verlag, Berlin.
Desor, J. A., Maller, O., and Turner, R. E. (1973). Taste in acceptance of sugars by human infants. *J. Comp. Physiol. Psychol.* **84**, 496–501.
Desor, J. A., Maller, O., and Andrews, K. (1975). Ingestive responses of human newborns to salty, sour and bitter stimuli. *J. Comp. Physiol. Psychol.* **89**, 966–970.
Desor, J. A., Maller, O., and Greene, L. S. (1977). Preference for sweet in humans: Infants, children and adults. In "Taste and Development: The Genesis of Sweet Preference" (J. M. Weiffenbach, ed.), pp. 161–172. U.S. Government Printing Office, Washington, D.C.
Ferrel, M. F., Mistretta, C. M., and Bradley, R. M. (1981). Development of chorda tympani taste responses in the rat. *J. Comp. Neurol.* **198**, 37–44.
Filer, L. J., Jr. (1978). Studies of taste preference in infancy and childhood. *Pediatr. Basics* **12**, 5–9.
Fomon, S. J., Thomas, L. N., and Filer, L. J. (1970). Acceptance of unsalted strained foods by normal infants. *J. Pediatr.* **76**, 242–246.
Fregly, M. S., and Fregly, M. J. (1982). The estimates of sodium intake by man. In "The Role of Salt in Cardiovascular Hypertension" (M. J. Fregly and M. R. Kare, eds.), pp. 3–17. Academic Press, New York.
Hill, D. L., and Almli, C. R. (1980). Ontogeny of chorda tympani responses to gustatory stimuli in the rat. *Brain Res.* **197**, 27–38.
Hill, D. L., Mistretta, C. M., and Bradley, R. M. (1982). Developmental changes in taste response characteristics of rat single chorda tympani fibers. *J. Neurosci.* **2**, 782–790.
Jensen, K. (1932). Differential reactions to taste and temperature stimuli in newborn infants. *Genet. Psychol. Monogr.* **12**, 363–479.
Maller, O., and Desor, J. A. (1973). Effects of taste on ingestion by human newborns. In "Fourth Symposium on Oral Sensation and Perception: Development of the Fetus and Infant" (J. F. Bosma, ed.), pp. 279–291. U.S. Government Printing Office, Washington, D.C.
Meneely, G. R., and Batterbee, H. D. (1976). Sodium and potassium. *Nutr. Rev.* **34**, 225–235.
Midkiff, E. E., and Bernstein, I. L. (1983). The influence of age and experience on salt preference of the rat. *Dev. Psychobiol.* **16**, 385–394.
Mistretta, C. M., and Bradley, R. M. (1983). Developmental changes in taste responses from glossopharyngeal nerve in sheep and comparisons with chorda tympani responses. *Dev. Brain Res.* **11**, 107–117.
Moe, K. E. (1986). The ontogeny of salt preference in rats. *Dev. Psychobiol.* (In press.)
Nowlis, G. H. (1973). Taste elicited tongue movements in human newborn infants: An approach to palatability. In "Fourth Symposium on Oral Sensation and Perception: Development of the Fetus and Infant" (J. F. Bosma, ed.), pp. 292–310. U.S. Government Printing Office, Washington, D.C.
Pangborn, R. M. (1970). Individual variation in affective responses to taste stimuli. *Psychon. Sci.* **21**, 125–126.
Steiner, J. E. (1973). The gustofacial response: Observation of normal and anencephalic new-

born infants. *In* "Fourth Symposium on Oral Sensation and Perception: Development of the Fetus and Infant" (J. F. Bosma, ed.), pp. 254–278. U.S. Government Printing Office, Washington, D.C.

Steiner, J. E. (1977). Facial expressions of the neonate infant indicating the hedonics of food-related chemical stimuli. *In* "Taste and Development: The Genesis of the Sweet Preference" (J. M. Weiffenbach, ed.), pp. 173–189. U.S. Government Printing Office, Washington, D.C.

Steiner, J. E. (1979). Human facial expressions in response to taste and smell stimulation. *Adv. Child Dev. Behav.* **13**, 257–295.

Vazquez, M., Pearson, P. B., and Beauchamp, G. K. (1982). Flavor preferences in malnourished Mexican infants. *Physiol. Behav.* **28**, 513–519.

Weiffenbach, J. M. (ed.) (1977). "Taste and Development: The Genesis of Sweet Preference." U.S. Government Printing Office, Washington, D.C.

Winer, B. J. (1971). "Statistical Principles in Experimental Design," 2nd ed. McGraw-Hill, New York.

Yamaguchi, S., and Kimizuka, A. (1979). Psychometric studies on the taste of monosodium glutamate. *In* "Glutamic Acid: Advances in Biochemistry and Physiology" (L. J. Filer, Jr., S. Garattini, M. R. Kare, W. A. Reynolds, and R. J. Wurtman, eds.), Raven Press, New York.

Yeung, D. L., Leung, M., and Pennell, M. D. (1984). Relationship between sodium intake in infancy and at 4 years of age. *Nutr. Res.* **4**, 553–560.

3

Effects of Dietary Protein on the Taste Preference for Amino Acids in Rats

KUNIO TORII, TORU MIMURA, AND YASUMI YUGARI

Life Science Laboratories
Central Research Laboratories
Ajinomoto Company, Inc.
Yokohama, Japan

I.	Introduction	45
II.	Taste Preference and Protein Intake in Rats during Growth	47
III.	Changes of Taste Preference in Rats Fed a Diet with or without L-Lysine Deficiency	52
IV.	Relation among Protein Intake, Taste Preference, and Genetic Predispositions	61
V.	Research Needs	63
VI.	Conclusion	65
	References	65

I. INTRODUCTION

The gustatory and anticipatory cephalic stimuli that are detected during a meal yield nutritional information and aid in the efficient digestion of food (Epstein and Teitelbaum, 1962; Grossman, 1967; Sarles et al., 1968; Novis et al., 1971; Steffens, 1976; Richardson et al., 1977; Naim et al., 1978; Brand et al., 1982). When animals detect that a food has a bitter and/or sour taste, they exercise caution in ingesting it since it may be toxic. They may stop ingesting and even vomit the food (Richter, 1953; Pfaffmann, 1970; Rozin and Kalat, 1971; Beauchamp and Maller, 1977). But foods having a familiar or pleasant taste are swallowed without caution. In general, foods having a sweet or sufficiently salty taste are palatable, but those that are sour and bitter are usually aversive (Rozin and Kalat, 1971). In addition, protein and polyribonucleotides, which are found in such foods as animal and fish meats, yield free forms of L-amino acids and 5′-ribonucleotides following autolysis; some of these L-amino acids and 5′-ribonucleotides are generally attractive not only to lower microorganisms but also to higher primates, including humans (Block and Bolling, 1951; Maeda et al., 1958; Kuninaka et al., 1964; Shimazono, 1964; Kirimura et al., 1969; Fernandez-Flores et al., 1970; Stadtman, 1972). Cephalic stimulation by both L-amino acids and nucleotides may be important for enhancing intake and allowing more efficient digestion and metabolism of food (Rogers and Leung, 1977).

However, it is not clear that the taste of every individual L-amino acid fits into any of the usual typical categories of taste such as sweet, salty, sour, or bitter (Yoshida et al., 1966; Yoshida and Saito, 1969; Schiffman and Engelhard, 1976). All essential L-amino acids except L-threonine (Thr) are generally bitter, whereas nonessential ones and Thr are sweet. The acidic L-amino acids, L-glutamic and L-aspartic acid (Glu and Asp), are sour, yet the sodium salts of both define a unique taste, which is called "umami" in Japanese. The term "umami" is usually translated as "delicious" or "savory" (Ikeda, 1912; Komata, 1976; Cagan, 1977; Yamaguchi and Kimizuka, 1979). In Japan, sea tangle and dried black mushrooms are used as condiments to season soup stocks (Kuninaka et al., 1964; Kuninaka, 1967). The former has large amounts of the ammonium salt of Glu, and the latter contains guanosine 5′-monophosphate (GMP). It is well known that stronger "umami" taste can be achieved when both of these components are added together, rather than alone (Yamaguchi et al., 1968; Yamaguchi and Kimizuka, 1979). There is, in effect, a taste synergism between the two (Cagan, 1977). The crystalline monosodium salt of Glu (MSG) with or without 5′-ribonucleotides, or both alone have been used worldwide as flavor enhancers. The synergistic enhancement of "umami" taste between MSG and GMP was defined quantitatively in sensory tests (Yamaguchi and Kimizuka, 1979; Rifkin and Bartoshuk, 1980), in nerve recordings from the chorda tympani of the rat (Sato and Akaike, 1965; Sato et al., 1970; Yamashita et al., 1973). and in studies of Glu binding to taste receptors of the bovine tongue (Cagan et al., 1979; Torii and Cagan, 1980). Reactivity to 5′-nucleotides was also reported in lower vertebrates, such as puffer fish (Hidaka and Kiyohara, 1975) and turbot (Mackie and Adron, 1978).

Protein is a major essential nutrient for the body, and it can also be metabolized to glucose and lipids. The taste produced by the L-amino acids and 5′-ribonucleotides released from foods by chewing may be a detection signal for foods that contain protein. The synergism between the two stimuli and the unique appealing taste property of both may be important for en-

3. Dietary Protein Effects on Taste Preference 47

hancing the palatability of these foods. Similar connections have been made between other tastes and food quality. For example, the sensation of saltiness may be related to foods containing salts, probably sodium chloride (NaCl), and sweetness to foods rich in energy sources, such as glucogenic L-amino acids or sugars. These gustatory stimuli from foods are probably important since they signal food quality and initiate digestion and metabolism (Naim *et al.*, 1978; Rogers and Leung, 1977; O'Hara *et al.*, 1979; Brand *et al.*, 1982). This chapter will discuss the manipulation of the taste preference for L-amino acids and NaCl in rats under various degrees of protein or essential L-amino acid restriction.

II. TASTE PREFERENCE AND PROTEIN INTAKE IN RATS DURING GROWTH

In the first series of experiments, taste preference for L-amino acids was determined in rats. Male Sprague–Dawley rats (strain S-D) were used throughout. Five individuals were housed together in large transparent polyacrylic-resin cages (30 × 30 × 80 cm) to which 15 drinking bottles were attached on a single face. These bottles could be filled with L-amino acid solutions in concentrations varying up to 500 mM. Animals were supplied with a 5% purified whole egg protein (PEP) diet *ad libitum*. For 1 week prior to the test, all 15 bottles were filled with water. After this, 10 of the 15 bottles were filled with varying concentrations (up to 500 mM or its saturation limit) of a single amino acid and the remaining 5 bottles with water. The position of these 15 bottles was randomized and never changed during the test period. After 12 days of testing, animals had differentially selected from among the concentration series. Table I lists the mean concentration that was ingested of each amino acid tested over the 12-day period plus the concentration that was most preferred. These data indicate that L-amino acids that are sweet and/or display an "umami" taste to humans were preferred by rats, but those generally regarded as bitter or sour by humans were not preferred by these rats. Overall, the observed preference ranking based on the intake data for rats tested against 20 amino acids was quite comparable to the preference measure of each in humans.

Second, taste preference in rats under various levels of protein restriction was determined. Five taste solutions were used: (1) 500 mM glycine (Gly) (sweet to humans), (2) 150 mM MSG (a mixed "umami" taste and slightly salty taste to humans), (3) 4.5 mM MSG with 4.5 mM GMP (eliciting an "umami" taste only to humans), (4) 150 mM NaCl, and (5) deionized distilled water. These were offered in a five-choice arrangement to rats (4 weeks of age, $N = 13$ in each group). Animals were housed in pairs and supplied a diet

TABLE I Mean Concentration Ingested and Most Preferred Concentration of Each L-Amino Acid Solution for Growing Rats Fed 5% Purified Whole Egg Protein Diet[a]

L-Amino acid offered	Concentration of solution tested		Essential L-amino acid for growing rats
	Mean ingested (mM)	Most preferable (mM)	
1. Histidine	15	200	Yes
2. Isoleucine	39	75	Yes
3. Leucine	20	75	Yes
4. Lysine–HCl	47	400	Yes
5. Methionine	4	150	Yes
6. Phenylalanine	2	45	Yes
7. Threonine	50	400	Yes
8. Tryptophan	6	25	Yes
9. Valine	11	75	Yes
10. Alanine	153	500	No
11. Arginine	19	50	No
12. Asparagine	153	200	No
13. Aspartate Na	53	200	No
14. Aspartic acid	6	27	No
15. Cysteine–HCl	72	100	No
16. Glutamate Na	74	150	No
17. Glutamic acid	1	12	No
18. Glutamine	212	200	No
19. Glycine	148	500	No
20. Proline	78	400	No
21. Serine	94	200	No
22. Tyrosine	1	1	No

[a] Male S-D rats, 4 weeks of age ($N = 5$), were supplied with 5% purified egg protein (PEP) diet and offered a choice of solutions in 15 drinking bottles. Ten of these were filled with varying concentrations of a specific amino acid. The concentrations varied stepwise up to 500 mM or to their solubility limit. Five bottles were filled with water. Tests were carried out for 12 days. Each value in the table was calculated from the data of days 2 through 12. Fourteen of these L-amino acids were later used at their most preferable concentration in taste preference tests. These 14 are numbered in this table as 1, 2, 3, 4, 5, 6, 7, 8, 9, 11, 16, 17, 18, and 19.

containing either 0% (nonprotein), 5, 10, 15, or 20% PEP for 34 days followed by protein restriction (0% protein) for 1 week.

The percentage intake of each taste solution is shown in Fig. 1. Water intake in all groups was negligibly small. A preference for NaCl was observed for a week after initial exposure to the solutions, regardless of the level of PEP in the diet. The preference for NaCl in the group of rats fed the

3. Dietary Protein Effects on Taste Preference 49

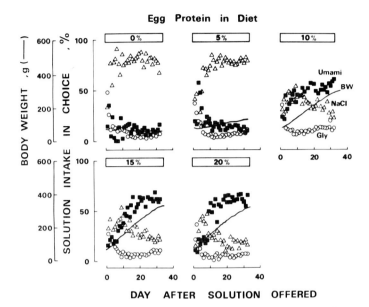

Fig. 1. Pattern of preference for NaCl, Gly, and MSG-containing ("umami") taste stimuli in weanling rats fed different levels of protein. Male S-D rats (4 weeks of age, $N = 13$/group except for the nonprotein diet group where $N = 6$) were supplied with a diet containing purified egg protein (PEP) at 0, 5, 10, 15, or 20% for 34 days. This feeding period was preceded by a 7-day period during which all animals were fed a 0% protein diet. Five taste solutions were offered in a multichoice arrangement: (1) 500 mM Gly, (2) 150 mM MSG, (3) 4.5 mM MSG with 4.5 mM GMP, (4) 150 mM NaCl, and (5) water. The figure displays the percentage intake based on the total volume intake of Gly (○), NaCl (△), and both solutions containing MSG (i.e., 2 + 3) (■). Water intake was negligibly small and is not noted on the figure. Mean body weight appears as a solid curve.

0 and 5% PEP diet was sustained and their growth was severely suppressed during the experiment. On the other hand, NaCl intake declined concomitantly with an increase in intake of both MSG-containing ("umami") solutions (i.e., solution Nos. 2 and 3) when 10, 15, and 20% PEP diets were offered to rats. These rats also grew normally. Moderate Gly intake was observed only at the onset of exposure to these stimuli regardless of experimental diets offered.

The total sodium intake from all drinking solutions in each group of rats is shown in Fig. 2. Animals eating the 0 and 5% PEP diet displayed maximal intake. Sodium intake decreased along with an increase in PEP in the diet. When the 10% PEP diet was offered to the rats, growth was slightly suppressed compared to that of animals fed the 15 or 20% PEP (Fig. 1); those animals fed the 10% PEP largely preferred solution No. 2 (150 mM MSG)

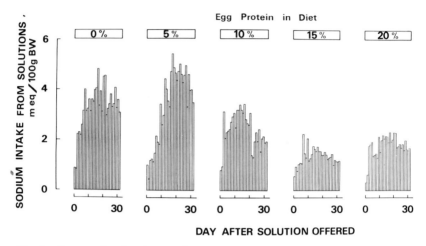

Fig. 2. Pattern of total sodium intake in weanling rats fed varying levels of purified egg protein (PEP). Sodium intake was calculated from all sodium ingested (i.e., solutions 2, 3, and 4) via solution as displayed in Fig. 1.

rather than solution No. 3 (4.5 mM MSG + 4.5 mM GMP). On the other hand, animals fed 15 and 20% PEP showed a marked preference for solution No. 3. These changes of taste preference lead one to postulate that development of a preference for taste solutions containing MSG may be related to adequate protein intake, but that rats under protein restriction prefer NaCl and/or Gly.

Another group of rats ($N = 8$) was supplied with the PEP diet wherein the PEP was increased stepwise, from 0% for 11 days to 5% for 6 days, to 10% for 10 days to 15% for 9 days, and to 20% for 18 days. During the entire experimental period, the animals were given a choice of the above five taste-test solutions. Prior to the testing, animals were maintained for a week on a nonprotein (0%) diet. The results (Fig. 3) indicated that the intake of NaCl was decreased whereas the intake of both MSG-containing solutions ("umami") was clearly elevated stepwise with protein concentration. Gly ingestion was notable only under the condition where a nonprotein diet was offered (Fig. 3). The phenomenon observed in Fig. 1 was, therefore, replicated. In addition, this study indicated that the alteration of solution preference lagged about 2 days behind the increases in protein concentration, but the alteration of preference was clearly dependent upon the PEP levels. This result suggested that the animals' change in taste preference was a reaction to a physiologic state brought about, over the 2-day period, by the increase in dietary protein.

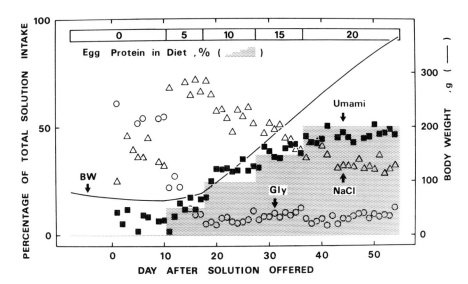

Fig. 3. Pattern of preference for NaCl, Gly, and MSG-containing ("umami") taste solutions in growing rats offered successively greater levels of purified egg protein (PEP). Male S-D rats (N = 8, 4 weeks of age) were supplied with diets containing varying levels of protein, the amount of which at any time can be seen by referring to the display at the top of the figure and by referring to the height of the shaded zone. Symbols for each solution and body weight are presented as in Fig. 1.

In general, the protein requirement for maximal growth in rats decreases with age (Hartsook and Mitchell, 1956; Forbes et al., 1958). If taste preference for MSG-containing solutions is coupled with the necessary levels of dietary protein required for normal growth, then the minimal PEP diet levels for the induction of this preference in adult rats should be lower than that for weanlings. Alterations of taste preference in rats 4, 8, and 12 weeks of age were observed when the diet contained 0, 2.5, 5, 10, or 15% PEP with taste stimuli offered concurrently in a choice situation for 4 weeks. The PEP requirement for maximal growth in these rats was estimated at 12.5% for animals that began the test at 4 weeks of age, 6.1% for those at 8 weeks of age, and 5.9% for those at 12 weeks of age. The PEP levels in the diet required for the induction of preference for MSG-containing solutions in rats, 4, 8, and 12 weeks of age, were 8.8, 3.8, and 2.5%, respectively (Fig. 4).

These observations suggest that taste preference may closely reflect the amount of dietary protein, and that intake of protein beyond its required level for normal growth could be coupled with a preference for MSG-containing taste solutions. Rats under severe protein deficiency preferred NaCl

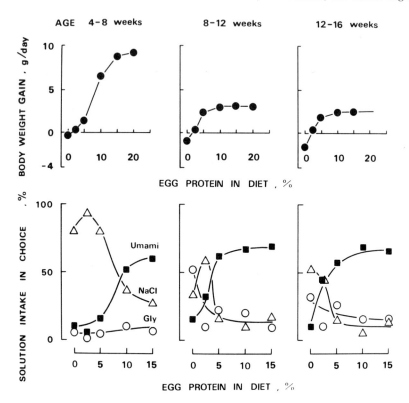

Fig. 4. Body weight gain and taste preference of rats of various ages each fed diets differing in level of purified egg protein (PEP). The same experimental procedure was used as for data in Fig. 1 except that egg protein content in the diet was 0, 2.5, 5, 10, and 15% for each subset ($N = 6$) of a group of rats of ages 4, 8, and 12 weeks. The top portion of the figure displays the average value for body weight gain (●). The bottom portion displays the percentage intake of each taste solution (the same symbols are used as in Fig. 1). The experimental time period was 4 weeks.

and sometimes Gly to the "umami" taste. NaCl may be selected to maintain electrolyte and body fluid balance. Preference for Gly may be indicative of its presumed ability to spare nitrogen and thereby minimize the negative nitrogen balance that the nonprotein diet induces.

III. CHANGES OF TASTE PREFERENCE IN RATS FED A DIET WITH OR WITHOUT L-LYSINE DEFICIENCY

Rats displayed a strong preference for MSG and MSG plus ribonucleotide solutions when diets contained PEP at levels of 10% or more. It is well known that PEP is an ideal protein source for growing rats. They deposit almost all of the dietary nitrogen as body protein (Block and Mitchell, 1946) when being fed a purified whole egg protein such as PEP. Changes of taste preference in rats fed diets deficient in an essential L-amino acid were next determined, and these data were compared with preferences obtained from animals fed various levels of PEP (Fig. 1).

The L-amino acid composition and nitrogen content of PEP, purified wheat gluten, and corn zein were assayed by complete hydrolysis (Table II). These materials were used as the protein sources with or without added essential L-amino acids (Mitchell and Beadles, 1950; Calhoun *et al.*, 1960; Rama-Rao *et al.*, 1959, 1961). The experimental diets contained 24.3% wheat gluten or 19.7% corn zein, each isonitrogenous to 20% PEP. Amino acids were added to each diet to bring the essential amino acid level of the wheat gluten and corn zein diets in line with those of the PEP diet. Only the amount of lysine was varied. In some cases, diets were prepared as lysine deficient ("Gluten − Lys," "Zein − Lys"); in other cases, they were lysine

TABLE II L-Amino Acid Composition of Purified Whole Egg Protein (PEP), Wheat Gluten, and Corn Zein[a]

	Concentration (% of 16 g nitrogen)		
L-Amino acid	PEP	Wheat gluten	Corn zein
Histidine	2.41	1.90	1.23
Isoleucine	5.72	3.23	3.91
Leucine	9.71	6.71	21.58
Lysine	7.39	1.48	0.05
Methionine	3.30	0.56	1.62
Phenylalanine	5.66	4.87	7.47
Threonine	4.83	2.46	2.81
Tryptophan	1.44	0.77	0.21
Valine	6.61	3.53	3.69
Alanine	6.05	2.50	10.98
Arginine	7.04	3.27	1.31
Aspartic acid	10.08	3.13	5.70
Cysteine	2.30	2.82	0.34
Glutamic acid	12.84	33.58	23.81
Glycine	3.53	3.31	1.18
Proline	4.28	13.95	1.15
Serine	7.76	4.87	5.48
Tyrosine	4.05	3.09	5.17

[a] Each protein sample was hydrolyzed and assayed for L-amino acid content using an automatic amino acid analyzer. Nitrogen content of each was determined using an automatic nitrogen analyzer (Micro-Kjeldahl method).

replete ("Gluten + Lys," "Zein + Lys") (Table III). Finally, every experimental diet was made isonitrogenous (final crude protein level in diets, 24%) and isocaloric with one other by altering the levels of L-glutamine (Gln) and corn starch (Table III). All zein diets without fortification were absolutely deficient in Lys, but the level of L-leucine (Leu) in zein (4.06%) was higher than that in 20% PEP diet (1.78%) (Tables II and III).

S-D rats, 4 weeks of age, were housed in pairs and supplied with these experimental diets along with the five taste solutions in a choice situation ($N = 6$ in each group). Preference for MSG-containing solutions was observed only in the cases where PEP and Gluten + Lys diets were used (Fig. 5). When suppression of growth occurred in rats fed Gluten − Lys or Zein − Lys diets, preference for NaCl and Gly solutions was observed. In the case of rats fed the Zein + Lys diet (Fig. 5C), animals still selected NaCl and Gly solutions, even if their growth was apparently normal. They did show some selection for the MSG-containing solutions, but this selection was not

TABLE III Specification of Experimental Diets[a]

		Content (% of diet)			
	PEP diet	Experimental diets			
Ingredient	(control)	Gluten − Lys	Gluten + Lys	Zein − Lys	Zein + Lys
PEP	20.0	—	—	—	—
Wheat gluten	—	24.3	24.3	—	—
Corn zein	—	—	—	19.7	19.7
L-Lys–HCl	—	—	1.4	—	1.7
L-Gln	4.7	1.7	0.7	1.3	—
L-Amino acids	—	3.3	3.3	3.5	3.5
Corn starch	61.1	56.5	56.1	61.3	60.9
Mineral mix	4.0	4.0	4.0	4.0	4.0
Vitamin mix	1.0	1.0	1.0	1.0	1.0
Choline chloride	0.2	0.2	0.2	0.2	0.2
Vitamin E	0.01	0.01	0.01	0.01	0.01
Cellulose	4.0	4.0	4.0	4.0	4.0
Corn oil	5.0	5.0	5.0	5.0	5.0
L-Lys in diet (%, w/w)	1.35	0.27	1.35	0.01	1.35

[a] These diets were prepared isocaloric and isonitrogenous with one another. The gluten and zein diets were fortified with any L-amino acids that were lower than those found in the purified whole egg protein (PEP) source (see Table II). The amount of Lys was controlled to either deficient levels ("−Lys") or fortified levels ("+Lys") for both the gluten and zein diets as indicated. L-Gln was added to the diets to make all of them isonitrogenous. The absolute L-Lys content in each experimental diet was calculated from the composition of protein (Table II) and the fortification of L-Lys–HCl and is noted at the bottom of this table.

robust. This result suggested that preference for MSG-containing solutions may occur only in those conditions without any deficit or excess of an essential L-amino acid.

Nevertheless, when these experimental diets were replaced with complete diets (from Gluten − Lys diet to Gluten + Lys, or from any zein diet to a PEP one), the growth rate of the rats accelerated, and they began to prefer the MSG-containing solutions in a pattern similar to that seen in the case of a 20% PEP diet offered to 4-week-old rats (Fig. 1). These facts suggest that animals can detect the supplementation of a deficient diet and, in this paradigm, alter their taste preference. When protein restriction was then introduced to each group of rats (as adults), both Gly and NaCl became preferred. It was postulated that preference for an "umami" taste might represent a parameter for nutritional status, particularly whether the protein balance is within normal limits, and that preference for NaCl or Gly over "umami" may be an indicator of marginal or severe amino acid restriction.

Additional experiments were performed to determine if the growing animals could detect Lys from among whole essential L-amino acid solutions in a choice situation when Gluten + Lys or Gluten − Lys diets were offered. Five S-D rats, 4 weeks of age, were housed together in a polyacrylic-resin cage (30 × 30 × 80 cm) to which 15 drinking bottles were attached in a row. Each drinking bottle was filled with one of the essential L-amino acid solutions (numbers 1–9 in Table I) in concentrations that were found to be preferable (Table I). The other bottles were filled with the nonessential L-amino acids numbered 11, 16, 17, 18, and 19 in Table I. The fifteenth bottle contained 4.5 mM MSG + 4.5 mM GMP. Of all of the L-amino acids, only Lys among the essential L-amino acids offered was consumed by the rats following the onset of their exposure, and only when the Gluten − Lys diet was offered. Lys intake was rarely recorded in the case of Gluten + Lys diet (Fig. 6). The amount of Lys ingested from both the solution and the diet, in rats fed either Gluten − Lys or (in another experiment) a Zein − Lys diet, was at similar levels to that found when the supplemented diets were used (Fig. 7). The replacement of these experimentally deficient diets by complete diets resulted in loss of Lys solution preference and appearance of a preference for L-amino acids with "umami" character, namely MSG, MSG + GMP, and glutamine, in that order of preference (Fig. 6). These observations are similar to those for essential L-amino acid deficiency or imbalance reported previously (Harper, 1959, 1964; Leung et al., 1968; Rogers and Harper, 1970) and clearly indicate that the rat can select a deficient essential L-amino acid from among 15 choices of amino acids and regulate the intake of each semiquantitatively to levels that slightly exceed but almost equal those required for body needs. Data from this study also support our hypothesis that taste preference for MSG-containing solutions and related "umami"

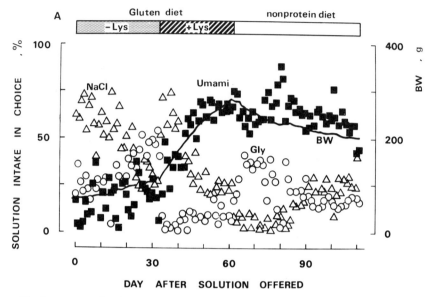

Fig. 5. Pattern of preference for NaCl, Gly, and MSG-containing ("umami") taste solutions in growing rats under Lys deficiency, Lys repletion, nonprotein diet, Leu excess, and purified egg protein (PEP) diet. Male S-D rats, 4 weeks of age ($N = 6$ per condition), were fed the isonitrogenous and isocaloric diets indicated in the figure. These diets are defined in Table III. Concurrently they were offered a choice of 5 taste solutions defined in Fig. 1. The diet supplied at any particular time for each group of animals is noted by referring to the display above the figures. Mean body weight and percentage intake of each solution are presented as they were in Fig. 1. (A) Rats fed the Gluten − Lys diet, followed by the Gluten + Lys diet followed by the nonprotein diet. (B) Rats fed the Gluten + Lys diet followed by the nonprotein diet. (C) Rats fed the Zein + Lys diet (Leu excess) followed by the PEP diet followed by the nonprotein diet.

tastes is related to the nutritional state of the animal. In particular, this preference appears to be dependent upon optimal balance of L-amino acids.

Arginine (Arg) ingestion was also found to be dependent on amino acid nutrition. The ingestion of L-Arg was elevated when rats were offered the Gluten + Lys or PEP diet, or when the animals were able to overcome the lysine-deficient diet by drinking a Lys solution (Fig. 6). But there were no cases during feeding of a deficient diet when Arg intake was observed, suggesting that the mechanism that dictates the preference for Arg has similar characteristics to that which induced preference for solutions with an "umami" taste character.

When protein restriction was introduced to each experimental group (by this time, the animals were adults), intake of Arg declined and a marked elevation in Thr intake (and subsequently in Gly) was observed. Note also

3. Dietary Protein Effects on Taste Preference 57

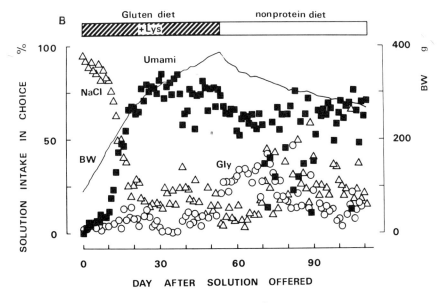

Fig. 5 (*Continued*)

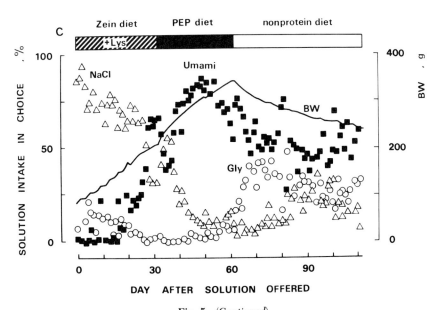

Fig. 5 (*Continued*)

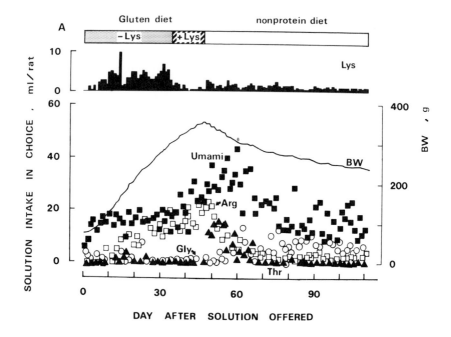

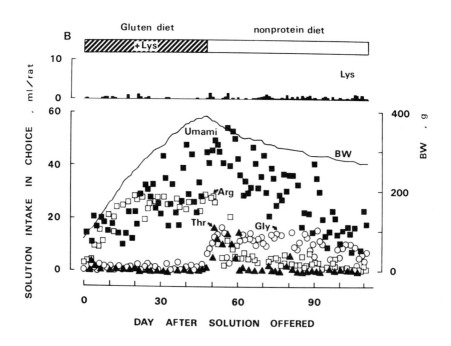

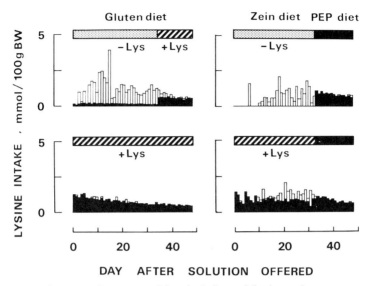

Fig. 7. Total amount of Lys ingested from both diet and drinking solution in growing rats under Lys deficiency or Lys repletion. The gluten diet is shown on the left; the zein diet (which includes a Leu excess) and PEP diet are on the right. Daily Lys intakes from diet (closed column) and solution (open column) were calculated from the data displayed in Fig. 6. The stippled horizontal bar represents a diet that was Lys deficient, the dashed bar represents a diet that was Lys replete, and the solid bar represents a complete PEP diet.

that preference for the "umami" taste in each group was sustained for more than 1 month under the complete protein restriction (Fig. 6). It was reported that Thr is able to spare endogenous nitrogen under dietary protein restriction (Yoshida and Moritoki, 1974). It was very interesting that rats that had been fed the Gluten − Lys diet and had, as a result, ingested Lys solutions, drank the Lys again under complete protein restriction (Fig. 6A). In contrast, those animals that had always been offered complete diets did not display this behavior toward Lys (Fig. 6B). These facts suggested that

Fig. 6. Pattern of preference for L-amino acids in growing rats under Lys deficiency, Lys repletion, or a nonprotein diet. Male S-D rats, 4 weeks of age ($N = 5$ per condition), were supplied with isonitrogenous and isocaloric diets (Table III) along with a multichoice of 14 solutions of L-amino acids (defined in the footnote to Table I) and one of 4.5 mM MSG + 4.5 mM GMP. The amount of the Lys solution ingested and the type of diet offered to a particular group of rats are noted by referring to the displays above the figures. Mean body weight is displayed as a solid curve and intake of each solution is noted: L-amino acids with an "umami" character (■); Arg (□); Thr (▲); and Gly (○). Intakes of other solutions were omitted since they were very small. (A) Rats fed the Gluten − Lys diet followed by the Gluten + Lys diet followed by the nonprotein diet. (B) Rats fed the Gluten + Lys diet followed by the nonprotein diet.

adult rats, which require much lower protein intake for maintenance than weanlings, may sustain a preference for MSG-containing solutions after sparing of endogenous protein by increased levels of plasma Thr. These increased plasma Thr levels could be induced both by the nonprotein diet and by increased preference for a Thr solution. In addition, these data show that presentation of a nonprotein diet to animals experienced with a Lys-deficient diet induces a subsequent Lys intake above that of animals not experienced with a deficient diet. It may be that the previous experience with the association of the taste of Lys and its beneficial consequences carries over to feeding of a nonprotein diet.

Rats of the S-D strain fed a 5% PEP diet for a long period of time consistently selected NaCl solutions. They began to prefer the MSG-containing solutions over NaCl solutions when their body weight reached 250 g (Fig. 8). This weight is reached by an 8-week-old rat under normal growth conditions. This fact suggests that a decrease in protein need, which accompanies growth, may be the most important factor in altering taste preference, even if this growth period is elongated due to a low-protein diet (Fig. 8). Of course, adequate amounts and quality of dietary protein during growth could minimize NaCl ingestion and prevent the excess intake of salt above that required for physiologic need. If these observations can therefore be extended to the human condition, one could postulate that those individuals who are in a good nutritional state and who have no genetic predisposition to

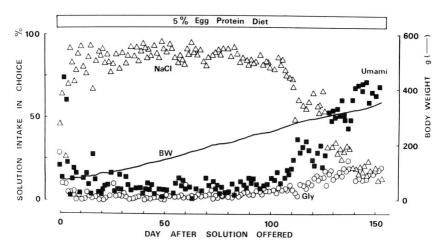

Fig. 8. Pattern of preference for NaCl, Gly, and MSG-containing ("umami") taste solutions in rats fed a low-protein diet (purified egg protein, PEP) long term. Male S-D rats, 4 weeks of age ($N = 6$), were fed a 5% PEP diet for 152 days and offered a choice of five taste solutions (defined in Fig. 1). The experimental procedure and expression of both mean body weight and intake of each taste solution were the same as for Fig. 1.

hypertension prefer foods having an "umami" character to those with salty character (Torii, 1980).

IV. RELATION AMONG PROTEIN INTAKE, TASTE PREFERENCE, AND GENETIC PREDISPOSITIONS

Individuals suffering from diabetes display a strong preference for sugars. This may reflect a glucose hunger in body tissues, muscles, and brain brought on by hypoinsulinemia. It has also been reported that people with essential hypertension show a stronger preference for salt, especially NaCl, compared to normotensives (Schechter et al., 1971, 1974; Henkin, 1974, 1980), reflected perhaps by elevated levels of sodium in saliva (Wotman et al., 1967). In addition, spontaneously hypertensive rats (SHR), which were developed from Wistar strain rats (Wistar/Kyoto) and are a typical animal model for human essential hypertension (Aoki, 1963; Aoki et al., 1972), displayed strong preferences for NaCl (Bartter et al., 1980; Henkin, 1980; Torii, 1980; Torii et al., 1983a,b), despite the fact that serum sodium levels are within normal limits (Torii, 1980). This observation may be related to the poor retention of sodium in the body. The dietary sodium requirement for normal growth in SHR is around 0.25%, much higher than that in normotensives (0.05%) (Torii, 1979, 1980) (Fig. 9). This fact is manifested in a sodium hunger (Fig. 9).

The previous series of studies indicated that rats under protein deficiency or deficit of the essential L-amino acid, Lys, displayed a preference for NaCl. However, they began to ingest MSG-containing taste solutions and other amino acid solutions having an "umami" character, and to decrease their total sodium intake when their nutritional state was corrected by increased protein availability or by the supplementation of the deficient amino acid (Figs. 1, 2, and 5). It was of interest, therefore, to determine if balanced amino acid intake could affect NaCl preference in SHR.

Male SHR, 4 weeks of age, were treated like the animals described in Section III and supplied with a diet containing either 0, 5, 10, 20, 30, or 40% PEP ($N = 6$ in each group), along with a choice of the same five taste solutions defined in Section II (500 mM Gly, 150 mM MSG, 4.5 mM MSG + 4.5 mM GMP, 150 mM NaCl, and water). The diets also contained an adequate amount of sodium for SHR. These amounts of dietary sodium were (w/w) 0.394% in the 0% PEP diet, 0.409% in the 5% PEP diet, 0.424% in the 10% PEP diet, 0.439% in 15% PEP diet, 0.454% in the 20% PEP diet, 0.484% in the 30% PEP diet, 0.514% in the 40% PEP diet, and 0.544% in the 50% PEP diet. The PEP requirement for maximal growth in SHR, 12.4%, is quite comparable to that for normotensive S-D strain rats (12.5%).

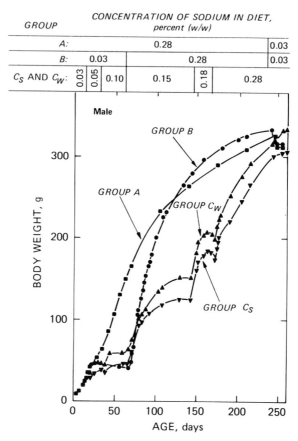

Fig. 9. Pattern of weanling SHR growth under dietary sodium restriction. Sibmated male SHR (F_2 generation) were used. Animals of Group A (■), control ($N = 10$; B (●) ($N = 11$); C_s (▼) ($N = 6$); and C_w (▲) ($N = 6$) were supplied *ad libitum* with deionized water as the only source of drinking water and diets containing varying quantities of sodium. The exact amount of sodium in the diet at any particular time for any particular group of animals is noted by referring to the display above the figure. Animals in Group C_s were nursed by a foster SHR (hypertensive) dam and those of Group C_w by a foster (normotensive) dam. Animals in Groups A and B were nursed by their own SHR dam. At 8 months of age (240 days), Groups A and B were provided with dietary sodium restriction (0.03% Na in the diet) for 10 days, but the magnitude of decreases in body weight in both groups was essentially the same. Body weight of Group C_w was significantly higher than that of Group C_s from weaning to the conclusion of the experiment. (From Torii, 1980.)

Results indicated that changes in taste preference in SHR were quite different from those in normotensive animals. The strong preference for NaCl in SHR was sustained despite the level of PEP in the diet. Preference for the MSG-containing solutions was induced, but this preference never overcame that for NaCl in animals fed diets containing more than 10% PEP (Fig. 10).

3. Dietary Protein Effects on Taste Preference

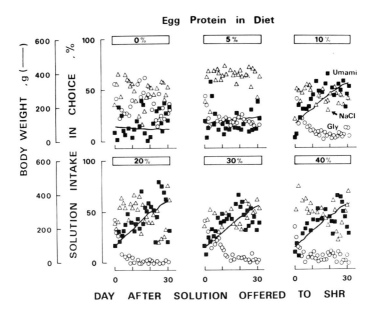

Fig. 10. Pattern of preference for NaCl, Gly, and MSG-containing ("umami") taste solutions in growing SHRs fed a diet containing various amounts of purified egg protein (PEP). Male SHRs, 4 weeks of age, were supplied with diets containing 0, 5, 10, 20, 30, and 40% PEP and a choice of five taste solutions (same as those defined in Fig. 1). The experimental procedures and expression of both mean body weight and intake of each solution were the same as explained for Fig. 1.

These facts suggested that adequate protein intake could elicit the preference for the MSG-containing solutions in SHR, but that there were no alterations in preference for NaCl no matter what level of PEP was offered. This preference for NaCl resulted in the animals consistently ingesting excess levels of sodium well above those required. This appetite for sodium is assumed to be linked to some genetic predisposition (Zinner *et al.*, 1971; Swaye *et al.*, 1972; Thomas, 1973; Torii, 1980). The pattern of taste solution intake in SHR is therefore not similar to that observed in normotensive rats under Lys deficiency or under cases of adequate protein intake. In the SHR case, NaCl preference is maintained even with adequate amino acids and dietary NaCl.

V. RESEARCH NEEDS

It may be that animals, including humans, can easily detect the amount and quality of protein ingested during a meal and use this information, via cephalic relays, to initiate digestion in the alimentary tract. They must ultimately have a means to determine whether the dietary protein intake level is sufficient for body needs. When the animal realizes that protein intake is deficient or imbalanced, the mechanisms of protein spairing as well as those that are needed for searching for the deficient materials come into play. In our studies, it was at this point that Thr and Gly ingestion was elevated (Fig. 6). Data from preference tests indicated that Arg and Gln ingestion was related to a relatively excessive intake of high-quality protein. This preference may be related to activation of urea formation. On the other hand, Thr and Gly ingestion spares endogenous protein (Yoshida and Moritoki, 1974). Animals under a normal nutritional state maintain preferences for solutions with an "umami" taste character over NaCl and Gly. Foods with an "umami" taste character contain high-quality animal protein that can be used for growth and maintenance. Data on the taste perception of MSG and its synergistic enhancement by some 5'-ribonucleotides in mammals (Sato and Akaike, 1965; Yamaguchi et al., 1968, 1971; Sato et al., 1970; Cagan et al., 1979; Yamaguchi and Kimizuka, 1979; Torii and Cagan, 1980) strongly suggest that "umami" stimuli could form a basic taste category. We hypothesize that the "umami" taste may represent a basic nutritional signal (Torii et al., 1983a), similar to saltiness for minerals (Dethier, 1980; Torii, 1980) and sweetness for an energy source.

Rats and humans generally show a taste preference for salt. Yet hypertension is often a complication of excess sodium intake. Some strains of animals can be made hypertensive by feeding high salt diets, others cannot. Still others, like the SHR, will become hypertensive even under salt restriction (Torii, 1980). Perhaps there is a human correlate to this phenomenon wherein some humans respond to a high salt diet by becoming hypertensive while others do not (Dethier, 1980; Torii, 1980). The present work would seem to suggest that hunger for salt might be decreased under adequate protein nutrition.

We also hypothesize the following: Since breast milk is an ideal food—isotonic and germ free, and containing high-quality protein and free forms of L-amino acids (notably a level of Glu sufficient to give it an "umami" taste) it is possible that every mammalian neonate may be "imprinted" during the nursing period, such that a normal nutritional state may become associated with a preference for the "umami" flavor. A recent report, for example, indicated that protein–calorie malnourished children in Mexico displayed greater preference for vegetable soup fortified with MSG following recovery after medical care (Vazquez et al., 1982).

Finally, the total supply of protein sources for humans is diminishing. This is epecially true of protein from animal sources. Almost all vegetable proteins are tasteless, and their L-amino acid composition is far from adequate. The adequate supplementation of both limiting L-amino acids and those that impart an "umami" taste to vegetable proteins could improve the nutritional state of the consumer (be it animal or human) as well as the palatability of the vegetable protein sources. In addition, under circumstances of adequate protein nutrition, our results suggest that the "umami" flavor will be preferred over saltiness. Consequently, protein resources now being wasted can be recovered.

VI. CONCLUSION

These series of studies led to the following hypotheses:

1. Taste preference is dependent upon the state of protein nutrition, which is defined by the level and quality of dietary protein. This preference parallels body need for protein, which declines with age.
2. Under adequate balanced protein nutrition, preference for the "umami" taste (MSG and MSG + GMP) is observed. Under marginal protein deficiency or when the dietary amounts of essential L-amino acids are imbalanced, preference for NaCl is induced. Under severe protein deficiencies, preference for both NaCl and Gly is induced.
3. Rats can select and regulate the ingestion of deficient L-amino acids alone or from among various solutions and then begin to elicit a preference for solutions with an "umami" taste.
4. Increase of Gln and Arg intakes occurred only when the rats were fed a completely fortified diet. However, both preferences disappeared when protein restriction was introduced, and preferences were never induced under L-amino acid imbalance, even if the crude protein level was high.
5. The elevation of Thr and subsequently Gly intakes may be an expression of the body's detection of a negative nitrogen balance.
6. Preference for NaCl in SHR reflects a genetic predisposition to hypertension and is independent of the dietary protein, be it adequate or not. This observation may have some medical implications.

ACKNOWLEDGMENTS

The suggestions and helpful discussions with J. Kirimura, T. Kobayashi, G. K. Beauchamp, and J. G. Brand are appreciated. The authors also acknowledge and appreciate the services of M. Konno and K. Hara in the care of animals throughout these studies and J. Blescia for typing the manuscript.

REFERENCES

Aoki, K. (1963). Experimental studies on the relationship between endocrine organs and hypertension in spontaneously hypertensive rats. I. Effect of hypophysectomy, adrenalectomy,

thyroidectomy, nephrectomy and sympathectomy on blood pressure. *Jpn. Heart J.* **4,** 443–461.
Aoki, K., Yamori, Y., Ooshima, A., and Okamoto, K. (1972). Effects of high or low sodium intake in spontaneously hypertensive rats. *Jpn. Circ. J.* **36,** 539–545.
Bartter, F. C., Fujita, T., Delea, C. S., and Kawasaki, T. (1980). On the role of sodium in human hypertension. *In* "Biological and Behavioral Aspects of NaCl Intake" (M. R. Kare, M. J. Fregly, and R. A. Bernard, eds.), pp. 341–344. Academic Press, New York.
Beauchamp, G. K., and Maller, O. (1977). The development of flavor preferences in humans. *In* "The Chemical Senses and Nutrition" (M. R. Kare and O. Maller, eds.), pp. 291–310. Academic Press, New York.
Block, J. R., and Bolling, D. (1951). "The Amino Acid Composition of Proteins and Food." Charles C Thomas, Springfield, Illinois.
Block, J. R., and Mitchell, H. H. (1946). The correlation of the amino acid composition of proteins with their nutritive value. *Nutr. Abstr. Rev.* **16,** 249–278.
Brand, J. G., Cagan, R. H., and Naim, M. (1982). Chemical senses in the release of gastric and pancreatic secretions. *Annu. Rev. Nutr.* **2,** 249–276.
Cagan, R. H. (1977). A framework for the mechanisms of action of special taste substances: The example of monosodium glutamate. *In* "The Chemical Senses and Nutrition" (M. R. Kare, and O. Maller, eds.), pp. 343–359. Academic Press, New York.
Cagan, R. H., Torii, K., and Kare, M. R. (1979). Biochemical studies of glutamate taste receptors: The synergistic taste effect of L-glutamate and 5'-ribonucleotides. *In* "Glutamic Acid: Advances in Biochemistry and Physiology" (L. J. Filer, Jr., S. J. Garattini, M. R. Kare, W. A. Reynolds, and R. J. Wurtman, eds.), pp. 1–9. Raven Press, New York.
Calhoun, W. K., Hepburn, F. N., and Bradley, W. B. (1960). The availability of lysine in wheat, flour, bread and gluten. *J. Nutr.* **70,** 337–347.
Dethier, V. G. (1980). Biological and behavioral aspects of salt intake: A summation. *In* "Biological and Behavioral Aspects of NaCl Intake" (M. R. Kare, M. J. Fregly, and R. A. Bernard, eds.), pp. 411–417, Academic Press, New York.
Epstein, A , and Teitelbaum, P. (1962). Regulation of food intake in the absence of taste, smell and other oropharyngeal sensations. *J. Comp. Physiol. Psychol.* **55,** 753–759.
Fernandez-Flores, E., Kline, D. A., Johnson, A. R., and Leber, B. L. (1970). Qualitative GLC analysis of free amino acids in fruits and juices. *J. Am. Oil. Chem. Soc.* **536,** 1203–1208.
Forbes, R. M., Vaughan, L., and Yohe, M. (1958). Dependence of biological value on protein concentration in the diet of the growing rat. *J. Nutr.* **64,** 291–302.
Grossman, M. I. (1967). Neural and hormonal stimulation of gastric secretion. *In* "Handbook of Physiology" (C. F. Code, ed.), Vol. 2, pp. 835–863. American Physiological Society, Washington, D.C.
Hartsook, E. W., and Mitchell, H. H. (1956). The effect of age on the protein and methionine requirements of the rat. *J. Nutr.* **60,** 173–195.
Harper, A. E. (1959). Sequence in which the amino acids of casein become limiting for the growth of the rat. *J. Nutr.* **67,** 109–122.
Harper, A. E. (1964). Amino acid toxicities and imbalances. *In* "Mammalian Protein Metabolism" (H. N. Munro and J. B. Allison, eds.), pp. 87–134. Academic Press, New York.
Henkin, R. I. (1974). Salt taste in patients with essential hypertension and with hypertension due to primary hyperaldosteronism. *J. Chron. Dis.* **27,** 235–244.
Henkin, R. I. (1980). Salt taste and salt preference in normal and hypertensive rats and humans. *In* "Biological and Behavioral Aspects of NaCl Intake" (M. R. Kare, M. J. Fregly, and R. A. Bernard, eds.), pp. 367–394. Academic Press, New York.
Hidaka, I., and Kiyohara, S. (1975). Taste responses to ribonucleotides and amino acids in fish.

In "Olfaction and Taste V" (D. A. Denton and J. P. Coghlan, eds.), pp. 147–151. Academic Press, New York.
Ikeda, K. (1912). On the taste of the salt of glutamic acid. *Int. Congr. Appl. Chem. 8th* **18**, 147 (abstract).
Kirimura, J., Shimizu, A., Kimizuka, A., Ninomiya, T., and Katsuya, N. (1969). The contribution of peptides and amino acids to the taste of foodstuffs. *J. Agric. Food Chem.* **17**, 689–695.
Komata, Y. (1976). In "Microbial Production of Nucleic Acid-Related Substances" (K. Ogata, S. Kinoshita, T. Tsunoda, and K. Aida, eds.), pp. 299–319. Wiley, New York.
Kuninaka, A. (1967). Flavor potentiators. In "Symposium on Foods: The Chemistry and Physiology of Flavors" (H. W. Schultz, E. A. Day, and L. M. Libbey, eds.), pp. 515–535. Avi, Westport, CN.
Kuninaka, A., Kibi, M., and Sakaguchi, K. (1964). History and development of flavor nucleotides. *Food Technol.* **18**, 287–293.
Leung, P. M. B., Rogers, Q. R., and Harper, A. E. (1968). Effect of amino acid imbalance on dietary choice in the rat. *J. Nutr.* **95**, 483–492.
Mackie, A. M., and Adron, J. W. (1978). Identification of inosine and inosine 5'-monophosphate as the gustatory feeding stimulants on the turbot, *Scophthalmus maximus*. *Comp. Biochem. Physiol.* **60(A)**, 79–83.
Maeda, S., Eguchi, S., and Sasaki, H. (1958). The content of free L-glutamic acid in various foods. *J. Home Econ. Jpn.* **9**, 163–167.
Mitchell, H. H., and Beadles, J. R. (1950). Biological values of six partially purified proteins for the adult albino rat. *J. Nutr.* **40**, 25–40.
Naim, M., Kare, M. R., and Merritt, A. M. (1978). Effects of oral stimulation on the cephalic phase of pancreatic exocrine secretion in dogs. *Physiol. Behav.* **20**, 563–570.
Novis, B. H., Banks, S. and Marks, I. N. (1971). The cephalic phase of pancreatic secretion in man. *Scand. J. Gastroenterol.* **6**, 417–421.
Ohara, I., Otsuka, S., and Yugari, Y. (1979). The influence of carrier of gustatory stimulation on the cephalic phase of canine pancreatic secretion. *J. Nutr.* **109**, 2098–2105.
Pfaffmann, C. (1970). Physiological and behavioral processes of the sense of taste. In "Taste and Smell in Vertebrates" (G. E. W. Wolstenholme and J. Knight, eds.), pp. 31–50. Churchill, London.
Rama Rao, P. B., Metta, V. C., and Johnson, B. C. (1959). The amino acid composition and the nutritive value of proteins. I. Essential amino acid requirements of the growing rat. *J. Nutr.* **69**, 387–391.
Rama Rao, P. B., Norton, H. W., and Johnson, B. C. (1961). The amino acid composition and nutritive value of proteins. IV. Phenylalanine, tyrosine, methionine and cystine requirements of the growing rat. *J. Nutr.* **73**, 38–42.
Richardson, C. T., Walsh, J. H., Cooper, K. A., Feldman, M., and Fordtran, J. S. (1977). Studies on the role of cephalic-vagal stimulation in the acid secretory response to eating in normal human subjects. *J. Clin. Invest.* **60**, 435–441.
Richter, C. P. (1953). Experimentally produced behavior reactions to food poisoning in wild and domestic rats. *Ann. N.Y. Acad. Sci.* **56**, 225–239.
Rifkin, B., and Bartoshuk, L. M. (1980). Taste synergism between monosodium glutamate and disodium 5'-guanylate. *Physiol. Behav.* **24**, 1169–1172.
Rogers, Q. R., and Harper, A. E. (1970). Selection of a solution containing histidine by rats fed a histidine-imbalanced diet. *J. Comp. Physiol. Psychol.* **72**, 66–73.
Rogers, Q. R., and Leung, P. M. B. (1977). The control of food intake: When and how are amino acids involved? In "The Chemical Senses and Nutrition" (M. R. Kare, and O. Maller, eds.), pp. 213–248. Academic Press, New York.

Rozin, P., and Kalat, J. W. (1971). Specific hungers and poison avoidance as adaptive specializations of learning. *Psychol. Rev.* **78**, 459–485.
Sarles, H., Dani, R., Prezelin, G., Souville, C., and Figarella, C. (1968). Cephalic phase of pancreatic secretion in man. *Gut* **9**, 214–221.
Sato, M., and Akaike, N. (1965). 5′-Riobonucleotides as gustatory stimuli in rats. Electrophysiological studies. *Jpn. J. Physiol.* **15**, 53–70.
Sato, M., Yamashita, S., and Ogawa, H. (1970). Potentiation of gustatory response to monosodium glutamate in rat chorda tympani fibers by addition of 5′-ribonucleotides. *Jpn. J. Physiol.* **20**, 444–464.
Schechter, P. J., Horwitz, D., and Henkin, R. I. (1971). Sodium chloride preference in essential hypertension. *J. Am. Med. Assoc.* **225**, 1311–1315.
Schechter, P. J., Horwitz, D., and Henkin, R. I. (1974). Salt preference in patients with untreated and treated essential hypertension. *Am. J. Med. Sci.* **267**, 320–326.
Schiffman, S. S., and Engelhard, H. H., III. (1976). Taste of dipeptides. *Physiol. Behav.* **17**, 523–535.
Shimazono, H. (1964). Distribution of 5′-ribonucleotides in foods and their application to foods. *Food. Technol.* **18**, 294–303.
Stadtman, F. H. (1972). Free amino acids in raw and processed tomato juices. *J. Food Sci.* **37**, 944–951.
Steffens, A. B. (1976). Influence of the oral cavity on insulin release in the rat. *Am. J. Physiol.* **230**, 1411–1415.
Swaye, P. S., Gifford, R. W., and Berrettoni, J. N. (1972). Dietary salt and essential hypertension. *Am. J. Cardiol.* **29**, 33–38.
Thomas, C. B. (1973). Genetic patterns of hypertension in man. *In* "Hypertension: Mechanisms and Management" (G. Onesti, K. E. Kim, and J. D. Moyer, eds.), pp. 67–73. Grune & Stratton, New York.
Torii, K. (1979). Growth, sodium balance and salt appetite of spontaneously hypertensive rats (SHR) under varying sodium supply. *Fed. Proc.* **38**, 1376.
Torii, K. (1980). Salt intake and hypertension in rats. *In* "Biological and Behavioral Aspects of NaCl Intake" (M. R. Kare, M. J. Fregly, and R. A. Bernard, eds.), pp. 345–366. Academic Press, New York.
Torii, K., and Cagan, R. H. (1980). Biochemical studies of taste sensation. IX. Enchancement of L-[^{3}H]glutamate binding to bovine taste papillae by 5′-ribonucleotides. *Biochim. Biophys. Acta* **627**, 313–323.
Torii, K., Mimura, T., and Yugari, Y. (1983a). Preference for Umami. Sweet and salty taste in rats fed diets containing various amounts and quality of protein. *Proc. Jpn. Symp. Taste Smell 17th*, pp. 89–92.
Torii, K., Mimura, T., and Yugari, Y. (1983b). Taste preference in spontaneously hypertensive rats (SHR) fed diets containing varied amounts of whole egg protein. *Proc. Jpn. Symp. Taste Smell 17th*, pp. 109–111.
Vazquez, M., Pearson, P. B., and Beauchamp, G. K. (1982). Flavor preferences in malnourished Mexican infants. *Physiol. Behav.* **28**, 513–519.
Wotman, S., Mandel, I. D., Thompson, R. H., Jr., and Laragh, J. H. (1967). Salivary electrolytes and salt taste thresholds in hypertension. *J. Chron. Dis.* **20**, 833–840.
Yamaguchi, S., and Kimizuka, A. (1979). Psychometric studies on the taste of monosodium glutamate. *In* "Glutamic Acid: Advances in Biochemistry and Physiology" (L. J. Filer, Jr., S. Garattini, M. R. Kare, W. A. Reynolds, and R. J. Wurtman, eds.), pp. 35–54. Raven Press, New York.
Yamaguchi, S., Yoshikawa, T., Ikeda, S., and Ninomiya, T. (1968). The synergistic taste effect

of monosodium glutamate and disodium 5'-guanylate. *J. Agric. Chem. Soc. Jpn.* **42**, 378–381.

Yamaguchi, S., Yoshikawa, T., Ikeda, S., and Ninomiya, T. (1971). Measurement of the relative taste intensity of some L-amino acids and 5'-nucleotides. *J. Food Sci.* **26**, 846–849.

Yamashita, S., Ogawa, H., and Sato, M. (1973). The enhancing action of 5'-ribonucleotide on rat gustatory nerve fiber responses to monosodium glutamate. *Jpn. J. Physiol.* **23**, 59–68.

Yoshida, A., and Moritoki, K. (1974). Nitrogen sparing action of methionine and threonine in rats receiving a protein-free diet. *Nutr. Reports Int.* **9**, 159–169.

Yoshida, M., and Saito, S. (1969). Multidimensional scaling of the taste of amino acids. *Jpn. Psychol. Res.* **11**, 149–166.

Yoshida, M., Ninomiya, T., Ikeda, S., Yamaguchi, S., Yoshikawa, T. and Ohara, M. (1966). Studies on the taste of amino acids. I. Determination of threshold values of various amino acids. *J. Agric. Chem. Soc. Jpn.* **40**, 295–299.

Zinner, S. H., Levy, P. S., and Kass, E. H. (1971). Familial aggregation of blood pressure in childhood. *N. Engl. J. Med.* **284**, 401–404.

4

Preference Threshold for Maltose Solutions in Rats Treated Chronically with the Components of an Oral Contraceptive

MELVIN J. FREGLY
Department of Physiology
University of Florida
College of Medicine
Gainesville, Florida

I.	Introduction	72
II.	Methods	72
	A. Experiment 1: Effect of Chronic Treatment with an Oral Contraceptive on Spontaneous Intake of NaCl and Maltose Solutions	73
	B. Experiment 2: Determination of the Preference Threshold for Maltose Solution	73
	C. Experiment 3: Preference Threshold Concentration for Maltose in Rats Treated with the Components of an Oral Contraceptive Drug	74
	D. Experiment 4: Effect Of Treatment with Ethynyl Estradiol on the Detection Threshold for Bitter and Acid Modalities of Taste in Ovariectomized Rats	75
	E. Experiment 5: Effect of Addition of Maltose to NaCl Solution on Intakes of NaCl Solution, Distilled Water, and Food .	75
III.	Results	76
	A. Experiment 1	76
	B. Experiment 2	77
	C. Experiment 3	78
	D. Experiment 4	79
	E. Experiment 5	80
IV.	Discussion	82
V.	Summary	85
	References	86

I. INTRODUCTION

Richter and Campbell (1940) were among the first to measure the preference threshold for glucose in rats by the two-bottle choice technique. In this experimental paradigm, one bottle contains distilled water, whereas the second contains glucose solution, the concentration of which is increased in small increments until a concentration is reached at which the rat ingests consistently more of its daily fluid intake from the glucose-containing bottle than from the water bottle (Richter and Campbell, 1940). This concentration was termed the "preference threshold" concentration by Richter (1939) and was 2 g/liter. Maximal intake occurred when the glucose concentration was about 110 g/liter. With increases in glucose beyond 110 g/liter, intake of glucose solution gradually decreased whereas simultaneous water intake increased. Preference–aversion curves such as that for glucose can be found only for solutions that are highly palatable to the rat, including salty (e.g., NaCl) and sweet (e.g., glucose, sucrose, and maltose) tastes. However, just as is the case for salts, not all sugars are equally palatable. Richter showed that glucose, maltose, and sucrose were more palatable, and therefore were ingested to a greater extent by rats, than galactose and lactose (Richter and Campbell, 1940; Richter, 1956).

Earlier studies from this laboratory reported both an appetite and a reduced preference threshold for NaCl solutions in rats treated chronically with an oral contraceptive preparation, as well as its constituents, ethynyl estradiol and norethynodrel, the estrogenic and progestational components, respectively (Fregly, 1972, 1973; Fregly and Newsome, 1980). To assess the effect of ethynyl estradiol and norethynodrel on the sweet modality of taste, an experiment was carried out to determine and compare preference thresholds for maltose in treated and control rats. Detection thresholds for other modalities of taste, including bitter (quinine) and acid (hydrochloric acid), were also determined in treated and control rats. Finally, the effect of a constant amount of maltose on the palatability of hypotonic and hypertonic concentrations of NaCl was determined. The results of these studies are described here.

II. METHODS

Female rats of the Blue Spruce Farms (Sprague–Dawley) strain weighing 200–300 g were used throughout these studies. The animals were maintained in individual stainless-steel hanging cages and were provided with finely powdered Purina Laboratory Chow and tap water during the periods prior to beginning any experiment. Water and food were freely available at all times. Food was provided in spill-resistant containers (Fregly, 1960), and water was provided in infant nursing bottles with cast aluminum spouts as described by Lazarow (1954). Temperature of the room was maintained at 26 ± 1°C, and illumination was from 7AM to 7PM.

The criterion of preference threshold used was that of Richter (1939), namely, the concentration of maltose solution at and above which simultaneous mean volume taken from the bottle containing maltose significantly exceeded that taken from the water bottle.

A. Experiment 1: Effect of Chronic Treatment with an Oral Contraceptive on Spontaneous Intake of NaCl and Maltose Solutions

Twelve rats were used in this experiment. Half received the oral contraceptive, Enovid (7.5 mg/kg food),* in their diet for 2 months prior to the experiment. Each rat was allowed to choose between distilled water and 0.25 M NaCl solution as drinking fluids. After a 2-day equilibration period, daily intakes of water, NaCl solution, and food were measured for 7 days. One week was allowed to elapse before beginning the next study. During this time the rats were each given two bottles of distilled water. At the end of this time, the rats were allowed to choose between distilled water and a maltose solution (8.0 g/liter) as drinking fluids. Intakes were measured daily for 5 days. The bottles containing both water and maltose were thoroughly washed and refilled daily.

The results of this experiment showed that oral contraceptive-treated rats manifested a strong appetite for maltose; however, it failed to differ from that of the control group. Since this concentration of maltose was well above preference threshold level for both groups of rats, it seemed important to determine both the preference or detection threshold for maltose in untreated rats and whether treatment with the components of an oral contraceptive altered their preference threshold concentration.

B. Experiment 2: Determination of the Preference Threshold for Maltose Solution

Five normal rats were used. Each rat, caged individually, was given a choice initially between two bottles of distilled water. The water ingested daily from each bottle was measured for 4 days. One of the bottles was then filled with maltose (1.0 g/liter), whereas the other contained distilled water. At 4-day intervals thereafter, the concentration of maltose in one bottle was

---

*Enovid tablets were used. Each contained 5 mg norethynodrel and 0.075 mg mestranol.

increased to 1.5, 2.0, 2.5, and 5.0 g/liter. Measurement of daily fluid intakes continued. After each measurement, the positions of the bottles on each cage were switched to avoid habit formation in the selection of drinking fluid.

C. Experiment 3: Preference Threshold Concentration for Maltose in Rats Treated with the Components of an Oral Contraceptive Drug

Twenty rats were divided randomly into four equal groups. Group 1 served as a control group and was implanted subcutaneously (sc) between the shoulder blades with an empty Silastic tube (602-231) while anesthetized with ether. Group 2 was implanted with a similar tube 20 mm long containing crystalline norethynodrel. Group 3 was similarly treated, except that the Silastic tube (602-281) was 10 mm long and contained crystalline ethynyl estradiol. The fourth group was implanted with both norethynodrel- and ethynyl estradiol-containing tubes. This method of administering steroids via Silastic implants allows diffusion into various media at a relatively constant rate over long periods of time (Fregly and Newsome, 1980). The amount of steroid released by each tube was determined by placing it in a vacuum desiccator for 72 hr prior to implantation and weighing it on an analytical balance. When removed from the rat, the tubes were again desiccated and weighed. The amount of steroid lost from the tube was determined by dividing the change in weight of the tube by the number of days the tubes were implanted and the mean body weight of the rats. Thus, the mean amount of norethynodrel lost from the tubes was 45 µg/day or 165 ± 8 µg/kg/day, whereas the mean amount of ethynyl estradiol lost was 9 µg/day or 36 ± 4 µg/kg/day.

The study began 7 weeks after implantation of the Silastic tubes. At this time each rat was given a choice between two bottles of distilled water as drinking fluids for 4 days. Water intakes were not measured during this period but were measured during the following 2 days. Subsequently, one bottle contained a maltose solution, whereas the second contained distilled water. The concentration of maltose was changed at the end of every second day. The following concentrations (grams/liter) of maltose were given in chronological sequence: 0.0125, 0.0250, 0.050, 0.075, 0.15, 0.20, 0.25, 0.30, 0.35, 0.40, 0.45, 0.50, 0.60, 0.75, 1.00, 1.75, 2.00, 2.50, 4.00, 5.00, and 6.00. Each concentration was made from a single stock solution by serial dilution. Preference threshold concentration for each rat was assessed as described in Experiment 2.

D. Experiment 4: Effect of Treatment with Ethynyl Estradiol on the Detection Threshold for Bitter and Acid Modalities of Taste in Ovariectomized Rats

Twenty-four rats were divided randomly into four equal groups. Three groups were bilaterally ovariectomized while anesthetized with pentobarbital (40 mg/kg, ip). The remaining group was anesthetized and sham operated, but the ovaries were not removed. While still anesthetized, the animals in one ovariectomized group had a 5-mm Silastic tube (602-285) containing ethynyl estradiol implanted subcutaneously. A second ovariectomized group was similarly treated, except that the length of the tube was 10 mm. The third ovariectomized group and the intact group were both implanted with empty Silastic tubes 10 mm in length. After a 10-day period for recovery, the rats were caged individually and given a choice between two bottles of distilled water to drink for 4 days. At the end of this time, the experiment began.

The first study was designed to determine whether the bitter taste modality, as assessed by detection threshold for quinine sulfate, was altered by ovariectomy and treatment with an estrogenic agent. A stock solution of 1×10^{-4} M was used to make the three concentrations used, i.e., 1, 5, and 10×10^{-6} M. Each concentration was given during a 2-day period. When the highest concentration had been available for 2 days, the rats were again given two bottles of distilled water.

Essentially the same methods used in Study 1 above were used in this study. The rats were also the same. The major difference between the two studies was that hydrochloric acid was used instead of quinine sulfate. As in the previous study, a 4-day period on distilled water preceded assessment of the drinking responses to graded increases in the hydrochloric acid concentration in one of the drinking bottles. The concentrations of hydrochloric acid used were 5, 8, 10, and 50×10^{-4} M. Each concentration was available for 2 days, as in Study 1.

The amount of ethynyl estradiol released from the 5-mm Silastic tube was 30 ± 3 µg/kg/day, whereas the amount released from the 10-mm tube was 72 ± 4 µg/kg/day.

E. Experiment 5: Effect of Addition of Maltose to NaCl Solution on Intakes of NaCl Solution, Distilled Water, and Food

The objective of this experiment was to determine whether maltose, a sugar highly palatable to rats, would affect their intake of NaCl. The latter is also highly palatable to rats.

Twelve normal rats were used. They were allowed choice between two bottles of distilled water for 4 days prior to beginning the experiment. At this time, the animals were divided randomly into two equal groups. Group 1 was then given a choice between a 0.15 M NaCl solution and distilled water, whereas Group 2 was given a choice between a 0.15 M NaCl solution containing 1% maltose and distilled water for 4 days. Fluid intakes were measured daily. Subsequently, at 4-day intervals the rats were offered the following NaCl solutions: 0.25, 0.075, and 0.35 M. Group 2 was always given the NaCl solution with 1% maltose added. Statistical analysis of the data in all experiments was carried out by a one-way analysis of variance where appropriate, or by means of Student's t test (Snedecor and Cochran, 1956). Significance was set at the 95% confidence limit.

III. RESULTS

A. Experiment 1

Chronic treatment with the oral contraceptive agent, Enovid, induced a striking appetite for 0.25 M NaCl solution (Table I). Although water intake was reduced below that of the control group, the decrease was not statistically significant, nor were there differences in food intake between groups. However, the ratio of NaCl solution ingested to total fluid (water + NaCl

TABLE I Effect of Chronic Treatment with the Oral Contraceptive, Enovid (7.5 mg/kg food), on Spontaneous Intake of NaCl (0.25 M) and Maltose (8.0 g/liter) Solutions by Female Rats

Treatment	Number of rats	Mean body weight (g)	Intake (ml or g/100 g body weight/day) of			Ratio (%) (test soln/ total fluid)
			Water	0.25 M NaCl or maltose	Food	
Study 1: NaCl solution (0.25 M)[a]						
Control	6	277 ± 7[b]	6.3 ± 0.8	5.4 ± 1.6	5.6 ± 0.2	42.6 ± 9.8
Enovid	6	231 ± 3	3.5 ± 0.8	16.0 ± 2.6[c]	6.3 ± 0.3	80.8 ± 5.6[d]
Study 2: Maltose solution (8.0 g/liter)[a]						
Control	6	277 ± 7	1.3 ± 0.2	13.2 ± 2.0	6.5 ± 0.1	89.6 ± 1.9
Enovid	6	229 ± 3	1.0 ± 0.1	16.2 ± 1.7	5.6 ± 0.2	94.0 ± 0.5[d]

[a] Intakes were measured for 7 (Study 1) and 5 (Study 2) days.
[b] One standard error of mean.
[c] Significantly different from control ($p < 0.01$).
[d] Significantly different from control ($p < 0.05$).

solution) ingested was significantly ($p < 0.01$) increased in the Enovid-treated group. In contrast, Enovid appeared to have no significant effect on intakes of maltose solution (8 g/liter or 0.8%), water, or food above that of the control group. However, the ratio of maltose solution ingested to total fluid ingested was significantly ($p < 0.05$) greater for the Enovid-treated group.

B. Experiment 2

When untreated rats were given a choice between maltose solution and distilled water to drink, they were able to distinguish between the two when the concentration of the maltose solution was between 1.0 and 1.5 g/liter (Fig. 1). The increase in intake of maltose solution was linearly related to the concentration of maltose in the solution. Simultaneous water intake de-

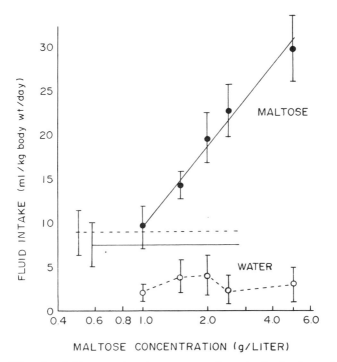

Fig. 1. Effect of graded increases in maltose concentration on intake of maltose solution in a two-bottle choice paradigm in which the second bottle was distilled water. One standard error is set off at each mean. The solid straight line represents the volume of water ingested during a 4-day control period from the bottle that contained maltose during the experiment. The dashed straight line represents the volume of water ingested during the control period from the bottle that contained water during the experiment. Five normal female rats were used.

creased below that observed during the control period prior to allowing access to maltose and remained at this low level for the duration of the experiment.

C. Experiment 3

Chronic treatment with norethynodrel, ethynyl estradiol, or the combination of the two drugs reduced the preference threshold for maltose solution below that of the untreated control group (Fig. 2). Table II shows the preference thresholds (grams/liter) of the four groups. All treatments reduced preference threshold by approximately 50%. Unfortunately, it is not possible to state whether the maximal concentration of maltose ingested by the treated rats differed significantly from that of the controls since concentrations sufficiently high to reduce the intake of maltose solution by both control and norethynodrel-treated groups were not achieved. However, for most concentrations above threshold, both groups receiving ethynyl estradiol in-

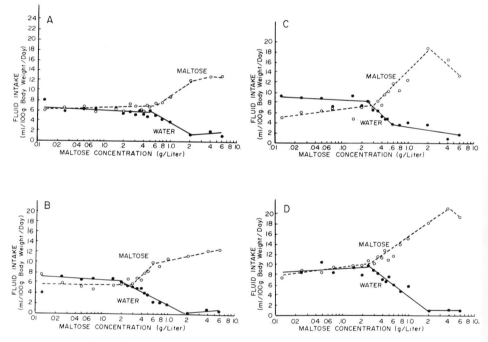

Fig. 2. The results of a study to determine the effect of norethynodrel (B), ethynyl estradiol (C), and a combination of the two drugs (D) on spontaneous intakes of distilled water and maltose solution. Data from control, untreated rats are shown (A). Preference threshold concentration for each group is shown in Table II.

TABLE II Preference Threshold Concentration for Maltose Solution in Female Rats Receiving Ethynyl Estradiol, Norethynodrel, and Both Combined

Treatment	Drug dosage (μg/kg/day)		Preference threshold conc. (g/liter)
	Ethynyl estradiol	Norethynodrel	
Control	—	—	1.30 ± 0.29[a]
Ethynyl estradiol	36	—	0.43 ± 0.05[b]
Norethynodrel	—	165	0.43 ± 0.09[b]
Ethynyl estradiol + norethynodrel	36	165	0.46 ± 0.06[b]

[a] One standard error of mean.
[b] Significantly different from control ($p < 0.01$).

gested more maltose solution than either the control or norethynodrel-treated groups.

D. Experiment 4

The threshold concentration for rejection of quinine sulfate solution by treated, ovariectomized rats was reduced by treatment with ethynyl estradiol but not by norethynodrel (Fig. 3). The concentration for rejection was 5×10^{-6} M for ethynyl estradiol-treated rats and 1×10^{-5} M for control and ovariectomized, untreated rats. When water was available in both bottles during the final 4 days of the study, water intakes did not match those measured at the beginning of the experiment for the ovariectomized group and for the ovariectomized group given the higher dose of ethynyl estradiol. There is no apparent reason for this difference in these two groups since the other two groups had similar water intakes at the beginning and end of the experiment.

The threshold concentration for rejection of hydrochloric acid solution by both estrogen-treated, ovariectomized groups was essentially unchanged by treatment (Fig. 4). The ovariectomized group given the lower dose of ethynyl estradiol increased their water intake and decreased their intake of hydrochloric acid solution at 1×10^{-3} M, but the difference between intakes of water and acid solution was not significant (Fig. 4C). The first significant difference between simultaneous intakes of water and acid solution occurred at 5×10^{-3} M. The ovariectomized group treated with the higher dose of estrogen detected the difference between water and acid solution when the concentration of the acid solution was 1×10^{-3} M (Fig. 4D). This is the same concentration at which intact control rats first detected a difference between the two solutions (Fig. 4A). In the case of the ovariectomized,

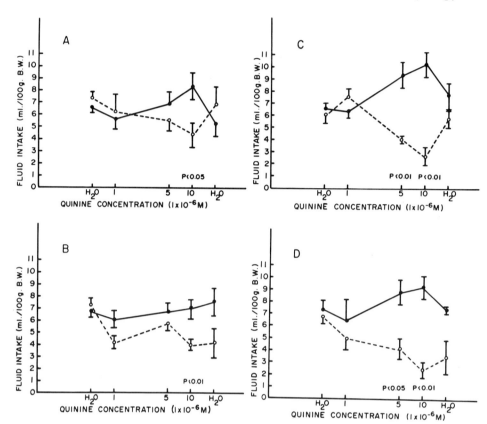

Fig. 3. The effect of ovariectomy and treatment with ethynyl estradiol (30 and 72 µg/kg/day) on intakes of water and quinine sulfate solutions is shown for each group. (A) Control; (B) ovariectomized; (C) ovariectomized + ethynyl estradiol (30 µg/kg/day); (D) ovariectomized + ethynyl estradiol (72 µg/kg/day). One standard error is set off at each mean. Detection threshold is indicated as the concentration at which a significant difference ($p < 0.05$) occurred between intakes of water and quinine solution offered simultaneously. (●) H_2O; (○) quinine.

untreated group, the first significant difference between water and acid solution was detected at a concentration of 8×10^{-4} M (Fig. 4B).

E. Experiment 5

When 1% maltose was added to NaCl solution, intake of the NaCl solution was greater than that ingested by the group given NaCl solution without maltose at all concentrations of NaCl offered except the highest (Fig. 5A).

4. Maltose Preference Threshold and Estrogen

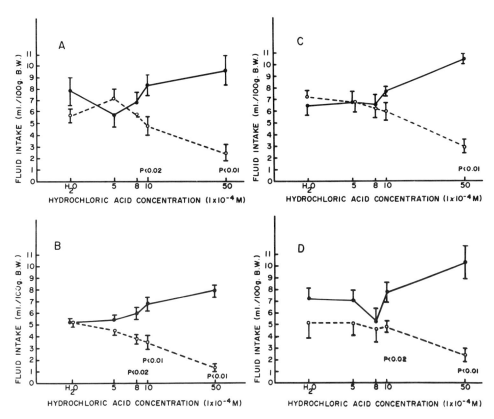

Fig. 4. The effect of ovariectomy and treatment with ethynyl estradiol (30 and 72 µg/kg/day) on intakes of water and hydrochloric acid is shown for each group. (A) Control; (B) ovariectomized; (C) ovariectomized + ethynyl estradiol (30 µg/kg/day); (D) ovariectomized + ethynyl estradiol (72 µg/kg/day). One standard error is set off at each mean. Detection threshold is indicated as in Fig. 3. (●) H_2O; (○) HCl.

Water intake of the NaCl + maltose-treated group (Group 2) was significantly less than that of the group given NaCl alone (Fig. 5A). The ratio of NaCl ingested to total fluid ingested was significantly higher for the group given NaCl + maltose at all concentrations except the highest (Fig. 5B). Urine output of the two groups failed to differ significantly throughout the study (Fig. 5C) whereas urinary sodium output of the NaCl + maltose-treated group tended to be higher than that of the group given NaCl alone (Fig. 5D). Urinary potassium output did not differ significantly between groups throughout the experiment.

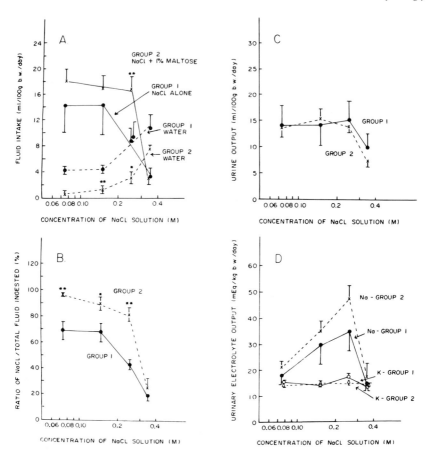

Fig. 5. The effect of addition of maltose (1%) to graded concentrations of NaCl solution on intakes of NaCl solution and water offered simultaneously. Group 1 was given choice between NaCl solution and distilled water, whereas Group 2 was given choice between the NaCl–maltose solution and distilled water. (A) Fluid intakes; (B) the ratio of NaCl ingested to total fluid (NaCl + water) ingested. (C) Urine output; (D) urinary electrolyte output. One standard error is set off at each mean.

IV. DISCUSSION

Rats treated chronically either with the oral contraceptive, Enovid, or its components, ethynyl estradiol and norethynodrel, developed a striking appetite for NaCl solution in that a hypertonic solution (0.25 M) was ingested in preference to water offered simultaneously (Table I). In fact, the amount of hypertonic NaCl solution ingested by the Enovid-treated group was about three times that ingested by controls (Table I).

When a choice was offered between water and maltose solution (8.0 g/liter), both the control and Enovid-treated groups ingested 13- to 16-fold more maltose solution than water (Table I). However, the maltose intake by the treated group did not significantly exceed that of the control group. This concentration of maltose is clearly above the preference threshold for both treated and control rats, judging by their 13- to 16-fold increase in intake of maltose solution above that of water offered simultaneously.

The results of the second experiment revealed that the preference threshold for maltose solution in our strain of rats is between 1.0 and 1.5 g/liter. This estimate of preference threshold is higher than that of Richter and Campbell (1940) (0.6 g/liter). However, a different diet and different sexes of animals were used. It is apparent, nonetheless, that the rats in Experiment 1 received maltose at a concentration well above preference threshold concentration.

The third experiment was then carried out to determine the effect of treatment with the components of an oral contraceptive, ethynyl estradiol and norethynodrel, on preference threshold for maltose (Fig. 2). The results of this study revealed a significant reduction in preference threshold for maltose solution in the groups treated with ethynyl estradiol, norethynodrel, or a combination of the two compounds (Table II). Judging from the data of both the ethynyl estradiol-treated group and the group treated with both ethynyl estradiol and norethynodrel, the concentration of maltose used in Experiment 1 would be expected to produce a maximal or near maximal intake of maltose solution in both groups. Differences between groups are less likely to be observed with such high concentrations. The results of Experiment 3 also suggest that administration of an estrogenic hormone increases the intake of maltose solution above that of either untreated or norethynodrel-treated groups (Fig. 2). These results are in agreement with those of Wade and Zucker (1969) that suggest that ovariectomy prior to the establishment of a preference for saccharin precludes its development whereas estrogen and progesterone replacement reinstates the preference in ovariectomized rats (Zucker, 1969). In contrast, Kenney and Redick (1980) reported that ovariectomy increased the intake of glucose solution above that of intact controls, and graded doses of estradiol, given to ovariectomized rats, decreased their glucose intake in a graded fashion toward that of intact controls. However, neither ovariectomy nor treatment with estradiol had any effect on intake of saccharin solutions.

Ovariectomy did not appear to affect the concentration at which the bitter taste of quinine sulfate could be detected (Fig. 3B). This finding is in agreement with the studies of Hirsch and Bronstein (1976). However, chronic treatment with ethynyl estradiol reduced by 50% the concentration that could be detected by the treated group when compared to either intact or ovariectomized controls. This suggests that treatment of rats with oral contraceptive agents or ethynyl estradiol reduces preference threshold for NaCl (salt) and maltose (sweet) solutions as well as reducing the detection threshold for quinine sulfate (bitter) solutions.

Dippel and Elias (1980) assessed sweetness preference of women using oral contraceptives having similar estrogenic potencies but either a high or a low progestin potency. They reported that a significantly greater number of women using oral contraceptives low in progestational potency showed a preference for very sweet solutions in comparison to the preferences of women using oral contraceptives high in progestational potency. They suggested that the progestational component may influence sweetness preference to a greater extent than the estrogenic component. In the case of rats, the progestational component of the oral contraceptive, Enovid, reduced the preference threshold for maltose but did not increase intake of maltose above that of controls, whereas treatment with ethynyl estradiol reduced preference threshold and increased intake of maltose at all concentrations above threshold. Although it is difficult to equate sweetness preference in humans with preference threshold concentration and sweetness preference in rats, the results suggest that humans and rats may differ in their responses to administration of estrogenic and progestational agents.

The results with hydrochloric acid are somewhat more difficult to interpret. They suggest that ovariectomy may have reduced the detection threshold for acid below that of intact controls and that treatment of ovariectomized rats with ethynyl estradiol may have returned detection threshold to that of intact controls. Treatment with ethynyl estradiol does not appear to affect detection threshold for acid, although this experiment should be repeated in an experimental paradigm in which both intact and ovariectomized rats receive ethynyl estradiol.

Hall and Hall (1966a,b, 1969) in a series of studies showed that addition of a number of different sugars to isotonic NaCl solution enhances its ingestion by rats. The objective of their studies was to increase the quantity of both fluid and electrolyte consumed to determine whether the severity of NaCl-induced hypertension was increased. Sucrose, fructose, maltose, and glucose, when added to isotonic saline, enhanced volume and electrolyte intake and the severity of the hypertension. Hall and Hall (1966a,b, 1969) did not study the effect of sugars on intake of hypertonic NaCl solutions. The results presented here indicate that maltose can enhance the intake of 0.25 M NaCl solution but has no effect on intake of 0.35 M. Fluid and electrolyte intake were increased at all concentrations of NaCl presented, except the highest, when maltose was added to the salt solution (Fig. 5). These results with maltose are similar to those reported by Young and Trafton (1964), who constructed "isohedons" to determine the concentrations of sucrose that had

4. Maltose Preference Threshold and Estrogen

to be added to various concentrations of NaCl solution to induce a rate of licking in 1-min test sessions equivalent to that observed with undiluted 1% or higher concentrations of sucrose. When a given concentration of sucrose was added to varying concentrations of NaCl solutions, a sharp reduction in licking rate occurred at concentrations between 2 and 4% (0.30 and 0.60 M NaCl solution). Similar results occurred in the study illustrated in Fig. 5 although maltose was used instead of sucrose. Thus, the question whether excessive ingestion of salt may be a function of the masking of its taste by excessive ingestion of sugars is of considerable interest and remains for further study.

V. SUMMARY

The preference threshold for maltose solution in untreated female rats, measured by the two-bottle choice paradigm, was 1.3 g/liter. Rats treated with the progestational agent, norethynodrel (165 µg/kg/day), with ethynyl estradiol (36 µg/kg/day), or with a combination of the two, had approximately a 65% reduction in preference threshold concentration and a greater intake of maltose solution than untreated controls at any concentration offered. The detection threshold concentration for quinine solution was unchanged by ovariectomy but was reduced significantly by treatment with ethynyl estradiol. On the other hand, ovariectomy reduced the detection threshold for acid whereas treatment with ethynyl estradiol returned it to that of intact controls. The results suggest that chronic treatment with an estrogenic agent reduces the preference threshold for both salt (NaCl) solution (as observed in earlier studies) and sweet (maltose) solution, as well as reducing the detection threshold for bitter (quinine sulfate) solution. Detection threshold concentration for acid was reduced by ovariectomy and returned to that of intact controls by treatment with ethynyl estradiol. Addition of maltose (1%) to NaCl solutions increased the volume of the NaCl solution ingested by intact rats at all concentrations of NaCl offered up to a concentration of 0.35 M. This observation has important implications regarding the masking of the taste of NaCl and for increasing the amount of salt ingested daily.

ACKNOWLEDGMENT

This work was supported by Grant HL 14526-13 from the National Heart, Lung, and Blood Institute, National Institutes of Health.

REFERENCES

Dippel, R. L., and Elias, J. W. (1980). Preferences for sweet in relationship to use of oral contraceptives and pregnancy. *Horm. Behav.* **14**, 1–6.
Fregly, M. J. (1960). A simple and accurate feeding device for rats. *J. Appl. Physiol.* **15**, 539.
Fregly, M. J. (1972). Effect of an oral contraceptive on spontaneous running activity, salt appetite, and blood pressure of ovariectomized rats. *J. Pharmacol. Exp. Ther.* **182**, 335–343.
Fregly, M. J. (1973). Effect of an oral contraceptive on NaCl appetite and preference threshold in rats. *Pharmacol. Biochem. Behav.* **1**, 61–65.
Fregly, M. J., and Newsome, D. G. (1980). Spontaneous NaCl appetite induced by administration of an oral contraceptive and its components to rats. *In* "Biological And Behavioral Aspects of Salt Intake" (M. R. Kare, M. J. Fregly, and R. A. Bernard, eds.), pp. 247–272. Academic Press, New York.
Hall, C. E., and Hall, O. (1966a). Comparative ability of certain sugars and honey to enhance saline polydipsia and salt hypertension. *Proc. Soc. Exp. Biol. Med.* **112**, 362–365.
Hall, C. E., and Hall, O. (1966b). Comparative effectiveness of glucose and sucrose in enhancement of hyperalimentation and salt hypertension. *Proc. Soc. Exp. Biol. Med.* **123**, 370–374.
Hall, C. E., and Hall, O. (1969). Interaction between desoxycorticosterone treatment, fluid intake, sodium consumption, blood pressure, and organ changes in rats drinking water, saline, or sucrose solution. *Can. J. Physiol. Pharmacol.* **47**, 81–86.
Hirsch, S. M., and Bronstein, P. M. (1976). Ovariectomy fails to affect rat's quinine aversion. *Physiol. Behav.* **16**, 375–377.
Kenney, N. J., and Redick, J. H. (1980). Effects of ovariectomy and subsequent estradiol replacement on intake of sweet solutions. *Physiol. Behav.* **24**, 807–809.
Lazarow, A. (1954). Methods for quantitative measurement of water intake. *Methods Med. Res.* **6**, 225–229.
Richter, C. P. (1939). Salt taste threshold of normal and adrenalectomized rats. *Endocrinology* **24**, 367–371.
Richter, C. P. (1956). Self-regulatory functions during gestation and lactation. In "Gestation" (C. A. Villee, ed.), pp. 11–93. Josiah Macy Jr. Foundation, New York.
Richter, C. P., and Campbell, K. H. (1940). Taste thresholds and taste preferences of rats for five common sugars *J. Nutr.* **20**, 31–46.
Snedecor, G. W., and Cochran, W. G. (1956). "Statistical Methods." Iowa State Univ. Press, Ames.
Wade, G. N., and Zucker, I. (1969). Taste preferences in female rats: Modifications by neonatal hormones, food deprivation, and prior experience. *Physiol. Behav.* **4**, 935–943.
Young, P. T., and Trafton, C. L. (1964). Psychophysical studies of taste preference and fluid intake. *In* "Thirst, First International Symposium on Thirst in the Regulation of Body Water" (M. J. Wayner, ed.), pp. 271–284. Pergamon, Oxford.
Zucker, I. (1969). Hormonal determinants of sex differences in saccharin preference, food intake and body weight. *Physiol. Behav.* **4**, 595–602.

5

The Chemical Senses and Nutrition in the Elderly

CLAIRE MURPHY
Department of Psychology
San Diego State University
San Diego, California

I.	Introduction: Nutritional Status in the Elderly	87
II.	Chemosensory Loss in the Elderly: Review of the Literature	89
III.	Chemosensory Preference and Biochemical Indexes in the Elderly	93
	A. Experiment 1	93
	B. Experiment 2	96
	C. Summary	101
IV.	Discussion and Conclusions	101
V.	Research Needs	102
	References	103

I. INTRODUCTION: NUTRITIONAL STATUS IN THE ELDERLY

Several recent papers have emphasized the deficiencies in dietary intake in the elderly population. Reviewing a number of nutritional surveys, Beauchene and Davis (1979) reported that a significant proportion of the elderly population consume less than two-thirds of the recommended dietary allowances (RDA) for nutrients such as calcium, iron, thiamine, riboflavin, niacin, vitamin A, and vitamin C. In many elderly people, caloric intake falls below the RDA.

In addition to decreased nutrient intake, biochemical studies have found nutritional deficiencies. Yearick *et al.* (1980) reported deficiencies in dietary intake for calcium, vitamin A, and thiamine in a significant proportion of a sample of elderly people, although with the exception of calcium, the *mean* nutrient intakes in their sample met the RDA for age and sex and were generally higher than the nutrient intakes reported in other studies of the elderly. Although there were fewer subjects in this study whose intake was less than two-thirds the RDA for specific nutrients, there was wide individual variation. Ten percent of the elderly in this study had inadequate dietary intake of protein. Biochemical studies of the same population showed 41% of the elderly to have total serum protein concentrations that were less than acceptable. Nine percent of the population had values indicative of protein deficiency (less than 6 g/dl). Jansen and Harrill (1977) also reported low serum protein concentrations. In their sample, 10% of the elderly had deficient values for serum protein, 26% had low values, and the remaining elderly had acceptable values, as defined by the Ten-State Survey. Their elderly subjects also showed a significant incidence of low values for serum albumin: 20% were low or deficient. Jansen and Harrill (1977) also related actual weighed amounts of food consumed to serum levels of protein and albumin. They reported significant correlations between dietary intake and serum levels for both protein ($r = 0.33$) and albumin ($r = 0.42$). Other studies, for example, Yearick et al. (1980), have found no significant correlation between dietary intake of protein and serum levels. It has been suggested that lower rates of absorption and utilization in the elderly may be responsible for the lack of a reliable relationship between these variables.

Establishing RDA for protein in the elderly has not been a simple matter. Some investigators have questioned the RDA for protein in the elderly, some because of the methodology used to arrive at acceptable levels (Young et al., 1976). Others have argued that because loss of lean body mass occurs with advancing age, the body has less need for dietary protein. Finally, higher protein intake has been recommended to compensate for reduced absorption and decreased organ function in the elderly.

Decreased activity levels in the elderly translate into decreased caloric intake if the body is to maintain energy balance. Under these circumstances, consuming adequate amounts of essential nutrients demands the choice of a diet high in nutrient content.

Many factors can work against the elderly person seeking to maintain a nutritionally adequate diet. Factors that have received little attention in the literature but that are undoubtedly important include income level, socioeconomic status, location in which meals are consumed, social context, knowledge about nutrition, multiple drug use, medical problems, and difficulties in marketing, food preparation, and mastication. In addition, problems with taste and smell loss may interfere with adequate intake. Problems of chemosensitivity in the elderly and their interrelationship with nutritional status will be the major focus of this chapter.

II. CHEMOSENSORY LOSS IN THE ELDERLY: REVIEW OF THE LITERATURE

Evidence is mounting to support the existence of a decline in chemosensory function with age. The vast majority of this evidence is in the form of threshold studies and studies of suprathreshold intensity function. The important question of whether or not there are alterations of chemosensory preference with increasing age has been left relatively untouched.

Threshold studies of the gustatory system have demonstrated modest increases in threshold with advancing age (see reviews by Murphy, 1979, 1986; Schiffman, 1979). Table I presents a brief summary of this literature. Although some authors have reported quality-specific changes with aging (e.g., Weiffenbach et al., 1982), none of the four basic tastes (sweet, sour, bitter, or salty) has escaped report of age-related decline by at least one investigator. There is currently some debate regarding both the magnitude of the decline in threshold sensitivity with age and the significance of threshold changes in the perception of real-world stimulation. In fact, there is a real lack of information on the rate of decline over the life span.

Several recent studies have investigated age-related changes in suprathreshold taste intensity perception (see Murphy, 1986, for a review). Enns et al. (1979) found no alteration in the slope of the psychophysical function for sucrose during the period from young adulthood to old age, although the slope generated by these age groups (1.12) was flatter than the slope for fifth graders (1.72). Bartoshuk et al. (1984) have reported stability of the shape of the psychophysical function in the elderly, although these authors have seen some flattening of the slope near threshold, which they suggest may be due to masking of the stimulus taste by a residual taste in the oral cavity caused by poor oral hygiene. Schiffman and Clark (1980), Schiffman et al. (1981), Hyde and Feller (1981), and Cowart (1983) all reported some flattening of the slopes of one or more taste intensity functions in elderly subjects. Schiffman and her colleagues investigated amino acids and artificial sweeteners and reported lower exponents for most functions generated by the elderly groups of subjects. Hyde and Feller reported decreased intensity of caffeine and citric acid for elderly subjects relative to young subjects, and no age-associated change in suprathreshold intensity scaling of sucrose and NaCl. In a life-span study, Cowart tested substances representative of the four basic taste qualities: sweet, sour, bitter, and salty. She noted that the taste functions of the oldest group were always flatter than those of any other group, but that only the quinine functions showed a consistent pattern of flattening across age groups. Magnitude estimates for salty stimuli were less for subjects over 35 than for those under 35. Age-

TABLE I Results of Existing Studies on Age-Associated Changes in Taste Threshold Sensitivity[a]

Reference	Sucrose	NaCl	Quinine	PROP	PTC	Acetic acid	Citric acid	HCl	Amino acids	Mean of 8
Balogh and Lelkes (1961)	▼									
Bouliere et al. (1958)	▼	▽								
Cooper et al. (1959)	▼	▼	◀							
Dye and Koziatek (1981)	▼		▼							
Glanville et al. (1964)		▼	▼	▼						
Grzegorczyk et al. (1979)					▼					
Harris and Kalmus (1949)	▼				▼					
Hinchcliffe (1958)						—				
Kalmus and Trotter (1962)	▼									
Kaplan et al. (1965)			—	—						
Moore et al. (1982)	▼							▼		
Murphy (1979)	▼									▼
Richter and Campbell (1940)								▼		
Schiffman et al. (1979)		▼	▼▼						▲	
Smith and Davies (1973)							—			
Weiffenbach et al. (1982)	—									

[a] —, Maintenance of function; ▼, decline in function with age; ▽, decline for males, but not for females; ▲, increase in function with age.

related decrements were also seen in those over 55 for bitter solutions and for elderly persons to concentrated sucrose solutions. More research is needed to resolve the apparent discrepancies between the results of these studies.

Studies of olfactory threshold across the life span have yielded close agreement that there is age-related decline, both for stimuli that are largely olfactory and for stimuli that are largely trigeminal (see reviews by Schiffman, 1979; Murphy, 1986). Table II summarizes the stimuli and results of these studies.

Recent work by Stevens *et al.* (1982, 1984) and Murphy (1983) demonstrated that there are age-related deficits in perception of the intensity of suprathreshold olfactory and trigeminal stimuli.

Schiffman (1977) demonstrated a decline in the elderly person's ability to identify, when blindfolded, foods which had been pureed and presented at the same temperature to both college students and residents of a retirement home. Murphy (1981, 1985) replicated this result and demonstrated that the decrement in identification was due more to an olfactory loss in the elderly than to a taste loss. Subsequently, Stevens *et al.* (1984) again showed that, relative to the loss of suprathreshold taste intensity, loss of suprathreshold olfactory intensity is greater in the elderly.

Given the growing body of evidence suggesting both threshold and suprathreshold changes in sensitivity of the chemical senses, the question arises: How is preference for chemosensory stimuli affected by age? There is a paucity of literature in this area, and what does exist sometimes presents the results of studies done on relatively small samples.

Laird and Breen (1939) reported an increased preference for tart taste over sweet taste in older subjects and, to a lesser extent, in females of all ages. Enns *et al.* (1979) found a similar effect when preferences for sucrose were measured for 21 fifth graders, 27 college students, and 12 elderly subjects. The elderly subjects and females in general showed a lesser preference for sweeter sucrose solutions than the children. In a sample of 618 children and 140 adults, Desor *et al.* (1975) demonstrated that 9–15 year olds preferred greater sweetness and saltiness than did adults 18–64 years old. No additional age effects emerged when preferences of adults 18–29 years old were compared to adults 45–64 years old. The proportions of the adult group falling into each of these age ranges were not reported. Dye and Koziatek (1981) measured pleasantness of suprathreshold aqueous solutions of sucrose for 79 diabetic and nondiabetic veterans. Preferences of men 41–65 years were compared with responses of men 65–88 years. Although the simple effect of age was not significant, the three-way interaction of age, patient group, and sucrose level was significant at the 0.0001 level, indicat-

TABLE II Results of Existing Studies on Age-Associated Changes in Olfactory Threshold Sensitivity et al.[a]

Reference	n-Butanol	Isoamyl acetate	Menthol	Phenol	Phenethyl alcohol	Town gas	Mean of 18	Mean of 8	Unidentified
Chalke and Dewhurst (1957)						▶			
Fordyce (1961)				▶▶					
Joyner (1961)									
Kimbrell and Furchtgott (1963)		*							
Minz (1968)	▶								
Murphy (1983)			▶						
Schiffman et al. (1976)					▶				
Strauss (1970)							▶	▶	
Venstrom and Amoore (1968)									▶

[a] ▶, Decline in function with age; *, trend toward a decline in function with age, but a nonsignificant result.

ing that age, patient group, and sucrose level combined to produce significant differences between means for preference.

Hence, the literature on age-associated changes in preferences is scant, but it does suggest the possibility that chemosensory preferences may be altered in later life. Although a single study dealing with flavor preferences in malnourished infants suggests the possibility of a link between preference and nutritional status at least in that age group (Vasquez et al., 1982), a survey of the current literature provided no quantitative data on the relationship between chemosensory preference and any measure of nutritional status in the elderly. The following research was designed to investigate the possibility of such a relationship.

III. CHEMOSENSORY PREFERENCE AND BIOCHEMICAL INDEXES IN THE ELDERLY

A. Experiment 1

An interest in the potential link between age-associated chemosensory loss and nutritional status in the elderly prompted the following study, which was designed as a first attempt to relate chemosensory preference in the elderly with biochemical indexes that have been used in biochemical studies of nutritional status in the elderly.

Experiment 1 was designed simply to investigate the effects of aging and the biochemical indexes serum albumin, total protein, and blood urea nitrogen (BUN) on the preference for casein hydrolysate, an amino acid mixture, presented in a soup base. The hypotheses were that (1) older subjects would rate high concentrations of the amino acid mixture as pleasanter than young subjects would, and (2) subjects with lower values for albumin and protein, and higher values for BUN would prefer higher concentrations of casein hydrolysate.

1. Method

a. Subjects. Subjects were 10 persons 18–26 years old and 16 persons over the age of 65. The young subjects were students and technicians. The elderly were recruited from senior centers and senior citizen's apartment complexes in an urban area.

b. Stimuli. The chemosensory stimuli consisted of an amino acid-deficient soup base to which were added the following concentrations of casein hydrolysate: 0, 1, 2, 3, 4, and 5% (w/v). Stimuli were prepared in deionized water, refrigerated, and heated before presentation.

c. Procedure. Subjects rated pleasantness of these chemosensory stimuli using a bipolar line scale described previously (Murphy, 1982). Marks to the left of the zero midpoint indicated unpleasantness. Marks to the right of zero indicated pleasantness. Distance from the zero midpoint represented degree of pleasantness or unpleasantness. Within the single, self-paced session, the six concentrations of casein hydrolysate were presented in random order. Since this experiment was part of a larger study, 12 other chemosensory stimuli were also judged in this session, which lasted approximately 45 min. Subjects rinsed before each stimulus and then sipped two teaspoonsful of the 30 ml provided in a disposable polyethylene bowl.

Blood was drawn from each subject, usually on the same day, and an independent laboratory provided an assay of serum total protein, albumin, and BUN.

Finally, each subject responded to a questionnaire regarding health and smoking status, drug usage, food preferences, etc.

d. Data Analyses. Each of the independent variables protein, albumin, and BUN was rendered discrete by taking a median to divide all values into high and low. In the few cases where values fell at the median, these values were not used in the analyses. Analyses of variance (ANOVAs) were then conducted on the biochemical indices.

2. Results

a. Biochemical Status. ANOVAs on the blood values for biochemical indexes showed significant differences between old and young subjects for protein, albumin, and BUN. The mean values for the elderly were 6.6, 4.0, and 23.9, respectively. For the young they were 7.0, 4.5, and 11.6, respectively. All differences between means were significant at the 0.05 level. Within the elderly group, 7 of the 16 had low serum protein levels, defined as less than 6.5 g/dl. One young subject had a protein level less than 6.5 g/dl.

b. Peak Preferred Concentration. Analyses of variance were conducted using the concentration chosen by the subject as his most preferred, referred to as the peak preferred concentration (PPC), as the dependent variable. One-way ANOVAs on the peak preferred concentration showed significant effects of age group, $F(1,24) = 19.03$, $p < 0.0002$; serum albumin, $F(1,22) = 16.26$, $p < 0.001$; and BUN, $F(1,24) = 12.81$, $p < 0.002$. Older subjects preferred higher concentrations of casein hydrolysate than did young subjects. In fact, as shown in Fig. 1, 80% of the young subjects chose the soup base with no added amino acids as the most pleasant. In contrast, the aver-

5. The Chemical Senses and Nutrition in the Elderly

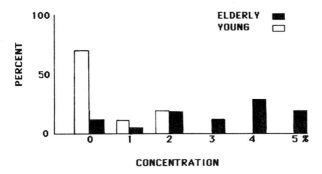

Fig. 1. The percentage of each age group in Experiment 1 who chose any given concentration of amino acids as the most preferred.

age concentration rated most pleasant by the elderly was 3.0% casein hydrolysate.

As indicated above, when the data were considered in terms of the biochemical indexes, it was clear that subjects with lower values for serum albumin and higher values for BUN chose higher concentrations of the amino acid mixture as being more pleasant. Figure 2 illustrates the effect of biochemical indices on chemosensory preference, showing peak preferred concentration as a function of albumin.

Age group and biochemical indexes were significantly correlated, regardless of whether the index was albumin ($r = -0.68$), protein ($r = -0.39$), or BUN ($r = 0.80$). A multiple regression, using peak preferred concentration as the dependent variable, with age group, albumin, protein, BUN, and gender as the independent variables showed serum albumin to be the best

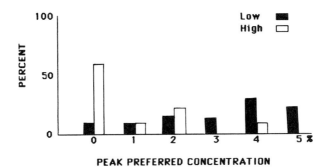

Fig. 2. The percentage of persons in Experiment 1 with values for serum albumin that were below the median (low) or above the median (high), who chose any given concentration of amino acids as the most preferred.

predictor of peak preferred concentration ($R = 0.72$, $R^2 = 0.51$). With albumin in the model, age would be the next best predictor, but it was not statistically significant, $F(1,23) = 3.12$, $0.1 > p > 0.05$. Since age and serum albumin were highly correlated ($r = -0.68$), the significant amount of shared variance probably accounts for this result.

B. Experiment 2

Experiment 1 suggested, in a small sample, the influence of age and blood indexes on the preference for amount of an amino acid mixture, added to a soup base. Experiment 2 was designed to further investigate these variables as well as to determine the effect of intensity on the pleasantness of casein hydrolysate. We sought to investigate the potential role of reduced chemosensory function in the elderly on their preferences. We reasoned that the differences in preference seen in the first experiment could be due simply to reduced input to the olfactory and taste systems of the elderly, and that their preferences for higher concentrations of amino acids might simply reflect a desire for greater intensity, rather than for the specific nutrients. If this were the case, then intensity should predict preference. The hypotheses were (1) that older subjects would prefer higher concentrations of the amino acid mixture, (2) that subjects of lower biochemical status would prefer higher concentrations of casein hydrolysate, and (3) that intensity would prove to be predictive of preference.

1. Method

a. Subjects. There were 40 participants in this study. Twenty were 18–26 years of age and 20 were 65 or more years of age. Each age group contained 7 males and 13 females. All were active, community-dwelling persons with no hospitalizations in the preceding 12 months. The young subjects were university students. The elderly subjects were recruited from a retired professor's club, a university-based education and enrichment program for older adults, and a senior center in an affluent area.

b. Stimuli. The chemosensory stimuli consisted of an amino acid-deficient soup base to which were added the following concentrations of casein hydrolysate: 0, 1, 2, 3, 4, and 5% (w/v). Stimuli were prepared in deionized water, refrigerated, and heated before presentation. Thirty-milliliter samples were presented in disposable polyethylene bowls. Subjects sipped two teaspoonsful of each sample.

The auditory stimuli were six sound pressure levels of a noise spectrum of 150–460 Hz presented via TDH-39 headphones, which had been calibrated to deliver the following: 70, 71, 72, 75, 80, 84 db.

c. **Procedure.** Subjects participated in two sessions each approximately 1.5 hr long. In the first, they were screened with a rapid modified auditory threshold procedure designed to ensure that all stimuli to be presented in the magnitude matching experiment were above threshold. In the same session, subjects also were tested on various measures of psychosocial and cognitive status, which will not be described here.

On the second day, subjects first had blood drawn at the university health center and then proceeded to the laboratory. The blood samples were later assayed for total serum protein, albumin, and BUN by an outside laboratory.

The method of magnitude matching, developed by Stevens and Marks (1980), was used to determine the intensity of both the auditory and the chemosensory stimuli. This method embodies elements of both magnitude estimation and cross-modality matching. See the paper by Stevens and Marks (1980) for an excellent discussion of the usefulness of this methodology in making comparisons of absolute intensity of stimuli between groups of subjects. Subjects assigned numbers on the same scale of magnitude to reflect perceived intensity of both the auditory stimuli and the chemosensory stimuli.

Subjects also rated pleasantness of both auditory and chemosensory stimuli using a bipolar line scale described previously (Murphy, 1982). Marks to the left of the zero midpoint indicated unpleasantness. Marks to the right of zero indicated pleasantness. Distance from the zero midpoint represented degree of pleasantness or unpleasantness.

Within the session, presentation of auditory and chemosensory stimuli was alternated. Within the sensory category, stimuli were randomly ordered. Subjects rinsed before and after each chemosensory stimulus. Subjects judged both intensity and pleasantness before proceeding to the next stimulus.

d. **Data Analyses.** The variables serum albumin, total protein, and BUN were rendered discrete, in the same manner as in Experiment 1, before the analyses were conducted. Because of the large number of values falling at the median for albumin (8), a third group was designated as middle values and the analyses were conducted using the three groups: low, middle, and high.

2. Results

a. **Biochemical Indexes.** One-way ANOVAs on the raw values for serum protein, albumin, and BUN, using age group as the independent variable in each analysis, showed age to be a significant predictor, regardless of the index of biochemical status. The mean values for protein, albumin, and BUN were 6.7, 4.2, and 15.9, respectively, for the elderly. The younger subjects had serum values of 7.2 for protein, 4.6 for albumin, and 12.8 for BUN. In this sample, 20% of the elderly subjects had low protein values (<6.5 g/dl). None of the young subjects had low protein values.

b. Peak Preferred Concentration. The concentration of casein hydrolysate that was rated the highest by each subject was labeled his peak preferred concentration (PPC). ANOVAs were performed on these data to determine which factors were significant in determining the peak preferred concentration. In each analysis, PPC was the dependent variable. In the first one-way ANOVA, age was the independent variable. Age group was shown to be a significant predictor of preferred concentration, $F(1,38) = 7.79$, $p < 0.01$. As in Experiment 1, older subjects chose higher concentrations as their most preferred. Figure 3 shows the distribution of their preferences over the concentration range. The younger subjects chose lower concentrations as their most preferred. In fact, as Fig. 3 illustrates, 75% of the young subjects chose the soup base with no added casein hydrolysate as the most preferred.

The biochemical indexes also proved to be predictive of peak preferred concentration. A one-way ANOVA with serum albumin as the independent variable demonstrated that those with lower albumin values preferred higher concentrations of the amino acid mixture than did those with higher values for albumin, $F(2,37) = 4.79$, $p < 0.02$. The one-way ANOVA investigating the influence of protein on peak preferred concentration was also statistically significant, $F(2,37) = 6.11$, $p < 0.01$. Subjects whose protein values fell below the median preferred higher concentrations of amino acids. Of the three biochemical indexes, only BUN did not prove to be significant at the 0.05 level of probability, $F(2,37) = 1.17$, $p > 0.1$. The trend was, however, in the predicted direction: Subjects with high values for BUN tended to prefer higher concentrations.

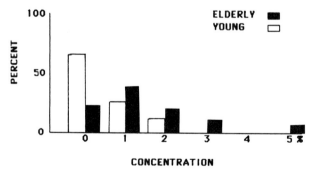

Fig. 3. The percentage of each age group in Experiment 2 who chose any given concentration of amino acids as the most preferred.

Since each of the three biochemical measures, protein, albumin, and BUN, played a role in predicting preferred concentration (although BUN was not significant at the 0.05 level in Experiment 2, it was a strong predictor in Experiment 1, where the range of BUN values was greater than in Experiment 2), it seemed that a composite of the three values should provide a better predictor than any one of the measures alone. To create a single biochemical index, each subject's standard (z) scores for the three blood measures were summed and weighted using standardized regression coefficients (0.338 for protein, 0.257 for albumin, and −0.237 for BUN). Since these blood measures were significantly affected by age (mean index z score for elderly = −0.41, for young = 0.41, $t(38)$ = 6.04, $p < 0.0001$), the index values were grouped above and below the appropriate age group medians, so that each subject was classed as having a high or low composite biochemical index according to his or her own age group.

A two-way ANOVA, testing the effect of age and biochemical index on preferred concentration of casein hydrolysate, showed both variables to be significant predictors of peak preferred concentration. Elderly participants preferred higher concentrations of amino acids, $F(1,36) = 10.20$, $p < 0.003$, $\eta^2_{partial} = 0.22$. Subjects with lower composite biochemical indexes also preferred higher concentrations, $F(1,36) = 12.59$, $p < 0.001$, $\eta^2_{partial} = 0.26$. Figure 4 shows peak preferred concentration as a function of age and composite biochemical index.

Finally, a stepwise multiple regression was performed on peak preferred concentration using age group, composite biochemical index, gender, and the subject's geometric mean chemosensory intensity rating as variables. The composite biochemical index accounted for the most variance, and so was the first variable to enter the model ($R = 0.46$), followed by age ($R = 0.62$). The remaining variables failed to improve the model and were not entered.

c. Intensity. To determine the effect of chemosensory intensity on the peak preferred concentration of amino acids, chemosensory intensity judgments were first normalized by auditory intensity judgments, using a procedure described by Stevens et al. (1982). The subject's geometric mean of the loudness estimates was computed. The geometric mean was then multiplied by a factor which made the geometric mean a constant across subjects. All of the magnitude estimates given by a subject in the experiment were then multiplied by his normalization factor. This operation effectively placed all subjects' judgments on the same number scale. The normalized data were used in the following analyses.

The grand geometric mean intensity was used to divide all subject geo-

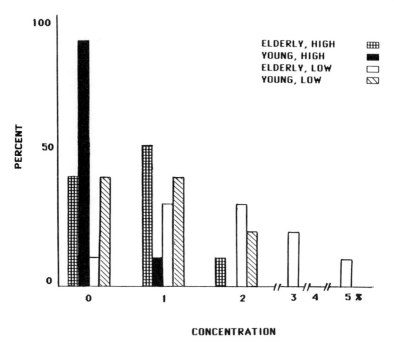

Fig. 4. The percentage of persons who chose any given concentration of amino acids as the most preferred, shown as a function of both age group and the composite biochemical index.

metric means into two groups: those above the grand geometric mean and those below the grand geometric mean.

ANOVA was run using high and low intensity as the independent variable and peak preferred concentration as the dependent variable. The effect of intensity on preference was statistically significant in a one-way ANOVA at the 0.008 level, $F(1,38) = 8.1$. There was, of course, a significant negative correlation between age and intensity, $r = -0.51$. Older subjects tended to have lower intensity estimates for the chemosensory stimuli. In fact, ANOVA with age as the independent variable and geometric mean intensity as the dependent variable was also statistically significant, $F(1,38) = 13.91$, $p < 0.0006$.

It should be appreciated that the stepwise multiple regression of the factors age group, gender, geometric mean intensity, albumin, protein, BUN, and composite biochemical index on the dependent variable peak preferred concentration showed only the composite biochemical index ($R = 0.46$) and age ($R = 0.62$) to enter the model. The remaining variables, including intensity, failed to improve the model for prediction of peak preferred concentration. Thus, it would appear that the effect of intensity on peak preferred concentration, seen in the one-way ANOVA, is due to age, which clearly predicts intensity.

C. Summary

Two of the initial hypotheses in Experiment 2 were confirmed: (1) older subjects preferred higher concentrations of the amino acid mixture, and (2) subjects with lower values for the combined biochemical index preferred higher concentrations of the amino acid mixture. The third hypothesis was not confirmed: intensity in itself did not predict the peak preferred concentration of amino acids. Thus, the present studies suggest individual differences in preference for a nutritionally significant chemosensory stimulus, and that these differences can be ascribed to both biochemical status and to age.

IV. DISCUSSION AND CONCLUSIONS

The results of these experiments together suggest the importance of further investigation of the link between the chemical senses and nutrition in the elderly.

The reader familiar with the animal nutrition literature may see the need for a study of threshold sensitivity to amino acids in subjects with relatively high and low values for albumin or total protein. Such a study would be analogous to animal experiments in which the preference threshold for a given nutrient is investigated in nutrient-deprived and nutrient-rich animals. The investigation would not be straightforward in the case of a study involving elderly and young human subjects with divergent values for serum albumin or total protein, simply because the thresholds for the chemical senses tend to rise with age. If there were a decrease in threshold for amino acids in elderly subjects with low values for the biochemical indexes, such an effect could be obscured by an age-associated increase in threshold, which would be suggested by the chemosensory literature. However, the appropriate experiment that also controls for age-related chemosensory decline is currently being designed. We hope to examine the age effects and the nutritional effects separately in future experiments. This was not possible in the present experiment since within the sample population, the elderly subjects tended to have the higher values for albumin and protein and the lower values for BUN.

In future experiments, we plan to address the question of dietary intake of protein directly by requiring 7-day diet records of all subjects. The not insignificant difficulties involved in nutritional assessment of the elderly prompted the use of biochemical indexes in the preliminary studies described above. The strength of these indexes in predicting peak preferred concentration suggests that more extensive investigation of the link between nutritional status and preference is warranted. By requiring subjects to record intake rather than to participate in dietary recall, we hope to overcome problems that can lead to inaccuracy. Decreased short-term memory, hearing loss, and poor communication skills have the potential for affecting the validity and reliability of the 24-hr recall method (Bowman and Rosenberg, 1982). We are particularly sensitive to the potential for these problems to occur within our elderly sample. Questions of validity and reliability also arise regarding the use of anthropometric measures of nutritional status within the elderly population. Triceps skinfold measurements for elderly samples are highly variable in the literature, particularly in females (Bowman and Rosenberg, 1982). Because of age-associated changes in lean body mass, total body fat, distribution of body fat and skin thickness, elasticity, and compressibility, the accuracy and interpretation of results are complicated.

Renal function has been shown to decrease with age. Hence, concern regarding the use of BUN as an indicator would seem justified. Of the three blood indicators measured in each person, BUN seems the least useful in providing information regarding nutritional status.

It should be understood that both elderly and young subjects in the present studies were generally not nutritionally deficient. We found no frank malnutrition in the subjects we sampled. High and low values for biochemical indexes refer to values above and below the median in the experiment, not to established values for assessment of nutritional status. We hope in future studies to sample from populations which show divergence in nutritional status that is of frank clinical significance. Though only limited conclusions can be drawn from the present experiments, it is suggestive that the present studies which focused on relative differences in biochemical indexes that generally fell within the normal range of variability were able to capture an effect on chemosensory preference.

The influence of other factors on dietary intake in the elderly cannot be ignored. In another study we are investigating factors such as socioeconomic status, level of education, income, location in which meals are consumed, etc. Demographic information may shed considerable light on the intake patterns of elderly persons.

V. RESEARCH NEEDS

Within the realm of chemosensory changes with age, there are several areas of research which have the potential for further developing our understanding. Closure needs to be achieved on the question of quality-specific, age-related changes in threshold for taste. The rate of change in taste threshold over the life span needs to be quantitatively defined for each quality. In both of these cases, there is a need for the development of methodologies that are both more sensitive and control completely for the subject's criterion.

More information is needed about the determinants of chemosensory preference, particularly in the elderly. Since cohort effects and other problems with cross-sectional studies are limiting, there is a real need for longitudinal data.

The link between nutrition and chemosensory preference needs to be explored further. The direction of this relationship also needs to be explored. Does nutritional status drive preference or does preference behavior ultimately affect nutritional status? It would not be surprising if the relationship were bidirectional; hence, determining the degree to which each factor affects the other represents a methodological challenge.

Furthermore, measures of chemosensory preference and sensitivity need to be directly related to dietary intake. Historically, research has proceeded without directly linking sensitivity to preference and dietary intake.

The effects of chronic disease and multiple drug therapy on both nutrition and the chemical senses deserve close attention.

ACKNOWLEDGMENTS

This research was supported by NIH Grant AG04085 from the National Institute on Aging. Experiment 1 was supported in part by funds from the Monell Chemical Senses Center. I am grateful to Dr. Gary K. Beauchamp for supplying the soup base and for helpful suggestions during the pilot study, and to Ms. Karen Knauff of Johnson and Johnson for supplying the casein hydrolysate. Dr. Sharon Greeley's comments on an earlier version of this manuscript were most helpful and very much appreciated. I thank Kelly Dozois, Elizabeth Konowal, Michele J. Reed, and Anne Woltjen for excellent technical assistance, and Kirsty Nunez and Jeanne Withee for expert computational assistance.

The research described above under Experiments 1 and 2 has been submitted for publication in full-length manuscript form.

REFERENCES

Balogh, K., and Lelkes, K. (1961). The tongue in old age. *Gerontologia Clinica.* **3** (Suppl.) 38–54.

Bartoshuk, L. M., Marks, L. E., Stevens, J. C., and Rifkin, B. (1984). Taste and aging. *Annu. Meeting Assoc. Chemoreception Sci., 6th, Sarasota, Florida.*

Beauchene, R. E., and Davis, T. A. (1979). The nutritional status of the aged in the U.S.A. *Age* **2**, 23.

Bouliere, F., Cendron, H., and Rapaport, A. (1958). Modification avec l'age des seuils gustatifs de perception et de reconnaissance aux saveurs salée et sucre, chez l'homme. *Gerontologia* **2**, 104–112.

Bowman, B. B., and Rosenberg, I. H. (1982). Assessment of the nutritional status of the elderly. *Am. J. Clin. Nutr.* **35**, 1142–1151.

Chalke, H. D., and Dewhurst, J. R. (1957). Accidental coal-gas poisoning. *Br. Med. J.* **2**, 915–917.

Cooper, R. M., Bilash, I., and Zubeck, J. P. (1959). The effect of age on taste sensitivity. *J. Gerontol.* **14**, 56–58.

Cowart, B. J. (1983). Direct scaling of the intensity of basic tastes: A life span study. *Annu. Meeting Assoc. Chemoreception Sci., 5th, Sarasota, Florida.*

Desor, J. A., Green, L. S., and Maller, O. (1975). Preferences for sweet and salty tastes in 9- to 15-year-old and adult humans. *Science* **190**, 686–687.

Dye, C. J., and Koziatek, D. A. (1981). Age and diabetes effects on threshold and hedonic perception of sucrose solutions. *J. Gerontol.* **36**, 310–315.

Enns, M. P., Van Itallie, T. B., and Grinker, J. A. (1979). Contributions of age, sex and degree of fatness on preferences and magnitude estimation for sucrose in humans. *Physiol. Behav.* **22**, 999–1003.

Fordyce, I. D. (1961). Olfaction tests. *Br. J. Ind. Med.* **18**, 213–215.

Glanville, E. V., Kaplan, A. R., and Fischer, R. (1964). Age, sex and taste sensitivity. *J. Gerontol.* **19**, 474–478.

Grzegorczyk, P. B., Jones, S. W., and Mistretta, C. M. (1979). Age-related differences in salt taste acuity. *J. Gerontol.* **34**, 834–840.

Harris, H., and Kalmus, H. (1949). The measurement of taste sensitivity to phenylthiourea (PTC). *Ann. Hum. Genet. (London)* **15**, 24–31.

Hinchcliffe, R. (1958). Clinical quantitative gustometry. *Acta Otolaryngol.* **49**, 453–466.

Hyde, R. J., and Feller, R. P. (1981). Age and sex effects on taste of sucrose, NaCl, citric acid and caffeine. *Neurobiol. Aging* **2**, 315–318.

Jansen, C., and Harrill, I. (1977). Intakes and serum levels of protein and iron for 70 elderly women. *Am. J. Clin. Nutr.* **30**, 1414–1422.

Joyner, R. E. (1963). Olfactory acuity in an industrial population. *J. Occup. Med.* **5**, 37–42.

Kalmus, H., and Trotter, W. R. (1962). Direct assessment of the effect of age on PTC sensitivity. *Ann. Hum. Genet.* **26**, 145–149.

Kaplan, A., Glanville, E., and Fischer, R. (1965). Cumulative effect of age and smoking on taste sensitivity in males and females. *J. Gerontol.* **20**, 334–337.

Kimbrell, G. M., and Furchtgott, E. (1963). The effect of aging on olfactory threshold. *J. Gerontol.* **18**, 364–365.

Laird, D. A., and Breen W. J. (1939). Sex and age alterations in taste preferences. *J. Am. Dietetic Assoc.* **15**, 549–550.

Minz, A. I. (1968). Condition of the nervous system in old men. *Z. Alternsforschung* **21**, 271–277.

Moore, L. M., Nielson, C. R., and Mistretta, C. M. (1982). Sucrose taste thresholds: Age-related differences. *J. Gerontol.* **37**, 64–69.

Murphy, C. (1979). The effects of age on taste sensitivity. *In* "Special Senses in Aging" (S. S. Han and D. H. Coons, eds.), pp. 21–33. University of Michigan Institute of Gerontology, Ann Arbor, Michigan.

Murphy, C. (1981). Effects of aging on chemosensory perception of blended foods. *Annu. Meeting Assoc. Chemoreception Sci., 3rd, Sarasota, Florida.*

Murphy, C. (1982). Effects of exposure and context on hedonics of olfactory–taste mixtures. *In* "Selected Sensory Methods: Problems and Approaches to Measuring Hedonics, ASTM

STP 773" (J. T. Kuznicki, R. A. Johnson, and A. F. Rutkiewic, eds.), pp. 60–70. American Society for Testing and Materials, Philadelphia, PA.

Murphy, C. (1983). Age-related effects on the threshold, psychophysical function, and pleasantness of menthol. *J. Gerontol.* **38**, 217–222.

Murphy, C. (1985). Cognitive and chemosensory influences on age-related changes in the ability to identify blended foods. *J. Gerontol.* **40**, 47–52.

Murphy, C. (1986). Taste and smell in the elderly. *In* "Clinical Measurement of Taste and Smell" (H. L. Meiselman and R. S. Rivlin, eds.), Macmillan, New York. (In press.)

Richter, C., and Campbell, K. (1940). Sucrose taste thresholds of rats and humans. *Am. J. Physiol.* **128**, 291–297.

Schiffman, S. S. (1977). Food recognition by the elderly. *J. Gerontol.* **32**, 586–592.

Schiffman, S. S. (1979). Changes in taste and smell with age: Psychophysical aspects. *In* "Sensory Systems and Communication in the Elderly" (J. M. Ordy and K. Brizzee, eds.), pp. 227–246. Raven Press, New York.

Schiffman, S. S., and Clark, T. B. (1980). Magnitude estimates of amino acids for young and elderly subjects. *Neurobiol. Aging* **1**, 81–91.

Schiffman, S. S., Moss, J., and Erickson, R. P. (1976). Thresholds of food odors in the elderly. *Exp. Aging Res.* **2**, 389–398.

Schiffman, S. S., Hornak, K., and Reilly, D. (1979). Increased taste thresholds of amino acids with age. *Am. J. Clin. Nutr.* **32**, 1622–1627.

Schiffman, S. S., Lindley, M. G., Clark, T. B., and Makins, C. (1981). Molecular mechanism of sweet taste: Relationship of hydrogen bonding to taste sensitivity in both young and elderly. *Neurobiol. Aging* **2**, 173–185.

Smith, S. E., and Davies, P. D. (1973). Quinine taste thresholds: A family study and a twin study. *Ann. Hum. Genet.* **37**, 227–232.

Stevens, J. C., and Marks, L. E. (1980). Cross-modality matching functions generated by magnitude estimation. *Percept. Psychophys.* **27**, 379–389.

Stevens, J. C., Plantinga, A., and Cain, W. S. (1982). Reduction of odor and nasal pungency associated with aging. *Neurobiol. Aging* **3**, 125–132.

Stevens, J. C., Bartoshuk, L. M., and Cain, W. S. (1984). Chemical senses and aging: Taste versus smell. *Chem. Senses* **9**, 167–179.

Strauss, E. L. (1970). A study on olfactory acuity. *Ann. Otol. Rhinol. Laryngol.* **79**, 95–104.

Vasquez, M., Pearson, P. B., and Beauchamp, G. K. (1982). Flavor preferences in malnourished Mexican infants. *Physiol. Behav.* **28**, 513–519.

Venstrom, D., and Amoore, J. E. (1968). Olfactory threshold in relation to age, sex or smoking. *J. Food Sci.* **33**, 264–265.

Weiffenbach, J. M., Baum, B. J., and Burghauser, R. (1982). Taste thresholds: Quality specific variation with human aging. *J. Gerontol.* **37**, 700–706.

Yearick, E. S., Wang, M. L., and Pisias, S. J. (1980). Nutritional status of the elderly: Dietary and biochemical findings. *J. Gerontol.* **35**, 663–671.

Young, V. R., Perera, W. D., Winterer, J. C., and Scrimshaw, N. S. (1976). Protein and amino acid requirements of the elderly. *In* "Nutrition and Aging" (M. Winick, ed.), pp. 77–118. Wiley, New York.

6

Micronutrients and Taste Stimulus Intake

SHARON GREELEY,* CHARLES N. STEWART,[†] AND
MARY BERTINO*
*Monell Chemical Senses Center
Philadelphia, Pennsylvania
[†]Department of Psychology
Franklin and Marshall College
Lancaster, Pennsylvania

I.	Introduction		108
II.	Effects of Deficiencies on Taste Preferences		110
	A.	B Vitamins	110
	B.	Zinc	111
	C.	Iron and Copper	113
	D.	The Case of D-Penicillamine	114
III.	Experimental Data		115
	A.	Experiment I: Severe Copper and Vitamin B_6 Depletions and Taste Behavior	115
	B.	Experiment II: Moderate Zinc and Vitamin B_6 Depletions and Taste Behavior	117
	C.	Experiment III: Severe Vitamin B_6 Depletion and NaCl Intake	120
	D.	Experiment IV: Severe Vitamin B_6 Depletion and KCl Intake	121
	E.	Experiment V: Severe Vitamin B_6 Depletion and Short-Term NaCl and KCl Intake	122
	F.	Discussion: Experiments I–V	123
IV.	Future Research		126
	References		126

I. INTRODUCTION

Micronutrients are food substances that are essential to normal metabolic function in animals. Generally, micronutrients are considered to belong to one of two categories: vitamins (organic compounds) or trace minerals (inorganic elements). Simple dietary inadequacy leads to what is called a primary micronutrient deficit. Conditioned deficiencies are secondary to some other cause as, for example, certain drug treatments, ingestion of radically imbalanced diets, some forms of pica, and malabsorption syndromes.

Dietary depletions of micronutrients are possible during all phases of the life cycle but fetal, suckling, and early growth are periods particularly sensitive to the development of nutrient deficits. Should B vitamin deficiencies be imposed during early development, neurogenesis may fail and severe brain damage ensue. Similarly, *in utero* zinc deficits may result in teratogenesis and copper deficits in neural disorders (National Research Council, 1978) with eventual death of the offspring.

Since adult animals often respond differently, it is important to understand the biological relationship between the micronutrient and the physiological state of the animal when the deficiency is imposed. For example, after development is complete, niacin deficiency coupled with limited tryptophan intake alters serotonin metabolism, leading to specific behavioral changes. Myelin degeneration, fatty liver, and anemia may accompany chronic riboflavin deficiencies in adult animals; anemia, hyperexcitability, and convulsions develop as a consequence of pyridoxine (vitamin B_6) deficiency (National Research Council, 1978). Zinc depletion of adult rats causes mild anorexia and reproductive failure in both males and females (National Research Council, 1978), whereas inadequate copper intake may result in anemia and cardiac arrhythmias (Klevay, 1984). Deficiencies can be detected clinically through gross observations such as stunted growth, as well as through precise biochemical reactions that measure single substrate transformations. The wide range of symptomology and biochemical measures reflects the diversity of the biofunctions associated with the micronutrients.

The B vitamins (e.g., niacin, riboflavin, thiamin, pyridoxine) are water soluble and therefore readily depleted from the body. In their active forms, B vitamins are essential to numerous enzyme reactions involving nitrogen and energy (Pike and Brown, 1975). Niacin, as nicotinamide, is a component of the coenzymes NAD and NADP. These coenzymes participate in many cellular metabolic processes, and NAD is required for ATP synthesis via oxidative phosphorylation. Riboflavin is a precursor of flavin coenzymes and is necessary for specific enzymes involved in oxidation–reduction reactions. Thiamin is a constituent of the coenzyme thiamin pyrophosphate, which takes part in reactions of intermediary metabolism, e.g., oxidative decarboxylations. Vitamin B_6 compounds are precursors of the coenzyme pyridoxal phosphate and, as such, are required for most reactions involving amino acids (Pike and Brown, 1975). The catalytic role of these micronutrients explains why B vitamin status is intricately involved with the utilization of dietary macronutrients (protein, fat, and carbohydrates), and probably contributes to growth failure and anorexia, which commonly accompany B vitamin deficiencies in animals and humans.

Copper and iron are trace elements that play a critical role in respiration. Iron is required for oxygen transport to cells and maintenance of adequate oxygen levels within muscle tissue. Copper is needed to facilitate iron absorption from the gut and iron mobilization from storage tissues. In addition, a copper-containing enzyme, ceruloplasmin, ensures the appropriate redox state for iron (Pike and Brown, 1975). Unlike copper, zinc apparently plays no direct role in oxygen metabolism or oxidation–reduction reactions. This is attributable to the stable valence state of zinc in biological systems. Zinc is important in synthetic and degradative processes involving nitrogen compounds (DNA, RNA, and protein) and hormone metabolism (insulin, nerve growth factor, and hormones of the pituitary–adrenal axis) (Pike and Brown, 1975). Zinc also serves to stabilize macromolecular structures within cells (Dinsdale and Williams, 1977) and the integrity of cell membranes (Chvapil, 1973).

While numerous bioactivities have been characterized and ascribed to various micronutrients, the role(s) that these dietary substances play in taste function are less clearly defined. We might assume that changes in taste behavior stem directly from cellular or biochemical events involving the micronutrient under investigation. Poor vitamin A status presents a classic example in which taste changes appear to be an extension of cellular lesions induced by the deficiency (Bernard et al., 1962; Bernard and Halpern, 1968). Rats, replete with vitamin A, prefer mildly salty solutions and avoid quinine. In contrast, vitamin A-deficient rats are indifferent to these salt solutions but will drink water containing quinine. Electrophysiological recordings made from the chorda tympani indicate that vitamin A deficiency results in reduced sensitivity to sodium chloride. The explanation for both the behavioral and electrophysiological observations rests with the cellular role that vitamin A plays in normal surface epithelia turnover. Prolonged absence of the vitamin induces metaplasia in which columnar epithelia are replaced by stratified epithelia. When this occurs in the oral cavity, the dorsal surface of the tongue becomes keratinized, as demonstrated by histological studies (Bernard and Halpern, 1968). It appears that both the aberrant taste behavior and the reduced gustatory neural response are due at least in part to limited accessibility of taste stimuli to the taste pore.

II. EFFECTS OF DEFICIENCIES ON TASTE PREFERENCES

A. B Vitamins

A deficiency of two B vitamins, thiamin and pyridoxine, alters tastant intake (Chan and Kare, 1979; Hastings and Van, 1981; Stewart and Bhavagan, 1982). An increased preference for 0.15 M NaCl has been reported in thiamin-deficient rats (Hastings and Van, 1981). Thiamin deficiency also influences adrenal gland function during sodium deficiency. Normally, during sodium deficiency the width of the zona glomerulosa of the adrenal glands and plasma and urinary aldosterone titers are increased. However, in thiamin deficiency, this response to sodium deficiency is severely depressed. Corticosterone is elevated by thiamin deficiency alone and unaffected by dietary sodium intake.

It has been postulated that the general metabolic stress produced by thiamin deficiency leads to ACTH release and hence corticosterone secretion, producing an increase in sodium appetite via this mechanism (Hastings and Van, 1981). Stressing the pituitary–adrenal system directly by chronic injection of ACTH causes a 50% reduction in plasma aldosterone and a threefold increase in plasma corticosterone (Aquilera et al., 1981). Daily injection of ACTH over 5 days produces a large increase in sodium chloride intake, which is eliminated by adrenalectomy (Weisenger et al., 1978). The inference is made that the increased sodium intake seen in thiamin deficiency is related to corticosterone release. The authors further suggest that since sodium appetite can be induced by other nutrient deficiencies (zinc, thiamin), it is not necessarily nutrient specific and may be part of a general stress syndrome (Hastings and Van, 1981). Since only sodium chloride has been studied in thiamin-deficient animals, the specificity of the effect on tastant intake cannot be ascertained. It should be noted, however, that the ACTH effect upon preference does appear to be sodium specific since intake of potassium, magnesium, and calcium chloride salts was unaffected (Weisenger et al., 1978). Perhaps there are other deficiencies (e.g., vitamin B_6, iron) that would fit into this model.

There are conflicting reports concerning the effects of pyridoxine deficiency on tastant preferences. One study used a pair-fed control procedure in which rats were fed control diets in amounts equivalent to what the anorexic, pyridoxine-deficient animals were consuming. In this study, pyridoxine deficiency has been reported to enhance the rats' preference for both 0.15 and 0.30 M NaCl (Chan and Kare, 1979). Enhanced KCl preference has also been reported (Stewart and Bhagavan, 1982). Preferences for other solutions such as sodium saccharin, hydrochloric acid, or quinine sulfate were unaffected (Chan and Kare, 1979). Whole chorda tympani responses to sodium chloride, sodium saccharin, hydrochloric acid, or quinine sulfate were unaltered by the deficiency (Chan and Cote, 1979). Deficient rats were found to have enlarged adrenal glands relative to body size but displayed normal plasma levels of sodium and zinc (Chan and Kare, 1979). These results contrast with earlier reports (Hsu et al., 1958) that pyridoxine-deficient rats had elevated serum sodium levels whereas rats treated with deoxypyridoxine (a vitamin B_6 antagonist) had reduced serum potassium levels. Because of these conflicting reports the role of plasma electrolyte levels in sodium and potassium intake is unresolved.

B. Zinc

It was reported that zinc deficient rats consumed more 0.3 M NaCl solution than water when compared to *ad libitum* or pair-fed zinc-replete rats (McConnell and Henkin, 1974). The authors proposed that the drinking behavior reflected changes in taste acuity due to zinc deficiency. These studies were extended, and the results indicated that zinc-depleted rats (but not pair-fed controls) increased their consumption of sucrose, quinine sulfate, and hydrochloric acid solutions in addition to sodium chloride (Catalanotto and Lacy, 1977). The sucrose data are especially interesting in light of the fact that zinc-deficient rats prefer high-carbohydrate diets to high-protein diets (Reeves and O'Dell, 1981). The data could indicate that zinc depletion alters taste function. However, it might be argued that the taste stimuli induced systemic changes because the tastants were offered to the animals for 96 (McConnell and Henkin, 1974) and 48 (Catalanotto and Lacy, 1977) hr. This period of time is sufficient for the tastant solutions to induce systemic effects, as for example, changes in water balance or sodium status. However, when water and fluids containing sodium chloride, quinine sulfate, or hydrochloric acid were presented in two-bottle tests for 1 hr, the aberrant drinking behavior was also evident (Catalanotto, 1979). Based on these findings, it was suggested that the effects of zinc deficiency on drinking behavior are associated with taste rather than postingestional (systemic) cues (Catalanatto, 1979).

There are several possible explanations for the changes in rats' ingestion of tastant solutions following zinc depletion. These include an impairment at the level of the taste receptor. It has been suggested that zinc deficiency may impair taste receptors through (1) reduced activity of zinc metalloenzymes localized in taste papillae, (2) depressed turnover of taste epithelial cells, (3) changes in the functional proteins of saliva, and (4) blockage in the taste pore due to parakeratosis. Other explanations include changes in neural metabolism, which accompany zinc deficiency (Wallwork et al., 1982), and alterations in the hormonal state of the animal (Jakinovich and Osborn, 1981).

Each is described briefly below with the exception of the last, which was discussed previously (see Section II,A).

Alkaline phosphatase (EC 3.1.3.1), a zinc metalloenzyme found in taste buds of many species, has been localized in membranes of bovine circumvallate papillae (Henkin, 1984). The enzyme functions maximally at pH 8–9 and minimally at pH 4.5, to hydrolyze or transfer phosphate groups from bioorganic molecules. A relationship between the specific activity of alkaline phosphatase and zinc concentration in taste buds was established (Law et al., 1983), but the manner by which enzyme activity might influence taste function is not clear.

Zinc is essential for normal cell turnover in all eukaryotic systems examined (Falchuk, 1979). The necessity of zinc in DNA (Rubin, 1972) and RNA (Terhune and Sandstead, 1972) synthesis is due to the fact that both DNA polymerase (EC 2.7.7.7) and RNA polymerase (EC 2.7.7.6) (Falchuk, 1979) are zinc-containing enzymes. Despite this requirement, there is increased rate of cell division in certain cell lines with zinc depletion. Epithelia of rat esophagus (Dinsdale and Williams, 1977), sheep tongue (Mann et al., 1974), and rat fetal neuroepithelia (Swenerton et al., 1969) showed increased mitotic figures during zinc deficiency. Chesters (1973) hypothesized that zinc plays a role in gene expression and proposed that zinc is essential for the transition from the resting phase (G_0) to the DNA replication phase (G_1) of the cell cycle. Zinc-depleted cells fail to enter the resting phase of cellular differentiation, with undifferentiated daughter cells no longer displaying the typical phenotype. If this hypothesis is correct, zinc-depleted taste bud receptor cells would no longer synthesize characteristic proteins and perhaps lose the specificity toward taste stimuli.

Saliva is the normal bathing medium for the oral cavity, and as such its quantity and composition can influence tastant–receptor activity (see Christensen, this volume). Data are available which indicate that concentrations of a histidine-rich salivary protein (Henkin et al., 1975) and a proline-rich protein of parotid saliva (Johnson and Alvares, 1984) are depressed during zinc deficiency. Zinc and saliva and their possible role in taste have recently been reviewed (see Henkin, 1984).

It had been observed that zinc-deficient rats exhibit hyperparakeratosis of the oral mucous membranes (Alvares and Meyer, 1968) and suggested that the pore area may become blocked, thus preventing tastant–receptor interactions (Catalanotto and Nanda, 1977). A morphological study found that the fungiform papillae of zinc-deficient rat tongues were acanthotic and parakeratotic (Calatanotto and Nanda, 1977). However, the pore area remained intact so that the differences in taste response did not appear to be due to blockage of the pore.

Adrenal hypertrophy and elevated blood corticosterone have long been established as consequences of zinc deficiency. This may have an influence on taste, perhaps by the general stress mechanism, proposed in the previous section (Hastings and Van, 1981).

Zinc deficiency may influence levels of neurotransmitters centrally. Wallwork *et al.* (1982) have demonstrated that concentrations of norepinephrine from whole brains of zinc-deficient rats are elevated. Brain dopamine was also increased, but this effect tended to be less consistent (Wallwork *et al.*, 1982). Brain catecholamines have been implicated in rat taste preference for sodium (Fitzsimons, 1979). Injection of norepinephrine into the third ventricle of the brain induced rats to increase their intake of saline, and, conversely, depletion of neural catecholamine stores prevented the sodium preference normally exhibited by control rats (Chiaraviglio and Taleisnik, 1969).

Peripheral neural taste response of the chorda tympani of zinc-deficient animals has been investigated in two studies. The first report found that zinc-deficient rats were less sensitive than their pair-fed, zinc-replete counterparts (Catalanotto and Frank, 1979). A second study found no difference between zinc-deficient and zinc-replete rats' chorda tympani responses to NaCl, sucrose, HCl, or quinine hydrochloride (Jakinovich and Osborn, 1978). This second study discussed possible sources of these discrepant results as being due either to differences in the degree of zinc depletion or to differences in surgical technique. In the second study, the myelin sheath was removed from the nerve before recordings were made; the first study left it intact.

C. Iron and Copper

Dietary iron deficiency has an effect on tastant preferences in rats. Forty-eight-hour, two-bottle testing indicated that iron-depleted rats had greater preferences for sodium chloride and potassium chloride solutions than iron-replete controls. Consumption of sodium saccharin, quinine sulfate, and hydrochloric acid solutions was unaffected (Chan *et al.*, 1983). It has been established that basal blood corticosterone levels of iron-deficient rats increased (Weinberg *et al.*, 1980; Dallman *et al.*, 1984). The specific ingestion of sodium and potassium fluids by iron-deficient rats may depend upon alterations in aldosterone or corticosterone secretion, as discussed previously.

In the case of copper, only toxicosis has been investigated. Rats injected intraperitoneally daily with cupric chloride showed increased 24-hr preferences for tartaric acid and quinine–HCl compared to saline-injected controls (Yamamoto, Kosugi, and Kawamura, 1978). Intakes of saline and sucrose solutions were unaffected. Electrophysiologic whole-nerve recordings from the chorda tympani in copper-toxic rats also showed no differences in response to NaCl, sucrose, tartaric acid, and quinine–HCl.

D. The Case of D-Penicillamine

Penicillamine is a pharmacologic agent used for the treatment of several human disorders (Gilman et al., 1980). It is also a substance which, depending on dosage, can secondarily induce certain nutrient deficiencies in humans and experimental animals. These secondary depletions include three micronutrients (Otomo et al., 1980) already associated with tastant preference changes: vitamin B_6, copper, and zinc. The mechanisms whereby the deficiencies develop appear to be unique for each of the nutrients. Vitamin B_6 and D-penicillamine presumably react *in vivo* to form a thiazolidine derivative, which is biologically inert. Copper and D-penicillamine undergo a redox reaction with the final product, a metal–ligand complex, rendering copper inactive. Zinc, on the other hand, reacts with D-penicillamine to form a stable metal chelate, but the chelated zinc is not available for normal bioactivity. It should also be noted that penicillamine is believed to form penicillamine disulfide *in vivo* (Scheinberg, 1981). The individual stoichiometric rates of each reaction, the chemical form of penicillamine, and circulating levels of vitamin B_6, copper, and zinc will greatly influence which of the micronutrients will be depleted at a given dose of the drug. For example, one might minimize the vitamin B_6 depletion response if the drug is presented as copper penicillamate (Scheinberg, 1981).

D-Penicillamine was reported to induce elevated consumption of 0.15 and 0.3 M sodium chloride solutions (Kare and Henkin, 1969; Henkin and Bradley, 1969); copper and zinc were reduced in the serum of the penicillamine-treated rats, and these physiologic measures were proposed to be associated with the increased consumption of salt solutions. It was hypothesized that penicillamine, a potent metal chelator, altered taste by virtue of copper and/or zinc depletion (Henkin and Bradley, 1969). Additional investigations showed that the D-penicillamine effect on salt solutions was reversed by discontinuation of the drug (Kare and Henkin, 1969) and by copper repletion (Henkin et al., 1968). Other studies have attempted to determine whether D-pencillamine treatment results in changes in whole-nerve chorda tympani response. One study replicated and extended the results of previous work; increased preference for high concentrations of saccharin, NH_4Cl, KCl, and NaCl were found in the D-penicillamine-treated rats (Zawalich, 1971). No differences were found between the groups, however, in activity recorded from the chorda tympani. A second study reported opposite effects upon preference testing; D-penicillamine reduced preferences for NaCl and sucrose (Ito, 1978). Recordings from the chorda tympani showed no significant differences between D-penicillamine-treated and control rats.

6. Micronutrients and Taste Stimulus Intake   115

III. EXPERIMENTAL DATA

Several questions arise in light of the reviewed data in the literature concerning taste behavior and micronutrient status.

1. Can tastant intake be altered by severe primary copper deficiency? And if so, is it influenced by concomitant severe vitamin B_6 deficiency?
2. Can the reported effects of zinc and vitamin B_6 deficiency on tastant intake during 1-hr and 24-hr exposures be observed in briefer tests?
3. Do moderate dietary depletions of vitamin B_6 and zinc also alter tastant intake?
4. Does severe vitamin B_6 deficiency in rats influence preference for hypotonic and hypertonic NaCl and KCl solutions?

The following section describes experiments designed to address these questions.

A. Experiment I: Severe Copper and Vitamin B_6 Depletions and Taste Behavior

1. Methods

Sixty-four male weanling Long–Evans rats were individually housed and fed a control diet* containing 68% corn starch and nutrients in sufficient quantities to meet growth requirements (National Research Council, 1978). While the rats were adapting to the control diet, they were exposed daily for 18 min to two drinking tubes: One contained distilled water, and the other, a 0.0062 M sodium saccharin solution prepared with distilled water (Cagan and Maller, 1974). These training sessions were conducted 4 hr into their light cycle (12 hr dark/12 hr light), and the position of the taste stimuli was alternated each test day.

After the rats attained body weights of 110 ± 5 g (SEM), the control diet was removed. The animals were randomly divided into four dietary groups and fed modified rations according to a 2-way factorial treatment design: group I, $-B_6$ $-Cu$; group II, $+B_6$ $-Cu$; group III, $-B_6$ $+Cu$; group IV,

---

*The basal diet formulation is as follows (grams/kilogram diet): vitamin-free casein, 200 g; DL-methionine, 30 g; corn starch, 683.95 g; corn oil, 50 g; fiber (cellulose), 30 g; choline bitartrate, 20 g; ethoxyquin, 0.01 g; modified AIN-76 mineral mix with cupric sulfate omitted, 35 g; AIN-76A vitamin mix with pyridoxine–HCl omitted, 10 g. Group I was fed the basal diet and groups II–IV were fed the basal diet with further modifications. Group II diet also contained 0.0072 g pyridoxine–HCl; group III diet was supplemented with 0.0126 g cupric sulfate; group IV (control) was supplemented with both 0.0072 g pyridoxine–HCl and 0.0126 g cupric sulfate. All salts were analytical reagent grade. The diets were purchased from Teklad (Madison, WI).

+B_6 +Cu. The rats were given their respective diets and deionized water *ad libitum*. Food intake, water consumption, and body weights were measured before each tastant intake test. Tastant intake tests were begun 3 days after initiation of the experimental diets. Four hours into the light cycle, food and water were removed from the cages. The rats were presented with two drinking tubes for a period of 18 min; one containing the taste stimulus dissolved in deionized water, the second containing deionized water. The position of the tastant tube was reversed each day.

The solutions tested were presented in the following order: sucrose, 0.02, 0.03, and 0.20 M; sodium chloride, 0.01, 0.15, and 0.30 M. The sucrose series was always followed by the salt series with each quality at each concentration being tested four times during the study. During the fourth week, all rats were tested with 1.28×10^{-6} M quinine sulfate solution for 2 consecutive days. Dietary treatment and testing continued for a total of 7 weeks. During the final week, short-term (18 min) and long-term (24 hr) tests were conducted with 1.28×10^{-6} M quinine sulfate.

2. Results

Intake for each tastant was expressed as grams fluid/100 g body weight and averaged across replication. Water intake during the test was less than 0.5 ml and is not presented here. A separate analysis of variance (ANOVA) was used to analyze each tastant. For example, a three-way ANOVA was used to analyze the NaCl data; there were 2 levels of dietary copper, 2 levels of dietary B_6, and 3 concentrations (repeated measures). Intake of sucrose and quinine sulfate was analyzed similarly.

Sodium chloride intake increased with vitamin B_6 deficiency [$F(1,63) = 146.7$; $p < 0.01$], whereas effects due to copper deficiency were not significant [$F(2,63) = 0.01$]. A significant interaction between vitamin B_6 deficiency and concentration [$F(1,120) = 32.5$; $p < 0.01$] (see Fig. 1) revealed that whereas the B_6 control group reduced their intake of the 0.3 M NaCl, the B_6-deficient groups did not ($p < 0.05$; Tukey).

Similar results were obtained with the sucrose analysis. Vitamin B_6-deficient rats consumed more sucrose per 100 g body weight [$F(1,63) = 39.5$; $p < 0.01$]. Copper deficiency did not influence intake of sucrose [$F(1,63) = 0.10$]. The single significant interaction was between vitamin B_6 deficiency and sucrose concentration ($F = 5.2$; df, 2,128). Post hoc tests (Tukey) showed that the vitamin B_6-deficient groups increased their intakes between 0.02 and 0.03 M sucrose whereas the controls did not ($p < 0.05$; see Fig. 2).

Intake of quinine sulfate was significantly depressed in copper-deficient [$F(1,49) = 4.4$; $p < 0.05$] and increased in vitamin B_6-deficient rats [$F(1,59) = 11.8$; $p < 0.01$]. Because significant effects were found with this tastant, the rats were exposed to the same concentration of quinine sulfate for 24 hr.

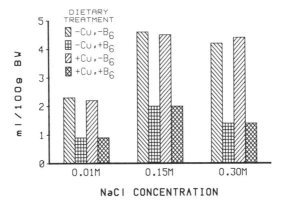

Fig. 1. Average intake of sodium chloride solutions during 18-min tests in copper- and/or vitamin B_6-deficient rats.

These results showed that the vitamin B_6-deficient rats consumed significantly less quinine sulfate [$F(1,57) = 4.6$; $p < 0.05$]. There was no copper effect (see Fig. 3).

B. Experiment II: Moderate Zinc and Vitamin B_6 Depletions and Taste Behavior

1. Methods

Sixty-four male weanling Long–Evans rats were individually housed and fed a control diet* containing nutrients in sufficient quantities to meet growth requirements (National Research Council, 1978). Rats were provided the control diet and trained as in Experiment I until they attained body weights of 185 ± 10 g (SEM). The animals were randomly divided into four treatment groups and fed the diet with the following nutrient modifications: group V, 3 μg Zn/3 μg vitamin B_6 per g food; group VI, 3 μg Zn/10 μg vitamin B_6 per g food; group VII, 12 μg Zn/3 μg vitamin B_6 per g food; group VIII, 12 μg Zn/10 μg vitamin B_6 per g food. The remainder of the

---

*The basal diet formulation is as follows (grams/kilogram diet): egg white solids, 200 g; corn starch, 689.78 g; corn oil, 50 g; fiber (cellulose), 30 g; choline bitartrate, 2 g; ethoxyquin, 0.01 g; modified AIN-76 mineral mix with NaCl omitted and 0.0032 g zinc carbonate, 35 g; modified AIN-76A vitamin mix with pyridoxine–HCl omitted, 10 g. Group V was fed the basal diet, and groups VI–VIII were fed the basal diet with further modifications. Group VI also contained pyridoxine–HCl, 0.0072 g; group VII was supplemented with 0.0182 g zinc carbonate; group VIII (control) was supplemented with both 0.0182 g zinc carbonate and 0.0072 g pyridoxine–HCl. All salts were analytical reagent grade. The diets were purchased from Teklad (Madison, WI).

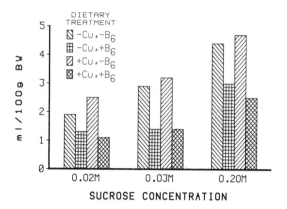

Fig. 2. Average intake of sucrose solutions during 18-min tests in copper- and/or vitamin B_6-deficient rats.

study, including the intake test procedure, was conducted as in Experiment I. Two concentrations of monosodium glutamate (MSG) (2.7×10^{-3} and $0.27\ M$) were also tested after 3 cycles of sucrose and sodium chloride were completed. Short-term tests for quinine ($1.28 \times 10^{-6}\ M$) were conducted between and following the two concentrations of MSG.

2. Results

Intake measures were expressed as grams fluid consumed/100 g body weight and analyzed separately with an ANOVA for each tastant. For each ANOVA, Zn content, B_6 content, and concentration (repeated measures) were the factors analyzed. The results of the sodium chloride analysis indicated significant effects due to moderate zinc depletion [$F(1,62) = 15.6$; $p <$

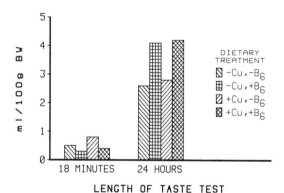

Fig. 3. Average intake of a quinine sulfate solution ($1.28 \times 10^{-6}\ M$) during 18-min and 24-hr tests in copper- and/or vitamin B_6-deficient rats.

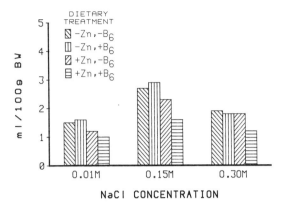

Fig. 4. Average intakes of sodium chloride solutions during 18-min tests in rats moderately deficient in zinc and/or vitamin B_6.

0.01]. There was no significant effect of moderate vitamin B_6 depletion $[F(1,62) = 2.2]$. Interactions of zinc and vitamin B_6 $[F(1,62) = 4.4; p < 0.05]$ as well as zinc and concentration $[F(2,124) = 3.8; p < 0.05]$ were found. Post hoc testing (Tukey) revealed that the zinc × B_6 and zinc × concentration interactions are due to small changes in intake at the low (0.01 M) and isotonic (0.15 M) concentrations: The presence of B_6 in the absence of zinc (group VI) increased NaCl intake, and the presence of zinc in the absence of B_6 (group VII) decreased the NaCl intake, when both groups were compared to Group V (3 μg B_6, 3 μg Zn) (see Fig. 4).

The partially zinc-depleted group consumed significantly more sucrose per 100 g body weight $[F(1,63) = 14.6; p < 0.01]$. There were no significant effects of moderate B_6 deficiency upon sucrose intake (see Fig. 5).

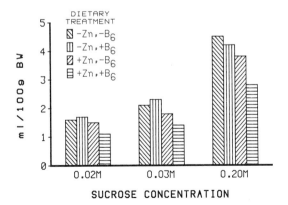

Fig. 5. Average intake of sucrose solutions during 18-min tests in rats moderately deficient in zinc and/or vitamin B_6.

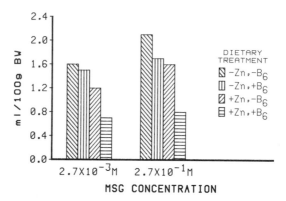

Fig. 6. Average intake of monosodium glutamate solutions in 18-min tests in rats moderately deficient in zinc and/or vitamin B_6.

Intake for monosodium glutamate is presented in Fig. 6. Moderately zinc-depleted rats consumed more MSG [$F(1,62) = 25.0$; $p < 0.01$] as did moderately vitamin B_6-deficient rats [$F(1,62) = 11.5$; $p < 0.01$]. MSG intake increased with concentration [$F(1,62) = 7.3$; $p < 0.01$]. There were no significant interactions. Moderate dietary zinc or vitamin B_6 deficiency had no effect on intake of quinine sulfate (data not shown).

C. Experiment III: Severe Vitamin B_6 Depletion and NaCl Intake

1. Methods

Fifty-two Sprague–Dawley albino male and female weanling rats were housed in individual stainless-steel cages under a 12-hr light/dark cycle (0900–2100 hr). At weaning (21 days of age), animals were separated into two groups, with each group containing approximately an equal number of males and females. Group 1 was fed a pyridoxine-deficient diet ($N = 25$), and the other group was fed a pyridoxine control diet ($N = 27$) *ad libitum* in ceramic food cups. The pyridoxine-deficient diet was composed of 640 g sucrose, 220 g casein, 40 g salt mixture No. 446, 50 g corn oil, and 10 g vitamin fortification mixture (ICN, Cleveland, OH; see Stewart and Bhagavan, 1982). The pyridoxine control diet contained the same ingredients with an additional 30 mg pyridoxine–HCl per kg diet. The animals were maintained on these diets for 8 weeks, and their body weights were recorded weekly. During the ninth week, rats began a two-bottle, 24-hr tastant intake test. Five different NaCl concentrations, 0.01, 0.05, 0.15, 0.30, and 0.45 M, were used. One NaCl tastant and distilled water were secured to the fronts of the cages in 125-ml glass bottles attached with drinking tubes. The testing lasted 10 days, and all rats received the ascending order of NaCl tastants twice. Bottles were weighed to the nearest 0.1 g every 24 hr, and their positions were alternated left to right, daily. The amount of NaCl and water intake per 24 hr was calculated per 100 g body weight for every 2 days. For each rat, the average number of milliliters/100 g body weight was calculated for each concentration of NaCl for both groups, to determine the group mean milliliter/100 g body weight.

2. Results

The data were analyzed by analysis of variance with the diet condition as one variable, sex as another, and concentration (repeated measures) as the third variable, yielding a 2 × 2 × 5 factorial design. The deficient groups drank significantly more of the saline $[F(1,48) = 26.45; p < 0.01]$ (see Fig. 7) and the female rats more than the male $[F(1,48) = 8.51; p < 0.01]$ (data not shown). A significant interaction between diet condition and NaCl concentration $[F(4,192) = 18.21; p < 0.01]$ was due to the elevated NaCl intake found in pyridoxine-deficient animals at the three highest concentrations (Fig. 7) (Tukey, $p < 0.05$).

D. Experiment IV: Severe Vitamin B_6 Depletion and KCl Intake

1. Method

This experiment was essentially a replication of Experiment III, with the exception that the tests were conducted with six different concentrations of

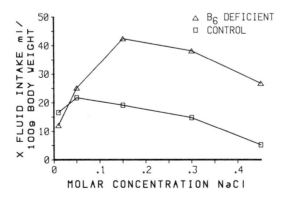

Fig. 7. Average intake of sodium chloride solutions in vitamin B_6-deficient rats as a function of concentration.

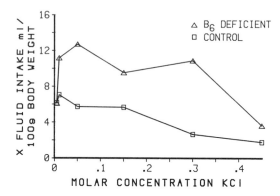

Fig. 8. Average intake of potassium chloride solutions in vitamin B_6-deficient rats as a function of concentration.

KCl (0.005, 0.01, 0.15, 0.30, and 0.45 M) and distilled water. The additional low concentration of KCl was added in light of the extremely strong preference shown for this substance in a previous study (Stewart and Bhagavan, 1982). The deficient group was made up of 8 females and 7 males, whereas the control group contained 6 females and 6 males.

2. Results

Analysis of variance (diet × sex × concentration) revealed that vitamin B_6-deficient rats consumed significantly more KCl [$F(1,23) = 5.10; p < 0.05$]. Concentration influenced intake with the greatest intake at 0.45 M KCl [$F(1,115) = 3.41; p < 0.01$]. Whereas no significant interactions were found, inspection of Fig. 8 indicates that at the lowest (0.005 M) and highest (0.45 M) concentration, the enhanced potassium intake is clearly attenuated in the deficient animals (Tukey, $p < 0.05$).

E. Experiment V: Severe Vitamin B_6 Depletion and Short-Term NaCl and KCl Intake

1. Methods

Fifty Sprague–Dawley rats (19 males and 31 females) were used in this experiment and maintained under conditions described previously. The pyridoxine-deficient diet was fed to 15 females and 11 males, whereas the remainder of the animals were fed the control diet. During the fourth week on the diet, animals were adapted to a schedule whereby they were deprived of food and water for 1 hr daily from 1200 to 1300 hr. Following this, they were given two-bottle drinking tests for 20 min with both bottles con-

6. Micronutrients and Taste Stimulus Intake

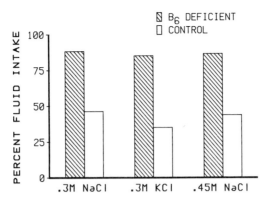

Fig. 9. Average percent preference of sodium chloride and potassium chloride solutions in vitamin B_6-deficient and control rats.

taining water, the positions of the bottles alternated after 10 min. After 6 days of adaptation to this schedule, they were given 6 days of a choice between 0.30 M NaCl and distilled water. Again, bottle positions were alternated after 10 min. Following this phase, the rats were given 1 day of testing with distilled water followed by 6 days of testing with 0.3 M KCl and distilled water, and then 6 days with 0.45 M NaCl and distilled water.

2. Results

Preference scores were computed [(gram tastant solution consumed/gram tastant solution + gram water consumed) × 100] and are shown in Fig. 9. Preference for the 0.30 M NaCl solution was averaged over 6 days of testing. The mean preference score was 88.3% ± 1.25 SEM for the deficient rats and 46.3% ± 2.28 SEM for controls. This difference was significant [$t(48)$ = 16.49; $p < 0.001$]. When the NaCl concentration was increased to 0.45 M, the deficient rats continued to show a strong preference for the electrolyte, with a mean preference of 86.6% ± 1.96 SEM as compared to 43.9% ± 2.11 SEM for controls [$t(47)$ = 14.83; $p < 0.001$]. Comparison of 0.30 M KCl intake between the two groups also revealed a significant difference [$t(47)$ = 13.76; $p < 0.001$], with a mean preference score of 85.2% ± 2.25 SEM for deficient rats and 35.1% ± 2.88 SEM for control rats.

F. Discussion: Experiments I–V

The results of Experiment I indicate that severe dietary copper deficiency does not alter intake of normally preferred stimuli, such as sucrose and sodium chloride, during short-term tests. However, there was a small but significant ($p < 0.05$) decrease in quinine sulfate consumption by these rats during short-term, but not long-term exposure. There is no doubt that these rats were copper deficient. Both serum and liver copper concentrations were severely decreased at the end of the experiment, and the depressed blood hemoglobin observed midway through the experiment indirectly confirmed the copper deficiency at that time (Greeley and Gniecko, unpublished observations). It remains to be seen whether or not rats dietarily depleted of copper at an earlier stage of development (e.g., *in utero*, during suckling, or at weaning) show any evidence of altered taste behavior toward normally preferred taste stimuli.

As observed in previous studies, zinc depletion is followed by increased intake of NaCl and sucrose solutions. This effect does not require a severe depletion. Marginal zinc status (Experiment II) was effective in increasing intake of 0.15 and 0.3 M NaCl, 0.0027 and 0.27 M MSG, and 0.2 M sucrose in 18-min tests. However, in contrast to previous studies, increased intake of 0.01 M NaCl, 0.02 M sucrose, 0.03 M sucrose, and 1.28×10^{-6} M quinine sulfate was not observed. This difference could have been due to the moderate degree of zinc depletion and/or to the short duration of the test. The fact that increased intake of NaCl, MSG, and sucrose was observed in short-term tests offers further evidence that intake changes seen with zinc depletion may reflect altered taste function.

Other interpretations are possible. Eighteen-minute tests lessen but by no means rule out a postingestional effect. It is possible that ingestion of sodium, sucrose, and MSG did provide some positive ingestional feedback, which became associated with the taste. It seems unlikely however, that two substances as different as sucrose and sodium would both provide positive systemic feedback. Another possible explanation could be that the zinc-deficient rats were showing enhanced neophilia, approaching and ingesting anything new. This strong preference for a novel tastant has been reported in zinc-deficient rats (Christensen *et al.*, 1974) as well as rats with other deficiencies (Rodgers, 1967). However, two pieces of evidence run counter to this latter interpretation. The first is that the enhanced intake was not shown to quinine sulfate, a new tastant. Second, the enhanced intake of sucrose and NaCl did not decrease across testing sessions as they normally should have done if the effects were due to neophilia alone. Finally, the suggestion that deficiencies produce a general metabolic stress leading to altered ingestive behavior must be raised as a possible cause of increased salt and perhaps sucrose intake.

The studies presented here confirm previous studies. Severe vitamin B_6 depletion leads to increased intake of NaCl, KCl, sucrose, and quinine sulfate solutions in 18-min tests and decreased intake of quinine sulfate solutions in 24-hr tests. Moderately deficient rats do not increase NaCl and sucrose intake. However, they do increase MSG intake. Preference thresholds for NaCl and KCl are not altered by pyridoxine deficiency in contrast to the findings with adrenalectomy, where a lowered preference threshold for sodium chloride is seen.

All of the possible explanations raised for zinc influences on intake must be considered for the case of vitamin B_6 depletion. It is possible that taste function in general is compromised by severe vitamin B_6 deficiency since increased intake of salts, sucrose, and quinine sulfate was seen. This appears unlikely since increased intake has not been shown with all tastant substances given during long-term tests of vitamin B_6-deficient rats (Chan and Kare, 1979).

Another mechanism which might be postulated to account for enhanced intake of tastants in vitamin B_6-deficient rats would be neophilia. Such a mechanism cannot, however, account for the prolonged increase in both potassium and sodium chloride intake. The initial neophilia should be lessened over time as the animals become familiar with the substances. Also, there are some novel substances for which deficient animals do not show an increased preference, such as Na saccharin and HCl, in 24-hr tests (Chan and Kare, 1979).

Particularly intriguing is the increased NaCl and KCl intake and preference observed in severely vitamin B_6-deficient rats. The adrenal glands enlarge with this deficiency, and preliminary data collected on pyridoxine-deficient rats have also shown increased potassium excretion as well as elevated aldosterone levels (Stewart et al., unpublished observations). It is possible that either corticosterone or aldosterone is influencing salt intake. Further supportive evidence that vitamin B_6-deficient rats may have hyperaldosteronism is the finding that the drug captopril (angiotensin converting enzyme inhibitor) reduces NaCl intake in deficient rats. This drug, which blocks the production of aldosterone by inhibiting the conversion of angiotensin I to II, increases NaCl preference in nondeficient rats (Fregly, 1980).

The effects on tastant intake seen with D-penicillamine are not seen in short-term tests with dietarily copper-depleted rats. These data contrast with reports that copper treatment reverses the copper-depleting effects of D-pencillamine on salt intake in long-term tests (Henkin et al., 1968). This can be interpreted in a number of ways. First, the effect of copper on taste perception could be small and become apparent only in long-term tests. Second, the increased salt intake of D-penicillamine-treated rats could be due to positive postingestional effects of salt, which increase intake only in long-term tests. Third, the effects of D-penicillamine could be due in part to other secondary deficiencies (such as vitamin B_6 and zinc) induced by the drug. Fourth, the time of the onset of copper depletion may be important. In one report, D-penicillamine and copper treatment were administered to weanling rats (Henkin et al., 1968); in the data presented here, copper-deficient diets were fed to 5-week-old rats. The mechanism underlying the taste changes reported earlier with D-penicillamine treatment remains to be elucidated.

IV. FUTURE RESEARCH

From the review and the experimental data presented, it is apparent that much work remains to be done before the mechanisms underlying micronutrient-induced changes in tastant ingestion are understood. Although dietarily induced changes in adrenal secretions are attractive as explanatory mechanisms, Denton's (1982) review of the effects of adrenal steroids on salt appetite raises the question of the role these steroids play at physiological rather than pharmacological levels. Nevertheless, it is important to be able to characterize the nature of the proposed stress, i.e., the ACTH–adrenal response to each of the micronutrient deficiencies including the other B vitamins, to search for similarities and differences. The salivary changes in sodium and potassium, induced by the deficiencies, have yet to be investigated, as have the possible changes in levels of central neurotransmitters. Finally, the mechanisms by which penicillamine induces changes in tastant ingestion need further study. In particular, the effects of zinc and pyridoxine repletion need to be examined in the penicillamine-treated animal.

ACKNOWLEDGMENTS

The authors wish to acknowledge the able technical assistance of Kathleen Gniecko, Susan L. Marra, and Suzanne VanVliet.

REFERENCES

Alvares, O., and Meyer, J. (1968). Regional differences in parakeratotic response to mild zinc deficiency. *Arch. Dermatol.* **98**, 191–202.

Aquilera, G., Fujita, K., and Catt, K. J. (1981). Mechanisms of inhibition of aldosterone secretion by adrenocorticotropin. *Endocrinology* **108**, 522–528.

Bernard, R. A., and Halpern, B. P. (1968). Taste changes in vitamin A deficiency. *J. Gen. Physiol.* **52**, 444–464.

Bernard, R. A., Halpern, B., and Kare, M. R. (1962). Effect of vitamin A deficiency on taste. *Proc. Soc. Exp. Biol. Med.* **108**, 784–786.

Cagan, R. H., and Maller, O. (1974). Taste of sugars: Brief exposure single-stimulus behavioral method. *J. Comp. Physiol. Psychol.* **87**, 47–55.

Catalanotto, F. (1979). Alterations of short-term tastant-containing fluid intake in zinc-deficient adult rats. *J. Nutr.* **109**, 1079–1085.

Catalanotto, F., and Frank, M. (1979). Depressed chorda tympani responsiveness in zinc deficient rats. *Abstr. Soc. Neurosci.* **5**, 126.

Catalanotto, F., and Lacy, P. (1977). Effects of zinc deficient diet upon fluid intake in the rat. *J. Nutr.* **107**, 436–442.

Catalanotto, F., and Nanda, R. (1977). The effects of feeding a zinc deficient diet on taste acuity and tongue epithelium in rats. *J. Oral Pathol.* **6**, 211–220.

Chan, M. M., and Cote, J. (1979). Fluid intakes and chorda tympani nerve responses in vitamin B-6 deficient rats. *Physiol. Behav.* **22**, 401–404.

Chan, M. M., and Kare, M. R. (1979). Effect of vitamin B_6-deficiency on preference for several taste solutions in the rat. *J. Nutr.* **109**, 339–344.

Chan, M., Brand, J., Ingle, D., and Kare, M. (1983). Feeding low iron diets affect fluid preference in growing rats. *Nutr. Res.* **3**, 511–518.

Chesters, J. (1973). Biochemical functions of zinc with emphasis on nucleic acid metabolism and cell division. *In* "Trace Element Metabolism in Animals" (W. Hoekstra, J. Suttie, H. Ganther, and W. Mertz, eds.), Vol. 2, pp. 39–50. University Press, Baltimore.

Christensen, C. M., Caldwell, D. F., and Oberleas, D. (1974). Establishment of a learned preference for a zinc containing solution by zinc-deficient rats. *J. Comp. Physiol. Psychol.* **87**, 415–421.

Chiaraviglio, E., and Taleisnik, S. (1969). Water and salt intake induced by hypothalamic implants of cholinergic and adrenergic agents. *Am. J. Physiol.* **216**, 1418–1422.

Chvapil, M. (1973). New aspects in the biological role of zinc. A stabilizer of macromolecules and biological membranes. *Life Sci.* **13**, 1041–1049.

Dallman, P., Refino, C., and Dallman, M. (1984). The pituitary-adrenal response to stress in the iron-deficient rat. *J. Nutr.* **114**, 1747–1753.

Denton, D. (1982). "The Hunger for Salt," pp. 301–309. Springer-Verlag, New York.

Dillman, E., Johnson, D., Martin, J., Mackler, B., and Finch, C. (1979). Catecholamine elevation in iron deficiency. *Am. J. Physiol.* **237**, R297–R300.

Dinsdale, D., and Williams, R. (1977). The enhancement by dietary zinc deficiency of the susceptibility of the rat duodenum to colchicine. *Br. J. Nutr* **37**, 135–142.

Falchuk, K. (1979). The role of zinc in the biochemistry of *Euglena gracilis* cell cycle. *In* "Trace Metals in Health and Disease" (N. Karasch, ed.), pp. 177–188. Raven Press, New York.

Fitzsimons, J. T. (1979). "The Physiology of Thirst and Sodium Appetite." Cambridge Univ. Press, Cambridge, MA.

Fregly, M. J. (1980). Effect of the angiotensin converting enzyme inhibitor captopril on NaCl appetite of rats. *J. Pharmacol. Exp. Ther.* **215**, 407–412.

Gilman, A. G., Goodman, L. S., and Gilman, A. (1980). "Goodman and Gilman's The Pharmacological Basis of Therapeutics," 6th ed. Macmillan, New York.

Hastings, M. M., and Van, J. L. (1981). Sodium deprivation during thiamin deficiency in rats: Hormonal, histological, and behavioral responses. *J. Nutr.* **111**, 1955–1963.

Henkin, R. (1984). Zinc in taste function: A critical review. *Biol. Trace Element Res.* **6**, 263–280.

Henkin, R., and Bradley, D. (1969). Regulation of taste acuity by thiols and metal ions. *Proc. Natl. Acad. Sci. U.S.A.* **62**, 30–37.

Henkin, R., Keiser, H. R., and Kare, M. R. (1968). The effects of D-penicillamine (D-pen) and of copper repletion on salt preference and fluid intake. *Fed. Proc.* **27**, 583.

Henkin, R., Lippoldt, R., Bilstad, J., and Edelhoch, H. (1975). A zinc protein isolated from human parotid saliva. *Proc. Natl. Acad. Sci. U.S.A.* **72**, 488–492.

Hsu, J. M., Davis, R. L., and Chow, B. F. (1958). Electrolyte imbalance in vitamin B6-deficient rats. *J. Biol. Chem.* **230**, 889–895.

Ito, H. (1978). Preference behavior and taste nerve responses in D-penicillamine treated rats. *Physiol. Behav.* **21**, 573–579.

Jakinovich, W., and Osborn, D. (1981). Zinc nutrition and salt preference in rats. *Am. J. Physiol.* **241**, R233–R239.

Johnson, D., and Alvares, O. (1984). Zinc deficiency-induced changes in rat parotid salivary proteins. *J. Nutr.* **114**, 1955–1964.

Kare, M. R., and Henkin, R. I. (1969). The effects of D-penicillamine on taste preference and volume intake of sodium chloride by the rat. *Proc. Soc. Exp. Biol. Med.* **131**, 559–565.

Klevay, L. M. (1984). The role of copper, zinc and other chemical elements in ischemic heart disease. In "Metabolism of Trace Metals in Man" (O. W. Rennert and W.-Y. Chan, eds.), Vol. I, pp. 129–157. CRC Press, Boca Raton, FL.

Law, J. S., Nelson, N., and Henkin, R. I. (1983). Zinc localization in taste bud membranes. *Biol. Trace Element Res.* **5**, 219–224.

McConnell, S., and Henkin, R. (1974). Altered preference for sodium chloride, anorexia and changes in plasma and urinary zinc in rats fed a zinc-deficient diet. *J. Nutr.* **104**, 1108–1114.

Mann, S., Fell, B., and Dalgarno, A. (1974). Observations on the bacterial flora and pathology of the tongue of sheep deficient in zinc. *Res. Vet. Sci.* **17**, 91–101.

National Research Council. Subcommittee on Laboratory Animal Nutrition. (1978). "Nutrient Requirements of Laboratory Animals," Third Revised Edition. National Academy of Sciences, Washington, D.C.

Otomo, S., Sasajima, M., Ohzeki, M., and Tanaka, I. (1980). Effects of D-penicillamine on vitamin B_6 and metal ions in rats. *Folia Pharmacol. Jpn.* **76**, 1–13.

Pike, R. L., and Brown, M. L. (1975). "Nutrition: An Integrated Approach," 2nd ed. Wiley, New York.

Reeves, P. G., and O'Dell, B. L. (1981). Short-term zinc deficiency in the rat and self-selection of dietary protein level. *J. Nutr.* **111**, 375–383.

Rodgers, W. L. (1967). Specificity of specific hungers. *J. Comp. Physiol. Psychol.* **64**, 49–58.

Rubin, H. (1972). Inhibition of DNA synthesis in animal cells by EDTA and its reversal by zinc. *Proc. Natl. Acad. Sci. U.S.A.* **69**, 712–716.

Scheinberg, I. (1981). Copper penicillamate for rheumatoid arthritis. *J. Rheumatol.* **8**, 178–179.

Stewart, C. N., and Bhagavan, H. N. (1982). Sodium and potassium chloride preferences in pyridoxine-deficient rats. *Proc. Soc. Exp. Biol. Med.* **170**, 15–18.

Swenerton, H., Shrader, R., and Hurley, L. (1969). Zinc-deficient embryo: Reduced thymidine incorporation. *Science* **166**, 1014–1015.

Terhune, M., and Sandstead, H. (1972). Decreased RNA polymerase activity in mammalian zinc deficiency. *Science* **177**, 68–69.

Wallwork, J., Botnen, J., and Sandstead, H. (1982). Influence of dietary zinc on rat brain catecholamines. *J. Nutr.* **112**, 514–519.

Weinberg, J., Dallman, P., and Levine, S. (1980). Iron deficiency during early development in the rat: Behavioral and physiological consequences. *Pharmacol. Biochem. Behav.* **12**, 493–502.

Weisenger, R. S., Denton, D. A., McKinley, M. J., and Nelson, J. F. (1978). ACTH-induced sodium appetite in the rat. *Pharmacol. Biochem. Behav.* **8**, 339–342.

Yamamoto, T., Kosugi, T., and Kawamura, Y. (1978). Taste preference and taste nerve responses of rats under copper toxicosis. *Pharmacol. Biochem. Behav.* **9**, 799–807.

Zawalich, W. (1971). Gustatory nerve discharge and preference behavior of penicillamine-treated rats. *Physiol. Behav.* **6**, 419–423.

7

Effect of Non-Insulin-Dependent Diabetes Mellitus on Gustation and Olfaction

LAWRENCE C. PERLMUTER,[*,†] DAVID M. NATHAN,[‡,§]
MALEKEH K. HAKAMI,[†] AND
HOWARD H. CHAUNCEY[*,†]
[*]Veterans Administration Outpatient Clinic
[†]Harvard School of Dental Medicine
[‡]Harvard Medical School
[§]Massachusetts General Hospital
Boston, Massachusetts

I.	Introduction	129
II.	Methods	131
	A. Subjects	131
	B. Gustatory Assessment	132
	C. Olfactory Assessment	133
III.	Results	133
	A. Gustatory Assessment	133
	B. Olfactory Assessment	135
	C. Gustatory Hedonics	135
	D. Olfactory Hedonics	135
	E. Relative Magnitude Judgments	136
IV.	Discussion	138
V.	Summary	141
	References	141

I. INTRODUCTION

The sensory modalities of taste and smell play a primary role in the control of food intake (Schiffman, 1983). In turn, gustation and olfaction can be altered by variables such as (1) certain local factors, including missing teeth, and prosthetic appliances for the replacement of missing teeth (Saxon and Etten, 1981), (2) systemic factors, including nutritional and metabolic imbalances, and (3) various diseases (Chauncey et al., 1981, 1984; Wayler et al., 1982, 1984). Alterations in these sensory modalities may modify food selection patterns and affect nutrient consumption. Other factors, such as dementia and depression can impair the identification and discriminability of common olfactory stimuli (Garg and Perlmuter, 1985), whereas age and sex are associated with variations in preference for and response to the quality and intensity of different gustatory (Schiffman, 1983) and olfactory stimuli (Doty et al., 1984).

Systemic diseases, including non-insulin-dependent diabetes mellitus (NIDDM), can affect gustatory and olfactory perception (Perlmuter et al., 1985). Lawson et al. (1979) reported that NIDDM was associated with higher taste thresholds (decreased sensitivity) for sucrose. Thus, individuals with NIDDM may ingest larger quantities of sucrose to produce a desired level of sweetness. This would lead to difficulty in adhering to dietary recommendations and contribute to greater obesity.

The importance of this observation is reflected by the high prevalence of dietary noncompliance demonstrated by patients with NIDDM. One-third of all adult diabetics ignore dietary restrictions, and an additional one-third show only moderate compliance with their recommended diets (Surwit et al., 1982).

Other factors may affect dietary compliance in NIDDM patients, the majority of whom are elderly. For example, depressed individuals may ingest food primarily as a means of increasing sensory input (Schiffman, 1983), a practice that could override normal food regulatory mechanisms. Since psychological depression levels increase with age, especially in NIDDM (Perlmuter et al., 1984), the elderly diabetic may increase food intake to alleviate depression. Thus, diabetics may utilize this mechanism to seek temporary relief from depression, at the cost of increased obesity.

In addition to the increased level of depression, central or peripheral neuropathies may also play a role in altering dietary regulation. The greater prevalence of peripheral neuropathies in persons with NIDDM is well recognized (Nathan et al., 1983). If central sensory function is affected in a manner similar to peripheral nerve function, this may modify gustatory and/or olfactory integrity, which might, in turn, lead to altered food intake. Furthermore, it has been suggested that insulin itself can affect the sense of taste (Saxon and Etten, 1981), thereby complicating this difficult situation.

Age-matched groups of diabetic and nondiabetic subjects were utilized to elucidate the relationships between diabetes, aging, depression, and the sensory modalities of olfaction and gustation. Participants were instructed to estimate and rate the gustatory magnitude of six solution concentrations of sucrose. Diabetes treatment modality (i.e., parenteral insulin, oral hypoglycemic agents, and diet) served as one major variable in this study, with a second being gender. Subjects in each of the treatment categories were separated according to their gender, with the design yielding 8 groups: male and female participants for the control group and each of the three treatment modalities. Olfactory sensitivity was also evaluated since it is a major contributor to flavor perception. Moreover, there is evidence (Settle, 1985) that diabetics may have an altered sense of smell. Therefore, magnitude ratings were collected on six concentrations of an olfactory stimulus, essence of orange.

The results indicated that the gustatory magnitude estimations were significantly greater for diabetic women than nondiabetic women. Diabetic and control males, however, performed in a similar manner and did not differ from the control females. Gustatory hedonic values were not dependent on diabetic status, but men and women exhibited significantly different response patterns, as the solution concentration was increased. Olfactory magnitude estimations were affected by disease status and solution concentration; diabetic subjects gave significantly higher intensity ratings than the controls. On the other hand, olfactory hedonic responses were affected only by concentration.

II. METHODS

A. Subjects

This study utilized 110 female and 107 male type II diabetics and 44 age-matched, nondiabetic controls (25 females and 19 males). The age range of these participants was from 55 to 74 years. In diabetics, the disease duration ranged from 1 to 20 years. Forty-three percent (93) of the diabetic subjects were treated with insulin, 24% (72) with oral agents, and 33% (52) with diet alone.

Information relating to a variety of different physiological and psychological parameters was collected. These included (1) depression, as indexed by the Zung Self-Rating Depression Survey (1965) to assess the affective and cognitive domains of depression, (2) weight, in kilograms, and (3) height, in meters. The two latter variables were measured to obtain an obesity index, which was calculated utilizing a body mass index: the weight divided by the height squared (Carson and Gormican, 1976).

A biothesiometer (Biomedical Instruments Co., Newbury, Ohio) measured vibratory threshold over the medial malleolus as an index of peripheral neuropathy. Measurements were made in triplicate (Gregg, 1951) and the arithmetic mean calculated. The degree of autonomic neuropathy was assessed by changes in electrocardiographic R–R intervals during deep inspiration and expiration, as described by Ewing et al. (1981).

Glycosylated hemoglobin (HbA_1c) levels, which are an index of chronic glucose control, were determined by high-performance liquid chromatography (Nathan et al., 1982), whereas plasma glucose levels were measured by an autoanalyzer.

A detailed medical history was also obtained, which included smoking habits, kind and frequency of medications used, and frequency of hypoglycemic reactions. A physical examination was performed, which evaluated tendon reflexes and measured supine and standing blood pressure in addition to pulse rate. Selected information pertaining to these data is included in this chapter, while a more complete description can be found in Perlmuter et al. (1984).

B. Gustatory Assessment

Intensity measurements were performed using a modification of the magnitude estimation procedure employed by Thompson et al. (1976). Six concentrations of sucrose solution served as the gustatory stimuli, and six concentrations of essence of orange served as the olfactory stimuli. Gustatory and olfactory estimation data were collected at separate times during the morning testing. Although these evaluations were part of a larger test battery, they comprised the first part of the test series for the subjects, all of whom had been fasting for the previous 12 hr.

Magnitude estimation scaling was undertaken using the sip method for stimulus presentation. Specifically, 5 ml of six sucrose solutions (0.031, 0.062, 0.125, 0.25, 0.50, and 1.00 M) were placed into paper cups and set before each subject. Participants were instructed to pick up and sip the test solution (0.125 M), hold it in their mouth for a few seconds, and then "spit it out." The sample cup containing the 0.125 M solution was presented at the beginning of the series as a standard. The subjects were informed that the standard had a magnitude of 10, based on a scale from 0 to 100. The participants were further instructed that a score of 0 should be assigned to the solution with no sweetness, and a maximum value of 100 could be assigned to the solution with the greatest degree of sweetness. The number assigned to each of the other cups was to be in proportion to the judgment of their relative sweetness.

Participants were periodically cautioned not to swallow the test solutions. They were also instructed to rinse their mouth twice with tap water before each tasting. An interstimulus interval of at least 30 sec was provided throughout.

C. Olfactory Assessment

In the assessment of olfactory stimulus intensity, subjects were instructed to rate the essence of orange odor intensity using an established procedure similar to that of Thompson et al. (1976). Six solution concentrations (0.4, 1.2, 3.7, 11, 33, and 100 mg%) of cold-pressed orange oil in light paraffin oil were used. Each stimulus solution was presented by uncorking the respective flask as it was placed beneath the subject's nose. Participants were not permitted to see the solutions during testing. The standard solution, presented at the start of testing, had an intermediate odor intensity (3.7% orange oil) and was arbitrarily defined as having a strength of 10, based on a scale from 0 to 100. The minimal interstimulus interval was 30 sec.

For each series of olfactory and gustatory magnitude estimation scalings—a total of six trials—one presentation for each of the six stimuli was given in a preestablished order. In all, each series was presented four times. Although presentation order for each of the six olfactory and six gustatory stimuli was identical for all subjects, half of the subjects performed olfactory testing prior to the gustatory evaluation, whereas for the remaining subjects this order was reversed. Stimulus presentation order in the initial gustatory trial was 1.0, 0.062, 0.031, 0.125, 0.25, and 0.500 M, whereas presentation order for the initial olfactory trial was 100, 1.2, 0.4, 3.7, 11, and 33 mg%. The data obtained from the first test series, containing one evaluation per subject for each of the six stimuli, are presented herein.

In addition to the magnitude estimations, hedonic evaluations were carried out for the gustatory and olfactory stimuli. After obtaining the magnitude estimation judgments for each of the stimuli, the subjects were asked to use a 12-in. scale, on which the midpoint was designated as being neutral. The left-hand side of the scale was used to indicate the extent of the subject's dislike, and the right side of the midpoint the extent of liking for each stimulus. Corresponding values were obtained from the back of the scale, which was not visible to the subject, and the experimenter was thus able to translate the qualitative ratings into a quantitative scale, ranging from 0 to 100.

III. RESULTS

Patients in the three diabetes treatment groups all performed similarly in response to the gustatory stimuli. Thus, for all comparisons, diabetic persons were segregated only according to their gender.

A. Gustatory Assessment

A comparison of the mean taste magnitude estimation values obtained from the first stimulus presentation series, for the control and diabetic

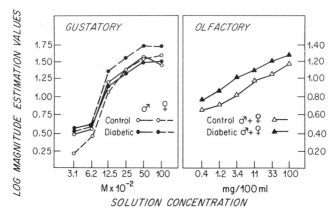

Fig. 1. Mean gustatory and olfactory magnitude estimation values for diabetic and control groups. Gustatory (six sucrose solution concentrations): (O——O) control males, (O- - -O) control females, (●——●) diabetic males, and (●- - -●) diabetic females. Olfactory (six essences of orange solution concentrations): (△) control males and females; (▲) diabetic males and females.

groups, was performed using a 2 × 2 × 6 analysis of variance (ANOVA). Diabetic versus control subjects as well as males versus females served as the between-subjects variables, whereas concentration of the sucrose solutions served as the within-subjects variable.

As expected, intensity ratings increased as a function of stimulus concentration, $F(5,1265) = 323.38$, $p < 0.001$, an effect which tended to level off at the next-to-highest concentration (Fig. 1). The main effects of groups, $F(3,253) = 2.78$, $p < 0.04$, and gender $F(1,253) = 23.53$, were statistically significant ($p < 0.001$) as was the Group × Gender interaction, $F(3,253) = 2.63$, $p < 0.05$. Examination of the data within the control group revealed that gender had no significant effect upon the magnitude estimation values. In the diabetic group, however, gender had a substantial impact on magnitude estimation of gustatory stimuli.

Female diabetic subjects yielded significantly higher overall mean magnitude estimation values for the six stimulus concentrations than did their male counterparts (ANOVA, $p < 0.05$). Moreover, the female diabetics showed significantly higher intensity ratings than the female controls (ANOVA, $p < 0.05$). The male diabetics showed no significant intergroup differences in their intensity rating of the sucrose concentrations, nor were they different from the control females.

Finally, the interaction of Concentration × Gender was found to be statistically significant: $F(5,1265) = 16.57$, $p < 0.001$, indicating a relatively sharper increase in rated intensity for females, as the stimulus concentration increased.

B. Olfactory Assessment

Since the data for the various diabetic treatment conditions were not significantly different, all the diabetic subjects were combined into one group for comparison with the control subjects. The differences in olfactory magnitude estimation values obtained for the diabetic versus control subjects, female versus male subjects, and the six stimulus concentrations were tested by a 2 × 2 × 6 between- and within-subjects analysis of variance. The intensity ratings increased as a function of the stimulus concentrations: $F(5,1265) = 60.47$, $p < 0.001$ (Fig. 1). More importantly, the main effect of groups was statistically significant: $F(3,253) = 3.11$, $P < 0.03$ and showed that the diabetic subjects generally gave higher intensity ratings for the stimulus concentrations than did the control subjects. Finally, in contradistinction to the gustatory testing, no significant gender effects or interactions with gender were observed.

C. Gustatory Hedonics

We examined taste preference as a function of increasing stimulus concentrations for diabetic and control subjects, as well as for males and females. As was done with the magnitude estimation values, the analysis was limited to the initial stimulus presentation series.

A repeated-measures, three-way ANOVA was conducted utilizing diabetic versus control groups and gender as the between-subjects variables, to test the effects of solution concentration on hedonics. The six stimulus concentrations served as the within-subjects variable.

The main effect of solution concentration, $F(5,1265) = 2.45$, $p < 0.03$, was statistically significant (Fig. 2). Whereas the main effect of gender was statistically significant $F(1,253) = 4.59$, $p < 0.03$, there was no effect of groups (control versus diabetes). There was a significant interaction of Concentration × Gender, $F(5,1265) = 4.11$, $p < 0.001$, which showed that hedonic ratings in females declined monotonically as concentration increased. By comparison, males showed an increase in hedonic rating as the stimulus concentration reached midrange, following which there was a decline.

D. Olfactory Hedonics

The subjects' hedonic ratings of the different olfactory stimulus (essence of orange) concentrations were examined by analysis of variance. The analytical procedures paralleled those employed for analysis of the gustatory data. The results (Fig. 2) indicated that increasing concentrations produced a greater hedonic rating ($F(5,1265) = 8.37$, $p < 0.001$, but no differences were observed for treatment or gender.

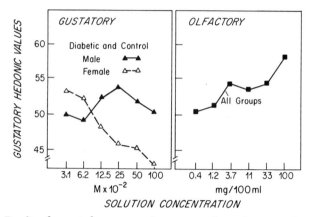

Fig. 2. Results of repeated measures, three-way analysis of variance for gustatory and olfactory hedonics in which groups (control and diabetic) and gender were the between-subjects variables, and the six stimulus concentrations were the within-subjects variable. (△) Control males and females; (▲) diabetic males and females.

E. Relative Magnitude Judgments

The described gustatory and olfactory magnitude estimates were based on absolute judgments made without arithmetic correction relative to the participants' intensity judgments of the standard stimulus. The following magnitude estimation analyses are based on transformed judgments of stimulus strength relative to the standard stimulus. When the gustatory standard stimulus was judged as part of the test series, magnitude judgments varied greatly within the diabetic subjects. An analysis of the distributions of these judgments was conducted comparing the diabetic males and diabetic females. Each judgment was placed into one of four categories: judgment of less than 10, judgment of exactly 10, judgment between 11 and 20, and judgment greater than 20. A statistical analysis of the number of subjects providing such judgments within each of the categories showed significant differences: $\chi^2 = 10.77$, $p < 0.05$. As seen in Table I, the effect was primarily due to the larger than expected number of women who judged the standard to have a value greater than 10.

Since some subjects, despite the precisely provided instructions, gave judgment values other than 10 to the standard stimulus, individual judgments for all subjects were calibrated with reference to their evaluation of the standard. If we denote the judgment of the ith stimulus as J_i and the judgment of the standard as J_3, the calibrating transformation may be expressed as

$$\ln J'_i = \ln J_i - \ln J_3 \qquad (1)$$

TABLE I Number of Subjects and Gustatory Magnitude Estimation Values for Diabetic Females and Males et al.[a]

	Magnitude estimation value			
	0–9	10	11–20	>20
Diabetic females	19	19	16	52
Diabetic males	19	28	29	31

[a] $\chi^2(3) = 10.77$, $p < 0.05$.

A mixed ANOVA was subsequently conducted using these transformed judgments, with gender and presence/absence of diabetes as the between-subjects variables and stimulus concentration as the within-subjects variable. For the gustatory judgments, there was a significant main effect of stimulus concentration, $F(4,1028) = 417.32$, $p < 0.0001$ and a significant Gender × Concentration interaction, $F(4,1028) = 5.75$, $p < 0.0001$. The main effect of concentration was due to an essentially monotonic increase in the judgment value as the concentration was increased, which was saturated at the two highest concentrations. The Gender × Concentration interaction resulted from the males, whose judgments of the first two concentrations tended to be larger than those given by females, but were smaller for the fourth, fifth, and sixth concentrations. That is, the males exhibited a lesser degree of variability per standard stimulus concentration than did the female subjects.

An analysis of the transformed olfactory judgments revealed the existence of a significant main effect only for concentration, $F(4,1028) = 64.11$, $p < 0.0001$. This effect was attributable to a monotonic increase in the estimated intensity as the concentration of the olfactory stimulus was increased. Neither diabetes versus control, gender, or any of the possible interactions was significant.

To examine the possible influence of the selected physiological and psychological variables on the taste and smell magnitude estimations, a series of multiple regression analyses were conducted. The dependent variable in each of the respective analyses was the mean taste or mean smell magnitude. The predictor variables entered into the regression included age, HbA$_1$c, R–R variation, gender, depression, smoking habits (past versus present versus never), orthostatic blood pressure changes, and fasting plasma glucose level.

When the mean gustatory magnitude estimations for the control subjects were evaluated, it was found that none of the variables that were entered into the regression equation had a significant influence. The analysis of the diabetic subjects showed that the gender variable was significant and indicated that females showed significantly higher magnitude estimations, $F(1,172) = 22.75$, $p < 0.007$, $r^2 = 0.12$. Moreover, higher levels of depression were associated with significantly higher magnitude estimation values, $F(2,171) = 15.54$, $p < 0.007$, $r^2 = 0.15$. In addition, greater orthostatic blood pressure changes, as measured by the drop in diastolic blood pressure when shifting from the lying to the standing position, were associated with higher taste-magnitude estimations $F(3,170) = 11.99$, $p < 0.007$, $r^2 = 0.17$.

A multiple regression analysis, comparing diabetic and control subjects, showed an overall effect by gender, $r^2 = 0.09$; depression, $r^2 = 0.12$, and a drop in diastolic pressure upon standing, $r^2 = 0.15$. Each of these variables was associated with higher taste-magnitude scores.

An analysis of the mean olfactory magnitude estimations for the control subjects indicated that higher HbA_1c values were associated with lower taste magnitude estimations, $F(1,30) = 7.23$, $p < 0.05$, $r^2 = 0.19$. In addition, diabetics and controls with higher levels of autonomic neuropathy, as measured by the R–R variation, provided lower olfactory magnitude ratings, $F(2,203) = 4.92$, $p < 0.05$, $r^2 = 0.05$. Neither smoking habits nor fasting plasma glucose levels were associated with mean gustatory or olfactory magnitude estimations.

IV. DISCUSSION

Although the diabetics, as a group, showed higher gustatory magnitude estimations than the control subjects, this effect was largely attributable to the diabetic women. By comparison, magnitude estimation values of control females tended to be lower than the values recorded for the control males; however, the latter difference was not statistically significant. The observed difference between diabetic and nondiabetics, with respect to their gustatory evaluations, requires closer examination. First, when the magnitude estimation values were adjusted with respect to the standard stimulus, the difference between diabetics and nondiabetics no longer appeared. The apparent discrepancy between the adjusted and nonadjusted analyses may be resolvable by examining the judgments of the diabetic females. That is, diabetic females tended to give higher estimations of the various stimulus concentrations including the standard. Thus, when the effects of this response bias were eliminated from the analysis, by use of adjusted scores, the difference between diabetics and nondiabetics was eliminated. Nonetheless, it remains to be determined which variables were responsible for the observed effects among diabetic women.

Parenthetically, it should be noted that higher depression levels were associated with higher mean taste magnitudes in diabetic females, $r = 0.35$, $p < 0.01$, whereas among control females the trend of this relationship was in the opposite direction, $r = -0.24$, $p < 0.15$. Although these relationships lack statistical robustness, they should not be dismissed a priori, given the previous findings that have shown that depression levels tend to be higher in diabetics. Moreover, Perlmuter et al. (1984) had observed that depression levels entered into significant relationships with a variety of cognitive measures among diabetics, and these effects tended to be strongest among female diabetics. Finally, when depression levels were correlated with mean taste magnitude in diabetic and control males, the respective relationships were clearly not significant: $r = 0.00$ and $r = 0.08$. Thus, depression levels appear to be related to altered mean gustatory evaluations only in diabetic females.

An examination of the diabetic females who provided the most distorted estimations of the standard stimulus revealed that they tended to have higher fasting plasma glucose values, higher HbA_1c values, and greater depression than diabetic males. In general, there were no systematic differences among the various subcategories of females with respect to these variables, one exception being that the females who judged the standard according to the instructions showed the best diabetes control, as indexed by the HbA_1c values.

It must be noted that all subjects were tested under fasting conditions, and it is known that such deprivation can alter gustatory judgments (Thompson and Campbell, 1977). Why such a regimen produces effects that are different in female diabetics relative to male diabetics can only be conjectured. An analysis of the hedonic evaluations did not further explicate the relationship between diabetic women and gustatory estimations. That is, although the gustatory hedonic data did show a significant gender effect, there was no differential effect due to diabetes. Dye and Koziatek (1981), studying aging males, failed to find any important effects of diabetes on hedonic ratings or thresholds with sucrose.

It may be argued that sensory evaluations involving sucrose as the test substance might yield ambiguous results in diabetic subjects because sucrose represents a "forbidden taste" and thus may create a psychodynamic stress in diabetic subjects. Another point of possible contention is that diabetics, as a group, may have relatively less experience with the taste of sucrose, and therefore their judgments might be distorted by a differential adaptation effect. Neither of these contentions is likely since the results showed that the diabetic individuals and controls differed with respect to their evaluation of the sucrose solutions, but more importantly, there was a Gender × Group interaction that negated these hypotheses.

In addition to the effects of gender and diabetes on magnitude estimations, there were significant effects associated with orthostatic-related changes in the diastolic blood pressure. Larger orthostatic decrements in diastolic blood pressure were associated with higher magnitude-estimation values. How this pathologically oriented change contributed to higher gustatory estimation judgments remains to be resolved.

In general, the gustatory hedonic evaluations indicated that the diabetic and control female subjects tended to prefer the weaker stimulus concentrations, whereas male participants generally showed a transient increase in their preference ratings, which peaked midrange in the series of stimulus concentrations. In contrast to the present findings, there is evidence that female rats tolerate higher glucose concentrations than male rats (Valenstein et al., 1967). In addition, among college age women, menstrual factors appear to alter sucrose hedonics (Weizenbaum et al., 1980). Neither finding, however, appears to adequately explain the gender-based hedonic effects observed in the present study.

The olfactory magnitude estimation data indicated that the diabetic subjects, male and female, rated the stimulus intensity as being significantly greater than did the control subjects. The magnitude estimation ratings were highest for the diabetic women and lowest for the control women, with the diabetic and control males being intermediate between the two female groups. This pattern, however, although similar to that noted for the gustatory-magnitude estimations, did not appear as a significant Gender × Group interaction. As with the gustatory estimations, adjustment of the olfactory estimations against the standard stimulus eliminated the effect of diabetes. An examination of the other variables that appeared to have an association with the magnitude estimation measures indicated that, for all subjects, high levels of autonomic neuropathy were associated with low olfactory magnitude estimations. Similarly, in the control subjects, high levels of hemoglobin A_1c were associated with low magnitude estimation values.

The olfactory hedonic measure did not distinguish between the diabetic and control subjects, nor did it show any difference between male and females. Only for the diabetic participants were the olfactory magnitude estimations and olfactory hedonic values found to be significantly interrelated, ($r = 0.16$, $p < 0.05$); in the control subjects, these measures appeared to be unrelated.

A comparison of gustatory and olfactory magnitude estimation responses indicated that these were significantly correlated within the diabetic ($r = 0.33$, $p < 0.05$) subjects and within the controls ($r = 0.28$, $p < 0.05$). Such cross-modality effects lend validity to the psychophysical procedures employed in the current investigation.

V. SUMMARY

There was evidence that the olfactory and gustatory magnitude estimations were elevated in diabetics; however, this effect was largely attributable to the diabetic women, who tended to give higher estimations than the males. An examination of the effects of variables other than gender and diabetes on magnitude estimations suggests that the evaluation of gustation and olfaction in the aged is complicated by a variety of factors. For example, as the levels of certain variables that reflect an increased degree of pathology changed, their influence tended to be more pronounced, or even limited to, one sensory modality. Moreover, the effects occasionally appeared to be limited to either the diabetics or controls. For example, increased depression was associated with elevated gustatory magnitude estimations, primarily in diabetic women. On the other hand, increased autonomic neuropathy was associated with decreased olfactory magnitude estimations among both diabetics and controls, whereas higher HbA_1c levels were associated with diminished olfactory and gustatory magnitude estimation values only among nondiabetics. Such intersensory comparisons, however, are valid only if a variety of stimuli within each modality are assessed. Although this study indicated that diabetes may alter certain aspects of sensory functioning, these findings must be tempered by the consideration that a variety of other coexistant factors may have multifactorial modulating effects.

ACKNOWLEDGMENTS

The authors would like to acknowledge the assistance of Kathy Flannery and Maria French. This work was supported by NIA Grant AGO 2300 and the Veterans Administration Medical Research Service.

REFERENCES

Carson, J. A., and Gormican, A. (1976). Disease-medication relationships in altered taste sensitivity. *J. Am. Diet. Assoc.* **68**, 550–553.

Chauncey, H. H., Kapur, K. K., Feller, R. P., and Wayler, A. H. (1981). Altered masticatory function and perceptual estimates of chewing experience. *Spec. Care Dent.* **1**, 250–255.

Chauncey, H. H., Muench, M. E., Kapur, K. K., and Wayler, A. H. (1984). The effect of the loss of teeth on diet and nutrition. *Int. Dent. J.* **34**, 98–104.

Doty, R. L., Shaman, P., Applebaum, S. L., Giberson, R., Siskorski, L., and Rosenberg, L. (1984). Smell identification ability: Changes with age. *Science* **226**, 1441–1443.

Dye, C. J., and Koziatek, D. A. (1981). Age and diabetes effects on threshold and hedonic perception of sucrose solutions. *J. Gerontol.* **36**, 310–315.

Ewing, D. J., Borsey, D. Q., Bellavere, F., and Clarke, B. F. (1981). Cardiac autonomic neuropathy in diabetes: Comparison of measures of R–R interval variation. *Diabetologia* **21,** 18–24.

Garg, M., and Perlmuter, L. C. (1985). Effects of depression and dementia on olfactory discrimination and identification. Unpublished manuscript, Harvard School of Dental Medicine.

Gregg, E. C., Jr. (1951). Absolute measurement of vibratory threshold. *AMA Arch. Neurol. Psychiat.* **66,** 403–411.

Lawson, W. B., Zeidler, A., and Rubenstein, A. (1979). Taste detection and preferences in diabetics and their relatives. *Psychosom. Med.* **41,** 219–227.

Nathan, D. M., Avezzano, E., and Palmer, J. L. (1982). Rapid method for eliminating labile glycosylated hemoglobin from the assay for hemoglobin A1. *Clin. Chem.* **28,** 512–515.

Nathan, D. M., Singer, D. E., Perlmuter, L., Harrington, C., Ginsberg, J., Katz, J., and Hakami, M. (1983). Hemoglobin A_1c influences prevalence of complications in elderly type II diabetics. *Diabetes* **32,** (Suppl. 1), 99A (Abstract).

Perlmuter, L. C., Katz, J., Ginsberg, J., Singer, D. E., Harrington, C., Hakami, M. K., and Nathan, D. M. (1984). Decreased cognitive function in aging non-insulin-dependent diabetic patients. *Am. J. Med.* **77,** 1043–1048.

Perlmuter, L. C., Nathan, D. M., and Hakami, M. K. (1985). Effects of diabetic treatment and gender on sensory functioning in the elderly. *In* "Handbook of Nutrition in the Aged" (R. Watson, ed.), CRC Press, Inc., Boca Raton, Florida.

Saxon, S. V., and Etten, M. J. (1981). Psychological aspects of nutrition in aging. *In* "Handbook of Geriatric Nutrition" (J. M. Hsu and R. L. Davis, eds.). Noyes Publications, Park Ridge, New Jersey.

Schiffman, S. S. (1983). Taste and smell in disease. Part 2. *N. Engl. J. Med.* **308,** 1337–1343.

Settle, R. G. (1985). Diabetes mellitus and the chemical senses. *In* "Clinical Measurement of Taste and Smell" (H. L. Meiselman and R. Rivlin, eds). Collamore Press, Lexington, Massachusetts.

Surwit, R. S., Scovern, A. W., and Feinglos, M. N. (1982). The role of behavior in diabetes care. *Diabetes Care* **5,** 337–342.

Thompson, D. A., and Campbell, R. G. (1977). Hunger in humans induced by 2-deoxy-D-glucose: Glucopruvic control of taste preference and food intake. *Science* **198,** 1065–1068.

Thompson, D. A., Moskowitz, H. R., and Campbell, R. G. (1976). Effects of body weight and food intake on pleasantness ratings for a sweet stimulus. *J. Appl. Physiol.* **41,** 77–83.

Valenstein, E. S., Cox, V. C., and Kakolewski, J. W. (1967). Further studies of sex differences in taste preferences with sweet solutions. *Psychol. Rep.* **20,** 1231–1234.

Wayler, A. H., Kapur, K. K., Feldman, R. S., and Chauncey, H. H. (1982). Effects of age and dentition status on measures of food acceptability. *J. Gerontol.* **37,** 294–299.

Wayler, A. H., Muench, M. E., Kapur, K. K., and Chauncey, H. H. (1984). Masticatory performance and food acceptability in persons with removable partial dentures, full dentures and intact natural dentition. *J. Gerontol.* **39,** 284–289.

Weizenbaum, F. A., Benson, B., Solomon, L., and Brehony, K. (1980). Relationship among reproductive variables. sucrose taste reactivity and feeding behavior in humans. *Physiol. Behav.* **24,** 1053–1056.

Zung, W. W. K. (1965). A self-rating depression scale. *Arch. Gen. Psychiat.* **12,** 63–82.

PART I

Discussion

Rolls: What is known about the way in which salivary flow and composition affect taste hedonics?

Christensen: This question has received little if any study. I don't, however, believe that marked effects will be observed. Many of the interactions that have been observed to date are near threshold level, with at least one exception for sour taste, where demonstrable effects are seen at suprathreshold. To the extent that hedonics are shaped by suprathreshold intensity functions, one would expect a change. But, by and large, salivary changes in taste perception are not observed at that level.

Fregly: Do you think that if sodium concentration in the body changes via a physiologic pathway that this might affect an individual's preference for it? As the adapting concentration of sodium shifts upward or downward, the point at which sodium becomes recognizable as salty also shifts. If one has a disease which lowers body sodium, such as Addison's disease (hypoaldosteronism), might one expect a shift in preference concentration for sodium, in addition to the change in detection threshold that has been described by Henkin et al. (J. Clin. Invest. **42,** 727, 1963).

Christensen: Changes in the perception of sodium as a function of salivary sodium concentration occur in a limited area near threshold. Regardless of where adaptation defines threshold, the intensity function is abnormal only near threshold, not over the entire intensity range. Preference probably will not change as a result of a small change in threshold, since it is mostly shaped by suprathreshold concentrations.

Maller: Dr. Christensen, is the information we have to date about the mechanism of adaptation limited by the fact that most all experiments that investigate the process use water as the diluent? That is, can we make generalizations to the "real" world of feeding behavior based on these observations?

Christensen: In all of the experiments I described where adaptation to saliva occurred, the method used a 2 min preadaptation period followed by the presentation of a taste solution followed by readaptation. One does not generally use these stringent processes in taste testing, and, of course, not in normal eating. It may be that the system exists in a state of partial adaptation, and that this is sufficient to change threshold. But it is not possible to achieve complete adaptation in normal situations. One can use dorsal flow methods, for example, but slight movements of the tongue can result in the adapted state being lost. The mechanism by which adaptation occurs has probably not been questioned since, no matter what method you use, you will find that sodium threshold occurs at a concentration above that found in saliva.

Yantis: You mentioned that caffeine can appear in saliva at concentrations above threshold. Does that mean that you could take a subthreshold level of quinine and have it be additive to the caffeine? Have you looked at this question?

Christensen: No, I have not yet investigated this. My evidence indicates, however, that salivary constituents alter taste stimuli concentrations. It is possible that the caffeine and urea in saliva could act in an additive way with bitter taste substances.

Houpt: I was wondering whether the fact that capsicum, the trigeminally active principal of hot peppers, causes the secretion of saliva may not be the answer to Dr. Paul Rosin's question as to why people like spicy-hot foods?

Christensen: Dr. Rozin entertained this notion, but I believe he has now abandoned it. Rozin determined salivary flow for people stimulated by capsicum. Interestingly, he found that individuals with a history of hot pepper consumption salivated less to capsicum than those who did not consume hot peppers, and he interpreted this as a desensitization to the capsicum stimulus.

Cagan: Dr. Beauchamp, is the phenomenon of failure of infants to thrive a familial one or is it just an occasional child in any family?

Beauchamp: I believe that it tends to cluster in families, which is why some investigators think it may be caused by a parental–child relationship.

Cagan: Dr. Murphy, is the variability in the data that you showed for menthol in the older group generally true for other stimuli that do not have trigeminal components?

Murphy: In general, a survey of the literature suggests that there is more variability in older groups.

Hegsted: I think variability in everything increases with age. Some of us lose hair, some don't! Drs. Greeley and Torii, it appears that every type of nutritional deficiency in rat increases their salt preference. Is that true and could you comment?

Greeley: There are a number of micronutrient deficiencies that will result in an increased preference for salt. Furthermore, I am not at all convinced that the effects are specific toward a particular taste quality.

Hegsted: If it is a general phenomenon, then it looks as though it is not related to the nutrient. There may be some factor related to growth rate, body size, or malaise that somehow leads to a liking for salt.

Torii: I would like to comment on this. In Japan, within the past four decades, ingestion of salt has decreased along with an increase in the ingestion of higher quality protein. A marginal deficiency in the nutritional state could lead to an inability to control either extracellular fluid volume or tonicity. Essential nutrients are constantly needed to maintain the active transport systems and the whole-body control of water. Every animal we have tested, especially young ones, when in a deficiency state, begin to drink salt water. This occurs not only for the micronutrient deficiencies mentioned by Dr. Greeley, but also for deficiencies of essential amino acids. But in the adult, macronutrient status of the diet does not seem to be of as immediate importance. The adult probably has reasonable stores of essential amino acids and when placed on either a deficient diet or low protein diet, they do not immediately show a preference for salt. Rats 4 weeks of age, for example, are still developing, and there are insufficient stores of essential amino acids. It is very easy to achieve a severe deficiency in a growing animal. But this question needs to be looked at as one not so much of age, but as one of growth or of body size. When a deficiency state is produced, the experimental animals are of a lower body weight than the controls, and what one usually does is compare taste preference between these two groups. But the control group may be an inapparopriate one since it is of a body size much greater than the experimentals. We have shown, for example, that when a rat reaches around 250 g, its taste preference is different after a challenge with a deficiency than if it were lighter than 250 g. The point in the growth of the rat at which one begins a nutritional deficiency is an important variable and one that, if taken into account, may explain much of the discrepancy in the literature. It is also best to obtain data from adult animals first, then go to the growing animal where severe deficiencies can be achieved rapidly.

Discussion

Maller: Is there any suggestion that the deficiencies you are creating in your animals have an impact on the nervous system in general rather than specifically affecting taste per se?

Greeley: My data do not show whether the deficiencies are causing alterations in the nervous system in general. It is true that in those animals where deficiencies were created we did observe growth deficits. One can see the alteration in taste behavior begin in states of only marginal deficiency, so I believe it is a very real phenomenon, but experiments have not been conducted which directly relate nutritional deficiencies to a change in a particular taste modality. We need to direct future experiments toward investigation of mechanisms rather than collection of behavioral data.

Hegsted: I did not mean to imply that the observation was not real. However, it may be unproductive to look at the taste system in search of a mechanism to explain how a particular nutrient works when most nutrient deficiencies alter salt preference.

Threatte: Dr. Greeley, what were the osmolalities of the sucrose solutions that were used in your tests?

Greeley: The millimolar concentrations were 10 and 20 mM then 200 mM sucrose.

Threatte: Did you see your larger differences at the highest concentration?

Greeley: It depends upon which deficiency. In vitamin B_6 deficiency there were significant effects at low and high concentrations of sucrose.

Threatte: If you gave your animals a choice between two sucrose solutions, one hypertonic, the other hypotonic, which do you believe these deficient animals would choose? I ask this because I am concerned that the results of the taste preference measures may be confounded by the animals' state of hydration.

Greeley: We have some information to suggest that these animals are dehydrated. This is based on data such as packed cell volume and water : food ratios. For example, serious vitamin B_6 deficiency leads to a reduced water : food ratio, far below anything I've ever seen before. They even look dehydrated. Some of the taste behaviors we are observing may, in fact, be linked with water metabolism. To answer your question is impossible without performing the experiment, but I would guess that the animals would choose the higher concentration of sucrose, only because it is a very compelling stimulus.

Fregly: Dr. Murphy, do you believe that part of the reason for the changes you reported in aging might be related to changes in renal function? Do you know this much detail about the medical history of the subjects?

Murphy: We have not asked our subjects to undergo a medical examination. In the first study, subjects were simply recruited, as it were, off the street. In the second study, subjects were required to be ambulatory, with no hospitalization within the last 12 months, and not institutionalized.

Fregly: This point was raised because of an earlier comment of Dr. Christensen's where she discussed the possibility that substances such as urea in the saliva may affect taste threshold. With progressive decline in renal function with age, it may be possible that taste function is affected by compounds that are not cleared efficiently from the body.

Maller: There seems to be a number of examples of discrepancies in results between studies using humans and those using animals as models for human nutrition. This, of course, raises the question as to what is the appropriate animal model for the human. In certain receptor studies, the catfish has become important; in gastrointestinal studies, it is the pig that serves as a model. I've assumed that the rat is a good model, but lately I've become wary of extending results from that animal to humans. Could anyone comment on this problem?

Fregly: I could certainly deliver a testimonial to the appropriateness of the rat as a model. But there are discrepancies between rat and human data. No animal will be the perfect model and, as we all know, we need to be careful in making extrapolations from our animal data to the human. One example that comes to mind is the effect of therapy on adrenalectomized rats compared with effects of therapy on humans with Addison's disease. Adrenalectomized rats display an increased preference for NaCl which can be brought toward normal by treatment with mineralocorticoid but not with glucocorticoid hormones. Addisonians, as reported by Henkin and co-workers (*J. Clin. Invest.* **42,** 727, 1963) display lower detection thresholds for all taste stimuli tested, including NaCl, yet their thresholds are normalized with glucocorticoids but not with mineralocorticoids. Since this is the only study that I have found regarding detection threshold in Addisonian patients, the results should be verified. If they are verified, we will have to recognize that the rat has some limitations as a model for the study of salt appetite in humans.

Beauchamp: As you know, adult humans dislike salty water solutions whereas most rats show a preference for isotonic saline solutions. On the other hand, humans show a marked preference for salty foods. Dr. Mary Bertino and I have offered rats food with various levels of NaCl and in short-term preference tests we have found no level of NaCl at which the food is preferred.

Fregly: Some years ago, I did a study with graded amounts of sodium in the diet, and then gave rats the opportunity to choose between isotonic saline and water to drink (*Am. J. Physiol.* **209,** 287, 1965). As the amount of sodium in the diet increased, the rats increased the amount of water, but not NaCl solution, that they ingested. Intake of NaCl solution remained constant despite increases in dietary sodium concentration. This suggested that the animals were regulating the amount of sodium through drinking, since they were ignoring the amount of sodium in the diet. It also suggested that only the sodium in solution was recognizable by the rat. When I increased the concentration of the NaCl solution offered, the rats drank less but maintained a constant intake (i.e., mEq/day) of sodium by way of the NaCl solution. I postulated that the rat controls its sodium intake by way of sodium in solution and not sodium in food.

Maller: What do you all think the effect of methodology is on the questions you are asking? Should this be a major consideration and how will it affect the direction of future research?

Fregly: We might ask what test in the rat might be comparable to a test in the human. I don't know the answer but it is a very basic question if we plan to use data from the rat to say something about the human.

Torii: I have a comment. I am always concerned about what is the most appropriate test solutions for a given deficiency state. Most researchers only offer a choice of some basic stimuli such as sucrose, NaCl, HCl, or quinine, but I think we need to establish which stimuli are appropriate for which test. Maybe the four basic ones need to be expanded to include others that may be more indicative of the animal's need under a particular test.

Fregly: I agree. We have now come to a threshold where we need to look at, for example, mixtures of stimuli. We have done a lot with individual substances in the past, but there are relatively few data on mixtures. I think P. T. Young is a pioneer in this regard with his studies of solutions that he termed isohedonic. He mixed concentrations of NaCl and sucrose and these were presented to rats for 10 min during which he measured their lick rate. He would, for example, begin with 1% sucrose and then add graded increments of NaCl to it and determine the lick rate of each mixture when presented to the rat. He constructed three-dimensional diagrams of this and, under these conditions, showed significant effects of one solute on another. I think this is an area we need to look at more carefully.

Christensen: The measure that you think affects taste function least in the physiological sense is preference. Yet, that is the measure used in the rat work. The most important data point in the animal literature is the so-called preference threshold. This may be the only point at which one can compare human and animal data.

Torii: I believe we must stop studying long-term taste preferences in isolation of the basic physiology of the animal. We did an experiment trying to link taste with physiological feedback, one that might remind some of you of the experiment of Dr. Naim and colleagues (*J. Nutr.* **107,** 1653, 1977). We continuously measured meal frequency and meal size with a strain gauge. If the diet contained quinine, rats found it aversive and tended to eat small meals at a high

Discussion

frequency. If the protein level was high, e.g., 20% egg protein, their meal pattern and meal size become like normal after a week. If we provided a nonprotein diet but added 0.2% saccharin to it, the meal patterns and size were normal initially, but a week later the meals were smaller and more frequent. I believe that the animal was looking for two types of information: (1) taste and (2) nutritional status that it received as a postingestional feedback. If the taste of a food is aversive, but its quality is good, it will hesitate at first, but later accept it. Postingestional reward has overridden taste, or perhaps the rat has habituated to the taste. If the food has an acceptable taste, it will be consumed normally until the postingestional consequences are evaluated. In the animal, the chemical senses and nutrition are coupled, and this is the best way to study them.

Greeley: In our copper–vitamin B_6 study, in collaboration with Dr. Mary Bertino, we did some electrophysiological recordings on a subset of these animals. Recall that with the copper-deficient animals, we saw little changes in taste behavior but with the vitamin B_6-deficient ones, we saw a large difference in taste behavior. I would like to ask Dr. Bertino to summarize the electrophysiological studies.

Bertino: The copper-deficient animals were more sensitive to NaCl, but this was not statistically significant. This did not match the behavioral data. But, on an individual basis, the five individual rats for which I have records demonstrated a decreased preference for NaCl.

Greeley: I think while this may have been an unfortunate choice of a subset of our animal groups, it points to the types of studies that must now be done. We must begin to look at the taste physiology of these animals under nutrient stress, not just at their taste behavior.

Rogers: Learned aversions have been studied for a number of years now. It has been clearly shown that not all deficiencies create learned aversions: for example, Cu doesn't but vitamin B_6 does. Kalat and Rozin defined the term "belongingness" for this, and in this context it can be applied to indicate that unless a deficiency causes a GI upset, it will not be coupled with a learned aversion. What you may be measuring is the capability of these animals to form a learned aversion, especially if these test solutions are novel taste stimuli. You may be asking the animal whether this deficiency causes a GI distress to the point that they will choose a novel stimulus over water?

Greeley: All the animals in the Cu–vitamin B_6 study were also tested against quinine. We did this on a short-term basis about midway through the study and on a long-term basis at the end of the study. The Cu-depleted animals showed more aversion to quinine than the controls during the first two short-term studies. On the third trial, the controls dropped down to a level comparable to the deficients. Long-term testing indicated that pyridoxine deficiency increased rats' aversion to quinine while copper-deficient rats' behavior was indistinguishable from that of control animals.

Rogers: I'd like to comment on two other things: first, amino acid deficiencies produce very strong learned aversions. I think taste and odors of the diets are used as a signal of those diets that have different postabsorptive consequences. Second, I'd like to comment on some work that Levkowski did, because it is relevant to this discussion on the relative role of nutrients in water versus nutrients in food. An animal will often associate the nutrient with the solid food and not the water. An example is with amino acid deficiency. If you put the amino acid in a diet, an animal will select the diet with complete nutrition. He will eat the complete diet to the exclusion of any other. If, now, you put the amino acid in solution, he will still choose it, but will do so in an approximate 2:1 ratio over water. The water is not the normal vehicle for amino acids. Lepkovsky asked the rat if milk is food or water. If he made them adipsic, they still drank the milk, but if it was diluted twofold with water, they ceased to drink it. So there is a point at which animals will not identify nutrients in water as food.

Maller: A similar effect was seen when saccharin was mixed with water or milk. Saccharin in water causes an increase in water intake; saccharin in milk does not lead to an increase in milk intake. Please, each of you make one more comment.

Beauchamp: I believe there is a large realm of cognitive factors involved in human food selection and choice that can never be reached by rat studies. So there is no animal model for those points. We must use the human.

Torii: When an animal is under an amino acid deficiency, he can detect the deficient amino acid in water clearly if given a two-choice test with water. I don't find that unusual. But if water is an inappropriate vehicle for nutritional information, could he detect amino acid from among 15 different amino acids and taste solutions presented simultaneously? It seems to me he should have a very hard time, but in fact, our data show that within 2 days he drinks almost exclusively from the tube containing the amino acid that is lacking from the diet.

Christensen: I hope two major points are remembered from the presentation and discussion on saliva. First, dietary variables do alter salivary characteristics and, thus, saliva is a plausible vehicle for dietary-induced changes in taste. Second, our understanding of the mechanisms by which saliva affects taste is only sketchy; therefore, research should be encouraged in this area.

Greeley: Fluid regulation should be considered as a variable in our taste experiments. The vitamin B_6-deficient rats in our experiments were hypodipsic if one considers daily water intake. Yet, when offered fluids containing preferred taste stimuli (repeated over many trials), short-term consumption of the aqueous stimuli increased. I suggest that this short-term behavior is taste related and the finding may be associated with the animal's state of hydration.

Murphy: I am concerned about individual differences and about taking a single measure at one point in time. Rose Marie Pangborn and her colleagues have shown that some stimuli will produce hedonic functions which are positively accelerated in some individuals, negatively accelerated in other individuals, and relatively flat in still other individuals (Pangborn, *Psychon. Sci.* **21**, 125, 1970; Lundgren et al., *Chem. Sen. Flav.* **3**, 249, 1978). Hence, it seems important that, for any substance in an hedonic study, the investigator be sensitive to the possibility of this type of individual variation before averaging over individual hedonic judgments to produce a "characteristic" group function. It is also important to consider that it is possible to alter hedonic responses to chemosensory stimuli through context and repeated exposure (Cain and Johnson, *Perception* **7**, 459, 1978; Murphy, "Selected Sensory Methods: Problems and Approaches to Measuring Hedonics," Am. Soc. Test. Materials, 1982; Riskey et al., *Percept. Psychophys.* **26**, 171, 1979; Stang, *Bull. Psychon. Soc.* **6**, 273, 1975). Hence, a single measurement may not be adequate.

Fregly: Perhaps I can highlight our discussion by telling you a relevant anecdote. We gave an exam question to medical students one year that had been given two or three years before. Some students were concerned that the question had been used previously. My response to them was "Yes indeed, the question is the same. It is the answer that has changed."

PART II

Effects of the Cephalic Phase on Digestion and Absorption

8

Intragastric Feeding of Fats

ISRAEL RAMIREZ
Monell Chemical Senses Center
Philadelphia, Pennsylvania

I.	Introduction	151
II.	Review of Intragastric Feeding	152
	A. Conditioning and Learning	152
	B. Digestion and Growth	153
	C. Appetite	155
	D. Fat	156
III.	Research Needs	162
	References	163

I. INTRODUCTION

When animals or humans are fed intragastrically, food enters the digestive tract without stimulating oral chemoreceptors. Comparison of the digestion of oral and intragastric meals thus provides a valuable method for studying the importance of cephalic phase responses. Intragastric feeding reliably alters such diverse variables as stomach emptying, insulin release, growth rate, and feed efficiency. As a result of these changes, intragastric feeding of fats actually seems to be aversive. A series of experiments are described in which rats were fed corn oil by mouth or by gastric catheter, and plasma levels of triglycerides were sampled 1–6 hr later. Intragastric feeding of oil resulted in a more rapid rise in plasma triglycerides, followed by a more rapid fall, than did oral feeding of the same amount of oil. Administration of an oily or sweet taste to the mouths of rats, immediately before intragastric feeding, prolonged the elevation in blood triglycerides 4 hr later. A sweet taste also decreased the rate of stomach emptying of an intragastric oil meal. Tastes were more effective in rats having previous experience with oil paired with that taste. The results show that cephalic phase responses influence the digestion of fat.

Intragastric feeding has often been used in physiological and behavioral research (e.g., Houpt, 1982; Hunt, 1956; McHugh and Moran, 1979), as well as in clinical work with humans (Stevens, 1982). In spite of its frequent use, there has been little attention given to the possibility that intragastric feeding might alter digestion. This lack of attention is surprising in light of the many ways intragastric feeding differs from normal oral feeding (e.g., taste, rate of ingestion, swallowing, saliva, presence of stomach tube or fistula, etc.).

Another reason for studying intragastric feeding is that it is a good way to analyze how taste and smell affect digestive responses. The most commonly used method of studying the role of taste and smell in digestion (cephalic phase responses) is sham feeding. In this method, animals or people are allowed to eat but the food is removed from the digestive tract before it is absorbed. This procedure allows one to observe the digestive responses induced by tasting, chewing, and swallowing food. Intragastric feeding is the opposite method, in which food is placed into the digestive tract (stomach) without its being tasted, chewed, or swallowed. Comparison of the digestion of oral and intragastric meals provides information about the consequences of bypassing the mouth. By itself, however, intragastric feeding does not enable one to establish a role for taste or smell. A slight modification of this method does allow one to implicate taste. In the modified approach, which I call taste plus intragastric feeding, the mouth is stimulated with a taste, and food is immediately put into the stomach. Digestion under this condition is then compared to that seen following intragastric feeding without oral stimulation. The power of this method will be illustrated below.

This chapter will briefly summarize previous studies on intragastric feeding of a variety of foods, and recent work in this laboratory on feeding of fat.

II. REVIEW OF INTRAGASTRIC FEEDING

A. Conditioning and Learning

It is relatively easy to reinforce operant behaviors (such as bar pressing) with oral food rewards, but it is very difficult to demonstrate rewarding effects of intragastric feeding (Cytawa *et al.*, 1972; Epstein, 1967; Snowdon, 1969). Interestingly, providing a taste in conjunction with intragastric feeding enhances the rewarding effects of this mode of feeding (Epstein, 1967; Holman, 1968), just as a taste may normalize digestion of an intragastric meal (see below). Another way to improve digestion and thus enhance the rewarding effect of an intragastric meal for a rat is using food that has already been partially digested by another rat (Deutsch, 1978; Puerto et al., 1976).

Under some conditions, intragastric feeding may be aversive. Pairing a taste with a treatment that produces malaise usually results in a reduced subsequent preference for that taste; this phenomenon is called learned taste aversion. If a particular treatment reduces the subsequent preference for the taste paired with it, it is commonly assumed that the treatment produces malaise or is somehow aversive (Deutsch, 1978). Rats that are given repeated exposures to a novel taste such as saccharin, immediately before intragastric oil feeding, subsequently showed lower saccharin preference than control rats (Deutsch, 1978; Ramirez, 1984). However, the magnitude of the change in taste preference is fairly small. For example, rats given a control treatment (intragastric water) paired with a taste of saccharin subsequently consumed 98% of their fluid from the saccharin bottle. Rats given intragastric oil consumed 88% of their fluid from the saccharin bottle; rats given intragastric lithium, which did not reduce food intake, consumed only 7% of their fluid from the saccharin bottle (Ramirez, 1984).

A significant complication is that the pairing of a taste with intragastric feeding can alter the digestive responses and perhaps the degree of aversion of an intragastric meal (see below). This factor might result in weaker taste aversions than might be expected for the degree of malaise produced without taste.

B. Digestion and Growth

Intragastric feeding of a glucose solution to rats results in a more rapid rise followed by a more rapid fall in blood glucose and insulin than does oral feeding of the same amount of glucose (Louis-Sylvestre, 1978). Oral ingestion of a liquid diet (Vivonex) produces a moderate rise in blood glucose and insulin, whereas intragastric infusion of the same amount of diet causes much greater elevations in blood glucose and insulin (Steffens, 1976). Analogous findings with fat feeding are discussed below. In each case, the differences between oral and intragastric feeding are substantial.

Some of these results may be due to changes in the rate of gastric emptying. Intragastric feeding has been reported to enhance stomach emptying about 10–30% in humans (Hunt, 1956; Malhotra, 1967) and as much as 40% in rats (Molina et al., 1977; Fig. 1). This difference even applies to simple solutions such as plain water or dilute saline (Hunt, 1956). Enhanced stomach emptying would be expected to produce the faster rise in glucose and insulin following intragastric carbohydrate meals.

Recent studies (Ramirez, 1986) indicate that the effects of intragastric feeding on gastric emptying are highly dependent on testing conditions.

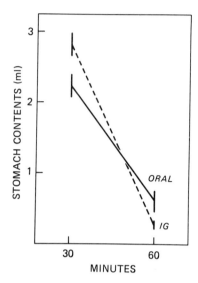

Fig. 1. Apparent rate of stomach emptying in rats fed 6 ml of 15% glucose solution. Nine rats were given glucose to drink (oral) or were fed by gastric catheter (IG), and stomachs were emptied by the same catheters 30 or 60 min later. Dilution by saliva and gastric secretions was assessed by a phenol red marker. Each rat was tested in each condition. The vertical bars are standard errors.

Intragastric feeding accelerated gastric emptying of phenol red in chronically food deprived rats (Fig. 1) but not in mildly food deprived rats. Direct assay of carbohydrates in the stomach gave different results than indirect assay via phenol red. The only consistent effect of intragastric feeding on gastric emptying of carbohydrate (glucose or glucose polymer) was to produce some dumping of gastric contents within 15 min. after the beginning of intragastric infusion; subsequent rate of emptying was unaffected by the method of feeding.

Two groups of researchers have used a variant of the taste plus intragastric feeding method in humans. One study indicated that tasting and chewing a meal did not greatly reduce the rate of stomach emptying (Schiller et al., 1980). However, the pH of the subjects' stomachs was kept constant by continuous intragastric titration; no data were reported on the effects of this manipulation. Malhotra (1967) reported that subjects who tasted a liquid food prior to intragastric feeding of the same food showed slower gastric emptying than did subjects who did not taste the food prior to intragastric feeding. Unfortunately, the study was conducted in a way that makes it difficult to determine whether the effect was due to tasting the food or to adding saliva to the food.

If the digestive abnormalities produced by intragastric feeding are physiologically significant, then one would expect long-term intragastric feeding to alter growth. Rats fed intragastrically by continuous intragastric infusion gained less weight than rats given the same diets orally (Young et al., 1982). This effect on body weight appeared with five different diets. On the other hand, two to four daily intragastric meals increased weight gain and/or fat deposition even when total caloric intake was about the same as the oral group in most (Cohn and Joseph, 1959; Kekwick and Pawan, 1966; Rothwell and Stock, 1978, 1979) but not all studies (Geliebter et al., 1984; Liu and Yin, 1974). This effect is not attributable to alteration of normal feeding patterns. Cox and Powley (1981) eliminated this problem by pair-feeding and did not observe substantial effects on total weight gain in normal rats. However, they found that normal rats fed intragastrically had elevated total carcass lipid. Cohn et al. (1957) showed that varying the number of intragastric meals from two to four per day or limiting oral intake to 2 hr/day did not alter body fat.

Enhanced weight gain and fat deposition without an increase in caloric intake implies a decrease in metabolic rate. LeBlanc et al. (1984) measured metabolic rate directly (oxygen consumption) in humans after an oral or intragastric meal. Resting metabolic rate was two to three times greater in people fed by mouth than people fed by gavage. Unfortunately, these investigators used different foods for the oral meal (sandwich, pie, and soft drink) and intragastric meal (liquid diet).

The abnormalities in growth rate associated with intragastric feeding are presumably due to a lack of oral stimulation, but there are few data at present to confirm this idea. Premature infants fed by gavage gain weight and mature faster when they are allowed to suck a pacifier during feedings (Bernbaum et al., 1983; Measel and Anderson, 1979). Furthermore, nonnutritive sucking decreases gastrointestinal transit time (Bernbaum et al., 1983).

C. Appetite

Intragastric feeding has been frequently used to study the importance of oral metering in the regulation of caloric intake. Animals fed intragastrically usually regulate caloric intake accurately (e.g., Booth, 1972; Geliebter, 1979; Geliebter et al., 1984; Houpt, 1982; Liu and Yin, 1974; Maggio and Koopmans, 1982; Ramirez and Friedman, 1983a; Rothwell and Stock, 1978, 1979) unless the loads are too large (Hansen et al., 1981). Interestingly, fats give variable results. In some studies, rats decrease food intake less than would be expected from the calories in the fat (Booth, 1972; Geliebter, 1979; Rothwell and Stock, 1978), whereas in other studies, rats decrease food intake enough to keep caloric intake constant (Geliebter et al., 1984; Liu and Yin, 1974; Maggio and Koopmans, 1982). A study in this laboratory suggests that intragastric feeding of oil can sometimes decrease food intake more than oral intake of the same amount of oil (Ramirez and Friedman, 1983). One factor contributing to this inconsistent literature is that suppression of food intake resulting from intragastric fat tends to increase with repeated testing (Geliebter et al., 1984; Ramirez, 1984).

Appetite studies involving intragastric feeding are very difficult to interpret. Since intragastric feeding can alter the rate of digestion of foods and the energetic efficiency of animals and humans as described above, it is hard to assign any meaning to the notion of accurate or inaccurate caloric intake regulation. If intragastric feeding produces severe malaise, as suggested by Deutsch (1978), then any decrease in food intake following an intragastric meal could be attributed to such malaise. Sawchenko et al. (1980) have suggested that gavage feeding can produce persistent alterations in the regulation of food intake. They force fed rats for 6 days, then tested their feeding response to a large dose of deoxyglucose 10 weeks later, and found that previously force fed rats ate less in response to the drug than control rats. If intragastric feeding does indeed disrupt normal regulatory mechanisms, then it is impossible to reach any conclusions about the regulation of food intake from studies of intragastric feeding until the nature of the disruption has been defined.

D. Fat

Most of the work conducted by the author has concentrated on vegetable oil because very little is known about possible cephalic phase responses to fats and fatty tastes. It seemed reasonable to predict that the taste of fats elicits some response that facilitates the digestion, absorption, or metabolism of fat. There is no specific taste for fats analogous to the sweet taste of most sugars. It is probable that organisms learn to associate tastes present in fatty foods with the presence of fat in those foods. Therefore, the role of learning has been investigated.

The work was facilitated by the finding that rats readily consume small amounts (1.5 ml) of vegetable oil, usually within 5 min after presentation, even when they have not been food deprived (Ramirez and Friedman, 1983b). Most of the experiments used a fat tolerance test, in which a small meal of corn oil was given either by mouth or intragastrically, and changes in blood levels of triglycerides were measured 1–6 hr later. This method is analogous to the carbohydrate tolerance test. In some experiments, blood levels of free glycerol were also examined because they are helpful in interpreting changes in triglyceride levels. Chronically implanted catheters were used to avoid the stress associated with gavage.

The first experiment demonstrated that rats allowed to consume oil orally showed an inverted U-shaped triglyceride tolerance curve, with blood triglycerides peaking at 2–4 hr after ingestion. When the rats were given oil intragastrically, plasma triglycerides were high within 1 hr, peaked at 2 hr, and by 6 hr were below those observed in the oral group (see Fig. 2). Blood glycerol levels showed a similar pattern (Ramirez, 1984).

It seemed possible that one factor contributing to the difference between oral and intragastric feeding is that with the latter method of feeding, oil was not emulsified with saliva. Therefore, the effects of feeding oil emulsified in water were examined (1:1 water:oil with 1% phosphatidylcholine). As may

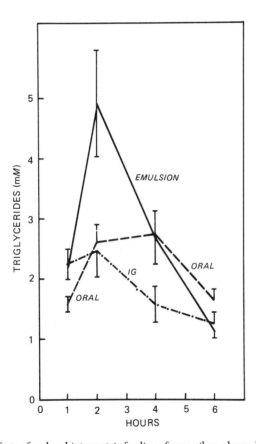

Fig. 2. The effects of oral and intragastric feeding of corn oil on plasma levels of triglycerides. Ten rats were given 1.5 ml corn oil by mouth (ORAL) or by stomach catheter (IG). Oil was also fed as a 50% oil-in-water emulsion stabilized with phosphatidylcholine (EMULSION). Each rat was tested in each condition. The vertical bars are standard errors. [Reproduced from Ramirez (1984) with permission.]

be seen in Fig. 2, emulsification exaggerated the abnormal triglyceride tolerance curve resulting from intragastric feeding.

The next experiment examined possible habituation to intragastric feeding. A group of rats was repeatedly offered 1.5 ml corn oil until they drank it readily. An oral fat tolerance test revealed an inverted U-shaped tolerance curve similar to that seen in the first experiment. When the rats were given oil intragastrically, blood triglyceride level showed a sharp peak at 2 hr and was significantly different from the oral curve at every time sampled (Fig. 3,

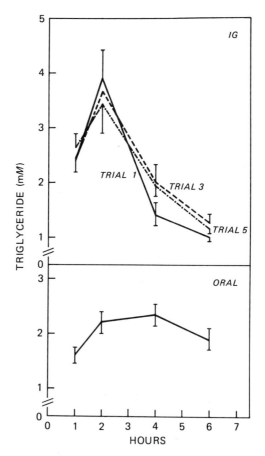

Fig. 3. Oral and intragastric triglyceride tolerance. The bottom panel shows plasma triglycerides 1–6 hr after rats ingested 1.5 ml oil by mouth. The top panel shows plasma triglycerides the first, third, and fifth times the rats were fed the same amount of oil via stomach catheter. There was no statistically significant trials effect. The vertical bars are standard errors. Each point is the mean of 10 rats.

8. Intragastric Feeding of Fats 159

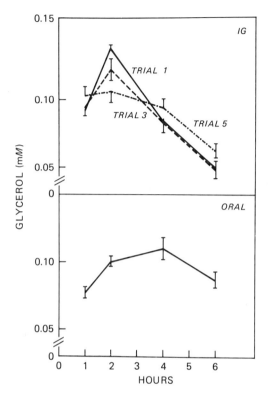

Fig. 4. Plasma glycerol during the same triglyceride tolerance tests as in Fig. 3. The bottom panel shows plasma glycerol 1–6 hr after the rats ingested 1.5 ml oil by mouth. The top panel shows plasma glycerol, the first, third, and fifth times the rats were fed the same amount of oil via stomach catheter. The trials effect is statistically significant ($p < 0.05$).

$p < 0.05$). The rats were given a series of five intragastric oil feedings, each of which was 3 days apart. Blood was sampled on the first, third, and fifth feedings. Blood triglycerides were similar on all three trials (Fig. 3), but free glycerol levels began to resemble those seen with oral feeding by the fifth trial (Fig. 4, trials effect, $p < 0.05$). Thus, learned adaptation to intragastric feeding in the absence of taste stimulation is slow.

The next series of eperiments examined whether intragastric fat tolerance could be modified by providing a taste along with the feeding (Ramirez, 1985). Rats were given either a small taste of fluid or nothing immediately prior to intragastric oil feeding. In some of the studies, this procedure was repeated up to three times on different days, and blood was sampled 1, 2, 4, and 6 hr after feeding on the first and third test days.

The results of several experiments showed that tastes that had not pre-

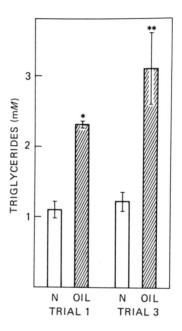

Fig. 5. The effects of oral stimulation on plasma triglycerides 4 hr after an intragastric meal of 1.5 ml corn oil. Rats were given either 0.3 ml 2% oil-in-water emulsion (OIL) or nothing (N) by cheek catheter immediately before intragastric feeding of oil. There were 5–8 rats/group. Vertical bars are standard errors. $*p < 0.01$; $**p < 0.001$.

viously been paired with fat ingestion altered fat tolerance slightly. However, tastes that had previously been paired with fat ingestion were more effective, particularly at 4 hr after ingestion. In one experiment, rats were given a taste of a 2% oil–water emulsion or nothing immediately before intragastric feeding. Each rat was given its assigned treatment combination on three trials, 3 days apart. Blood was sampled on the first and third trials. As may be seen in Fig. 5, a taste of oil appeared to increase plasma triglycerides at 4 hr, particularly on the third trial. Analysis of variance on data from all blood samplings indicated that the taste by trials by hours interaction is statistically significant ($p < 0.05$). Thus, the taste of oil becomes more effective with experience.

The taste employed did not have to be a "fatty" taste. Another experiment was conducted similar to the previous one, but employed 0.1% Na-saccharin as the taste stimulus instead of oil–water emulsion. Some of the results are shown in Fig. 6. The two left bars show 4 hr plasma triglycerides for rats given an intragastric meal of 1.5 ml corn oil preceded by nothing or by 0.3 ml of saccharin solution on Trial 1. The next two bars show plasma triglycerides

8. Intragastric Feeding of Fats 161

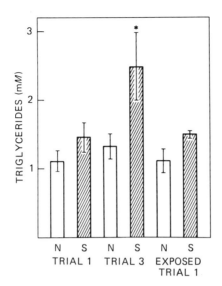

Fig. 6. The effects of oral stimulation on plasma triglycerides 4 hr after an intragastric meal of corn oil. The experiment is similar to that described in Fig. 5 except that the rats were given oral saccharin (S) or nothing (N) immediately before intragastric feeding. The two bars on the extreme right (EXPOSED) show the results for the first intragastric feeding for rats that had previously been permitted to drink the saccharin *ad libitum* for 2 days. Vertical bars are standard errors. $*p < 0.05$.

for the same rats on Trial 3. The effect of repeated exposure to saccharin with intragastric oil can be seen even more clearly than in the previous experiment. An additional experiment was conducted to determine whether this learning is attributable to mere exposure to saccharin alone. Rats were allowed to drink 0.1% saccharin *ad libitum* for 2 days prior to testing the effects of saccharin on fat tolerance. The two bars on the right-hand side of Fig. 6 (labeled Exposed Trial 1) show that these rats had the same triglyceride values the first time saccharin was paired with oil as in the previous experiment. Thus, pairing of the taste with intragastric feeding seems to be crucial for the learning observed in the previous experiments.

Stomach emptying was next examined because it provides a more direct measure of digestion than the triglyceride tolerance tests described above. The procedures were similar to those used in the previous experiments except that the rats were sacrificed at the appropriate time instead of bled (Ramirez, 1985). One hour after intragastric feeding, the stomachs and duodenums were dissected out and their contents promptly put into a bottle containing isopropanol:heptane:sulfuric acid (Dole, 1956) to stop hydrolysis of triglycerides. Total fat was measured gravimetrically and free fatty acids measured by titration (Dole, 1956). Saccharin was used as the taste solution because it is more convenient to use than oil or oil–water emulsion.

Stomachs of rats fed oil without any taste contained 0.48–0.65 g of fat, depending on the experiment (1.38 g was fed intragastrically). On the first trial, a taste of saccharin had no detectable effect on the amount of fat recovered from the stomach (0.53 g). On the third trial, however, 0.74–0.92 g of fat was recovered from the stomachs of rats given a taste of saccharin. This reduction in gastric emptying on Trial 3 was statistically significant in two separate experiments (both $p < 0.05$).

It might be assumed that intragastric feeding somehow causes the stomach to dump its contents into the duodenum and that oral stimulation may prevent or reduce this. If so, one could expect to find that the fat content of the duodenum would be inversely proportional to the stomach contents. This is not the case. The saccharin-tasting rats had a slightly lower fat content on the first trial but a slightly higher fat content on the third trial (interaction significant, $p < 0.05$).

Since the oil fed to the rats contains very little free fatty acids, any such acids present in the stomachs should be the result of hydrolysis of fat in the stomach by the enzyme lingual lipase (Hamosh, 1979). A taste of saccharin had no significant effect on the amount of free fatty acids found in the stomachs. If the effects of oral stimulation were mediated by lingual lipase in the saliva, one ought to find that oral stimulation increases the amount of free fatty acids present in the stomach. This was not the case.

The available data suggest that taste affects fat digestion as follows: A taste which has been previously paired with fat ingestion slows gastric emptying. The slow emptying rate permits more efficient digestion of the fat, which results in elevated blood triglycerides 4 hr after the meal. It does not seem probable that the more efficient digestion is the cause of the slow gastric emptying because tastes have not significantly altered blood triglycerides prior to 4 hr in any experiment conducted up to this time.

III. RESEARCH NEEDS

Intragastric delivery of nutrients profoundly alters the physiologic response to nutrients when compared to oral ingestion. Intragastric feeding differs from oral feeding with regard to such diverse variables as stomach emptying, insulin release, growth rate and behavioral reinforcement (see above). It is not at all surprising that intragastric feeding can sometimes be shown to be aversive. It is surprising, however, that animals fed intragastrically sometimes regulate caloric intake accurately despite abnormal digestion. Indeed, rats are capable of regulating caloric intake by feeding themselves intragastrically (Epstein, 1967; Snowdon, 1969). Thus, not only can animals regulate caloric intake in the absence of oral sensory cues but they can regulate caloric intake when normal digestive cues are severely disrupted.

The experiments described above provide the first evidence that a sensory stimulus in the mouth can alter fat metabolism or digestion in adult rats. Previous studies have shown that sucking by human infants during intragastric feeding accelerates growth and maturation (Bernbaum et al., 1983), apparently because sucking increases the amount of lingual lipase swallowed. Lingual lipase is an enzyme found in saliva which is important in the digestion of fats by infants (Hamosh, 1979). The results of the experiments described above are not readily attributable to lingual lipase because oral stimulation did not seem to alter the rate of hydrolysis of triglycerides in the stomach. However, more studies of the possible role of lingual lipase are needed.

It has been shown previously that oral stimulation can facilitate the acquisition of a behavioral response reinforced by intragastric feeding (Epstein, 1967; Holman, 1968). This finding had been interpreted from a motivational–sensory perspective (Epstein, 1967; Holman, 1968). It now seems probable that the enhanced learning is due to altered digestion of the food. Studies of this issue might provide clues about the physiologic mechanisms for reinforcement by food.

Taste is probably one of several factors responsible for the differences between oral and intragastric feeding. Other factors (e.g., saliva, esophageal stimulation, temperature, etc.) may contribute to the differences between oral and intragastric feeding. A better understanding of these factors could improve care of human patients as well as enrich our understanding of the relationships between the chemical senses and nutrition.

ACKNOWLEDGMENTS

This work was supported by National Institutes of Health Grant AM 34375-01. I thank M. Tordoff for critical remarks on the manuscript and R. Threatte for advice.

REFERENCES

Bernbaum, J. C., Pereira, G. R., Watkins, J. B., and Peckham, G. J. (1983). Nonnutritive sucking during gavage feeding enhances growth and maturation in premature infants. *Pediatrics* **71**, 41–45.
Booth, D. A. (1972). Postabsorptively induced suppression of appetite and the energostatic control of feeding. *Physiol. Behav.* **9**, 199–202.

Cohn, C., and Joseph, D. (1959). Changes in body composition attendant on forced feeding. *Am. J. Physiol.* **196,** 965–968.

Cohn, C., Joseph, D., and Shapiro, E. (1957). Effect of diet on body composition. I. The production of increased body fat without overweight ("nonobese obesity") by force-feeding the normal rat. *Metabolism* **6,** 381–387.

Cox, J. E., and Powley, T. L. (1981). Intragastric pair feeding fails to prevent VMH obesity or hyperinsulinemia. *Am. J. Physiol.* **240,** E566–E572.

Cytawa, J., Luszawska, D., Scheonborn, R., and Zajac, M. (1972). Lack of instrumental conditioning with intragastric reinforcement in normal experimental procedure. *Acta Neurobiol. Exp.* **32,** 767–772.

Deutsch, J. A. (1978). The stomach in food satiation and the regulation of appetite. *Prog. Neurobiol.* **10,** 135–153.

Dole, V. P. (1956). A relation between non-esterified fatty acids in plasma and the metabolism of glucose. *J. Clin. Invest.* **35,** 150–154.

Epstein, A. N. (1967). Feeding without oropharyngeal sensations. *In* "The Chemical Senses and Nutrition" (M. R. Kare and O. Maller, eds.), pp. 263–280. The Johns Hopkins Press, Baltimore, Maryland.

Geliebter, A. (1979). Effects of equicaloric loads of protein, fat and carbohydrate on food intake in the rat and man. *Physiol. Behav.* **22,** 267–273.

Geliebter, A., Liang, J. T., and Van Itallie, T. B. (1984). Effects of repeated isocaloric macronutrient loads on daily food intake of rats. *Am. J. Physiol.* **274,** R387–R392.

Hamosh, M. A. (1979). A review on fat digestion in the newborn: Role of lingual lipase and prejejunal digestion. *Pediatr. Res.* **13,** 615–622.

Hansen, B. C., Jen, K-L. C., and Brown, N. (1981). Regulation of food intake and body weight in rhesus monkeys. *In* "The Body Weight Regulatory System: Normal and Disturbed Mechanisms" (L. A. Cioffi, W. P. T. James, and T. B. Van Itallie, eds.), pp. 69–72. Raven Press, New York.

Holman, G. L. (1968). Intragastric reinforcement effect. *J. Comp. Physiol. Psychol.* **69,** 432–441.

Houpt, K. A. (1982). Gastrointestinal factors in hunger and satiety. *Neurosci. Biobehav. Rev.* **6,** 145–164.

Hunt, J. N. (1956). Some properties of an alimentary osmoreceptor mechanism. *J. Physiol. (London)* **132,** 267–288.

Kekwick, A., and Pawan, G. L. S. (1966). The effect of feeding pattern on fat deposition in mice. *Metabolism* **15,** 173–180.

LeBlanc, J., Cabanac, M., and Samson, P. (1984). Reduced postparandial heat production with gavage as compared with meal feeding in human subjects. *Am. J. Physiol.* **246,** E95–E101.

Liu, C. M., and Yin, T. H. (1974). Caloric compensation to gastric loads in rats with hypothalamic hyperphagia. *Physiol. Behav.* **13,** 231–238.

Louis-Sylvestre, J. (1978). Relationship between two stages of prandial insulin release in rats. *Am. J. Physiol.* **235,** E103–E111.

McHugh, P. R., and Moran, T. H. (1979). Calories and gastric emptying: A regulatory capacity with implications for feeding. *Am. J. Physiol.* **236,** R254–R260.

Maggio, C. A., and Koopmans, H. S. (1982). Food intake after intragastric meals of short-, medium-, or long-chain triglycerides. *Physiol. Behav.* **28,** 921–926.

Malhotra, S. L. (1967). Effect of saliva on gastric emptying. *Am. J. Physiol.* **213,** 169–173.

Measel, C. P., and Anderson, G. C. (1979). Nonnutritive sucking during tube feeding: Effect on clinical course in premature infants. *JOGN Nurs.* **8,** 265–272.

Molina, F., Thiel, T., Deutsch, J. A., and Puerto, A. (1977). Comparison between some

digestive processes after eating and gastric loading in rats. *Pharm. Biochem. Behav.* **7**, 347–350.
Puerto, A., Deutsch, J. A., Molina, F., and Roll, P. (1976). Rapid discrimination of rewarding nutrient by the upper gastrointestinal tract. *Science* **192**, 485–487.
Ramirez, I. (1984). Behavioral and physiological consequences of intragastric oil feeding in rats. *Physiol. Behav.* **33**, 421–426.
Ramirez, I. (1985). Oral stimulation alters digestion of intragastric oil meals in rats. *Am. J. Physiol.* **248**, R459–R463.
Ramirez, I. (1986). Intragastric feeding differentially affects apparent rate of gastric emptying of phenol red and carbohydrate. *Physiol. Behav.* (In press.)
Ramirez, I., and Friedman, M. I. (1983a). Suppression of food intake by intragastric glucose in rats with impaired glucose tolerance. *Physiol. Behav.* **31**, 39–43.
Ramirez, I., and Friedman, M. I. (1983b). Food intake and blood fuels after oil consumption: Differential effects in normal and diabetic rats. *Physiol. Behav.* **31**, 847–850.
Rothwell, N. J., and Stock, M. J. (1978). A paradox in the control of energy intake in the rat. *Nature (London)* **273**, 146–147.
Rothwell, N. J., and Stock, M. J. (1979). Regulation of energy balance in two models of reversible obesity in the rat. *J. Comp. Physiol. Psychol.* **93**, 1024–1034.
Sawchenko, P. E., Gold, R. M., and Bisson, B. (1980). Depriving or pair feeding intact controls induces some persisting regulatory deficits similar to those of the recovered lateral rat. *J. Comp. Physiol. Psychol.* **94**, 128–144.
Schiller, L. R., Feldman, M., and Richardson, C. T. (1980). Effect of sham feeding on gastric emptying. *Gastroenterology* **78**, 1472–1475.
Snowdon, C. T. (1969). Motivation, regulation and the control of meal parameters with oral and intragastric feeding. *J. Comp. Physiol. Psychol.* **69**, 91–100.
Steffens, A. B. (1976). Influence of the oral cavity on insulin release in the rat. *Am. J. Physiol.* **230**, 1411–1415.
Stevens, P. J. d'E. (1982). Stoma care. *Clin. Gastroenterol.* **11**, 345–350.
Young, E. A., Cioletti, L. A., Traylor, J. B., and Balderas, V. (1982). Gastrointestinal response to oral versus gastric feeding of defined formula diets. *Am. J. Clin. Nutr.* **35**, 715–726.

9

The Stomach and Satiety

PAUL R. McHUGH AND TIMOTHY H. MORAN
Department of Psychiatry and Behavioral Sciences
The Johns Hopkins University School of Medicine
Baltimore, Maryland

I.	Introduction	167
II.	Gastric Emptying of Liquids	168
III.	Intestinal Control of Gastric Emptying	170
IV.	The Two Phases of Gastric Emptying	171
V.	The Stomach and Glucose Consumption	173
VI.	The Stomach and Chow Intake	175
VII.	Cholecystokinin and Gastric Distention	176
VIII.	Conclusions	178
	References	179

I. INTRODUCTION

A role for the stomach in both hunger and satiety has long had many champions (Cannon and Washburn, 1912; Carlson, 1916). Our particular interest in the stomach developed from our studies of the regulation of food intake by rhesus monkeys responding to a nutrient infusion into the gastric lumen.

Monkeys given an infusion of liquid nutrients equivalent to some fraction of their daily caloric intake reduced the chow they ate as their food an equivalent number of calories. We demonstrated this phenomenon in several different experimental settings, but the results were always the same. The infusions of liquid nutrients into the stomach just before the daily feeding period produced, in normal monkeys and as well in hypothalamic obese monkeys, fractional reductions in food intake equivalent to the calories in the infusion (McHugh et al., 1975a,b). The effect could be demonstrated with carbohydrates, fats, proteins, or a mixture of all three (McHugh and Moran, 1978).

This phenomenon carries several implications. One is a conceptual problem tied to the word "satiety." Satiety and sated tend to imply an all-or-none phenomenon. In fact, as shown from these results in which a portion of the expected chow intake was prevented by a preload, satiety, defined as the inhibition of food intake produced by food itself, must be a graded phenomenon. It must derive from coordinated actions of several mechanisms, none of which is well understood, and yet together they produce precisely modulated inhibitions of feeding in a variety of settings.

We review here some of our studies on nutrient-based physiological actions of the stomach. We chose to focus on the stomach because our preload appeared to act on food intake before all of it had left the stomach (McHugh and Moran, 1978). It seemed likely that some aspects of gastric functioning must be active in these effects and that a better understanding of gastric physiology might reveal mechanisms of importance in some forms of satiety.

II. GASTRIC EMPTYING OF LIQUIDS

The mode of our studies depended on unanesthetized, cage-confined rhesus monkeys fitted with soft leather vests to which were attached multiflexible stainless-steel cables in order to protect the Silastic cannulas that we employed to infuse solutions into the stomach, duodenum, and venus blood in a variety of experiments. The monkeys were kept in separate cages in temperature-controlled rooms with a cycle of 12 hr of light and 12 hr of darkness. The animals had *ad libitum* access to water, but were on a schedule by which they were provided chow pellets for 4 hr each day from 11 AM to 3 PM. Most of the experiments that we report here were done in the morning, 16–20 hr after the previous days feeding. This overnight fasting and time interval for feeding meant that the animals began each experiment with an empty stomach and were on each occasion comparably hungry. The monkeys learned quickly to accommodate to this schedule of food intake. They were taking amounts of chow equivalent to that taken by monkeys offered food *ad libitum* (Hamilton et al., 1976). Therefore, although fasted at the time of each experiment, they were not famished or in any way undernourished. All were gaining weight slowly.

In these gastric emptying studies, we applied our modification of the serial test-meal method using phenol red as a marker (Hunt and Spurrell, 1951). A saline or nutrient solution (test meal) with phenol red is infused through the cannula into the monkeys' stomachs and is left in place for a varying period (emptying time). At the end of this time, the contents of the stomach are removed, their volumes measured, and the dilution of the original meal by gastric juice or saliva documented from the colorimetric dilution of phenol red. From such calculations, the amount emptied from the stomach into the duodenum during the emptying time is estimated. By varying the emptying time on subsequent days, the pattern of delivery of a given nutrient solution can be displayed from 5 to 240 min after filling the stomach.

The results of these experiments were intriguing. The stomach emptied nonnutrient physiological saline in an exponential fashion such that larger volumes emptied more rapidly, and the half-time for emptying of saline was quite constant, averaging about 14 min in our monkeys. The emptying of glucose, however, was quite different. In fact, it took on a biphasic form. Upon filling of the stomach, there was a brief "rush" of a portion into the small intestine; the volume passed in this rush varied with the initial volume. However, after a minute or two, the stomach ceased to empty rapidly, but took on a slow and intermittent emptying pattern, but with some regulatory feature such that the caloric load delivered to the intestine over time was constant. For the monkey, the stomach settled down to deliver approximately 0.4 kcal/min. The glucose-emptying displayed by our method is shown for volume emptied in Fig. 1. As can be seen, with increasing glucose concentration the emptying of volume was slowed. But converting into calories emptied, as in Fig. 2, for three glucose concentrations (0.2, 0.5, and 1.0 kcal/ml), the emptying of calories was constant. We demonstrated a similar

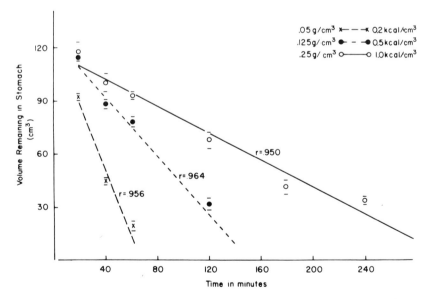

Fig. 1. Gastric emptying of three different glucose meals (0.2, 0.5, 1.0 kcal/ml). [Taken with permission from McHugh and Moran (1979).]

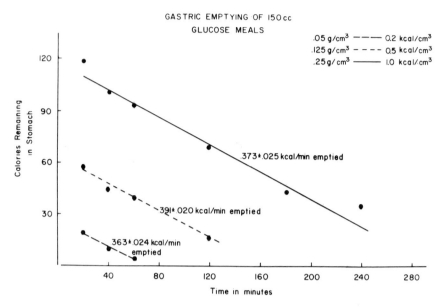

Fig. 2. Gastric emptying of same glucose meals as in Fig. 1, but displayed as calories emptied over time. [Taken with permission from McHugh and Moran (1979).]

rate of delivery of casein hydrolysate and medium-chain triglycerides (McHugh and Moran, 1979).

III. INTESTINAL CONTROL OF GASTRIC EMPTYING

The biphasic aspect of nutrient emptying, that is, the initial rush followed by the slow intermittent pattern with the regulation of caloric delivery, suggested that the control represented by the second phase of gastric emptying derived from the entrance of nutrients into the small intestine. This point has been proposed by many other investigators, who have used such terms as the "duodenal brake" (Shahidullah et al., 1975). We sought to quantitate the control of gastric emptying from beyond the pylorus. To do so, we took advantage of the characteristics of the stomach's emptying of saline. We proposed to infuse glucose directly into the small intestine while the stomach was filled with saline and observe whether such postpyloric glucose might delay the emptying of saline from the stomach for a period of time that might vary with the amount of glucose in the intestine.

For this experiment we threaded a thin Silastic cannula through the larger bore gastric cannula. The tip of this inserted cannula could be carried by gastric activity through the pylorus. We checked the thin cannula's placement in the postpyloric intestine by infusing phenol red through it and then washing the stomach through the larger cannula with saline. Colorless returns from the stomach indicated that the thin cannula had passed the pylorus. With such a preparation, we infused glucose into the intestine and filled the stomach with saline. Repeated emptying and refilling the stomach with saline permitted the demonstration of an inhibition of gastric emptying, the duration of which could be correlated with the amount of glucose placed postpylorically (Fig. 3).

Results of this experiment demonstrated that the duration of gastric inhibition varied directly with the glucose load (and not with volume concentration or osmolarity). As shown in Fig. 4, the inhibition was approximately 2.5 min/kcal in the intestine. This was an intriguing result because 2.5 min/kcal is the reciprocal of the 0.4 kcal/min emptying rate assumed in the second phase of gastric emptying. This result led us to appreciate that the slow-emptying phase of gastric delivery of glucose into the intestine is the steady-state result of the interrelationship between gastric delivery of glucose and inhibition on the stomach produced by glucose in the intestine (McHugh et al., 1982).

IV. THE TWO PHASES OF GASTRIC EMPTYING

Thus, the gastric emptying of liquids has two phases that occur sequentially with nutrient solutions (Fig. 5). The initial filling of the stomach produces a period of rapid, exponential emptying. This phase of emptying is rapid if the volume of the test meal is large, and it persists until the stomach empties if the test meal is osmotically neutral and nonnutrient.

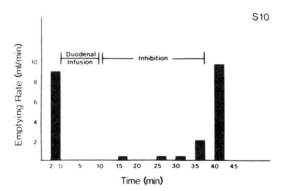

Fig. 3. Method for demonstrating gastric inhibition of saline emptying after duodenal glucose. [Taken with permission from McHugh et al. (1982).]

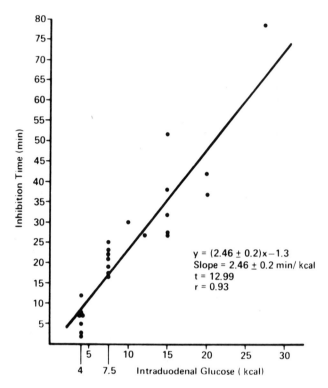

Fig. 4. Dose–response duration of gastric inhibition to duodenal glucose. [Taken with permission from McHugh *et al.* (1982).]

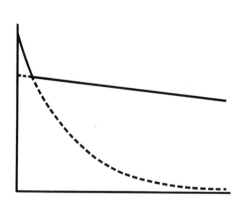

Fig. 5. Two phases of gastric emptying, diagrammed.

With a nutrient test meal, however, this initial phase will be interrupted after some of the test meal has entered the intestine. With this interruption, the second phase of gastric emptying begins and persists until this nutrient solution is emptied. This phase consists of intermittent, brief periods of emptying interspersed with periods of little or no emptying. The stomach delivers the nutrient liquid to the intestine in intermittent gushes.

There is, furthermore, a relationship between the amount of nutrient delivered with each gush and the duration of the following nonemptying period. The greater the nutrient amount delivered in the gush, either in nutrient concentration or in gush volume, the longer the intergush period will be. When gastric emptying is examined by a technique such as the serial test-meal method, which measures a remnant volume of a meal in the stomach over time, then a linear pattern emerges, smoothing out the intermittent features of this phase of gastric emptying to reveal the regulation in the delivery of glucose to the intestine (McHugh *et al.*, 1982; Brener *et al.*, 1983).

For each of these phases of gastric emptying, we need to illuminate the mechanisms that control it. The rapid initial phase, with its dependence on the volume of fluid filling the stomach, would seem to be an issue of gastric muscular tone and might be affected by such features as receptive relaxation of the stomach and by sympathetic and parasympathetic tone.

For the second phase, we also know little about the mechanism. The nature of reception in the intestine must be complicated in order to produce caloric regulation for different nutrients in the stomach. The transmission of intestinal inhibitions can be both neural and humoral. The control on the emptying must depend on pressures and resistance generated in the antrum, pylorus, and receptive duodenum (Weisbrodt *et al.*, 1969). All of these mechanisms require further study.

V. THE STOMACH AND GLUCOSE CONSUMPTION

We have, however, carried out a series of experiments demonstrating in the monkey how aspects of the control of food intake might derive from these physiological actions of the stomach as it managed its contents.

The first and simplest is a demonstration that in a fasted animal the consumption of a glucose solution is profoundly affected by the state of the stomach. If a monkey was offered a glucose solution at 0.5 kcal/ml, it consumed a large volume, 200–500 ml in less than 10 min, but then became sated and ceased consumption. This glucose solution, which had been swallowed, was managed by the stomach as described above. A portion promptly entered the small intestine and a portion was retained in the stomach to be delivered slowly over time. If the portion in the stomach was removed, however, there was a prompt behavioral response by the monkey. The monkey returned to the glucose solution and consumed once again a considerable volume. The removal of gastric contents relieved its satiety.

The intriguing point, however, derives from the regulation of gastric emptying. Removing the gastric contents did not change the amount of glucose that had passed through the pylorus. As a result, the postpyloric glucose was comparable when the stomach was emptied or left undisturbed after a bout of glucose consumption. Yet the behavior was quite different. The monkey reconsumed glucose, refilling its stomach in the one situation, and took little or none if the original glucose load in the stomach was left in place. If the gastric contents were undisturbed, then further consumption was minimal. Yet, over the time of study the amount of glucose that passed through the pylorus was the same under the two conditions.

This is one demonstration of the importance of gastric fullness in provoking satiety. It is demonstrated most clearly in Fig. 6. Here are compared 15-min periods during which the monkey was offered glucose to consume. The solid bars represent its behavior when the stomach was emptied after each 15-min period; the open bars, the consumption when the stomach was undisturbed. The amount of consumption was obviously greater with the stomach emptied. But it is also clear that with time and the accumulation of postpyloric glucose, there was a steady decline in the amount taken at each bout after the stomach was emptied. It seems evident that the postpyloric glucose was gradually inhibiting the consumption, but that the state of the

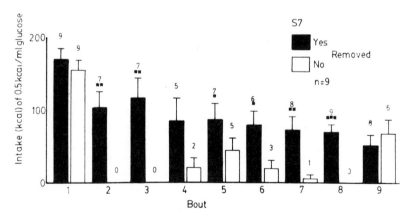

Fig. 6. Glucose consumption by same monkey when the stomach is emptied after each 15-min period of drinking (solid bars) compared to consumption when the stomach is not emptied.

stomach—empty or full—had a much more powerful and prompt effect on the behavior.

Under physiological conditions, a fasted, hungry animal will for a considerable time demonstrate in its feeding a dependence on the stomach. The amount of glucose that the stomach passes will have a lesser and only slowly increasing effect. The influence of the postpyloric nutrient only gradually becomes apparent and is at first modest in degree (Wirth and McHugh, 1983).

VI. THE STOMACH AND CHOW INTAKE

We could show this same relationship of gastric distention to feeding in another fashion. When a load of glucose was infused through an intragastric cannula into the stomach, some passed into the intestine and some remained in the stomach. The relative influences of these compartments on feeding could be displayed when such an infusion preceded the monkey's daily feeding period. These monkeys were eating chow that they obtained by pulling a lever. The lever activity could be recorded cumulatively, and thus the consumption of chow over time displayed.

If the infused glucose load was left in the stomach, there was then an inhibition of chow intake that persisted through the 4 hr of feeding and resulted in a reduction in the total chow eaten. This reduction was quite close in calories to the glucose calories given in the preload.

If, however, after some 20 min of filling the stomach with glucose, the glucose in the stomach was withdrawn just before the monkeys were offered chow to eat, then despite the small amount of glucose that passed into the intestine (about 18 kcal) during the 20 min, there was little or no effect on chow consumption. This small amount of intraintestinal glucose was apparently not discerned by these hungry monkeys.

If, however, after removing the glucose from the stomach, the stomach was refilled with a saline load (so that the gastrointestinal tract had saline in the stomach and glucose in the intestine), then the chow intake was inhibited for some 20–40 min. This inhibition, however, disappeared after the 20–40 mins; normal feeding began, and the animal ate enough over the next hours to catch up to controls (Fig. 7).

We interpret these results as indicating that the glucose that can pass into the intestine from our gastric infusions has little effect on feeding in a fasted animal with an empty stomach. But if that stomach is full, as by a glucose load in the stomach or by a saline load retained by the inhibition on gastric emptying from glucose in the intestine, then for the duration of that disten-

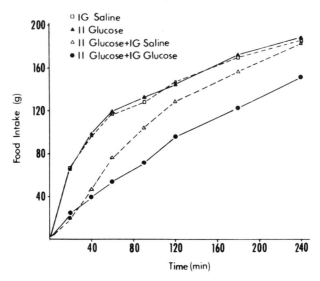

Fig. 7. Cumulative record of chow consumption under 4 different conditions: (□) saline in stomach; (▲) glucose in stomach empty; (△) glucose in intestine, saline in stomach; (●) glucose in both stomach and intestine. II, Intraintestinal; IG, intragastric. [Taken with permission from McHugh (1979).]

tion, feeding will be inhibited (short-lived for saline, persistent for gastric glucose) (McHugh, 1979).

Finally, these experiments with glucose and chow consumption display a satiety function that is not so much a long-term restriction of caloric intake—i.e., a resetting of caloric need—as it is a short-term mealtime satiety. Meals are the natural sequences of feeding in nonruminants. The distensible stomach is a structure that seems designed for the reception and storage of a quantity of nutrient. As it fills, it has mechanisms that promote its retention of food. When it is full, it has a means of interrupting further filling. Cannon and Washburn (1912) proposed that the stomach is linked to the behavioral mechanisms that start and stop feeding. Our observations indicate that the physiological interactions of the stomach and intestine, as they manage nutrients, can generate signals that control the progress of a bout of feeding.

VII. CHOLECYSTOKININ AND GASTRIC DISTENTION

As this concept emerged, it became clear that there are numerous sites in which a physiological element can be examined more carefully. Thus, the comprehensive view needs to be reinforced and presumably modified by the details.

We have focused attention on the intestinal hormone cholecystokinin (CCK) for several reasons. It has been proposed as a satiety factor, although no mechanism has been discerned. It is released from the intestine with the entrance of food from the stomach and has been shown to inhibit gastric emptying. We sought to demonstrate how these effects could be placed in a sequence related to behavior.

The first point is that CCK in low doses has a remarkably rapid onset and termination of its inhibitory action on the stomach (Fig. 8). An intravenous infusion of CCK inhibits the emptying of saline within a minute of the start of the infusion and, in like fashion, the inhibition disappears within a minute of its cessation. This result with exogenous CCK implies that endogenous CCK can act as one of several intestinal hormones to control the delivery of nutrients to the intestine. In fact, such a dynamic physiology can be fitted into a steady-state balance, controlling the delivery of nutrients to the intestine as CCK rises and falls during the passage and absorption of nutrients in the intestine.

The second observation is derived from receptor autoradiography employing ^{125}I-labeled Bolton–Hunter CCK-33 to seek a receptor mechanism for CCK in the gut. In rats, monkeys, and humans, we have defined a specific and restricted site for CCK reception in the circular muscle of the pylorus. This site is precisely the area where a contraction of the muscle could have the dynamic actions described above. The receptor auto-

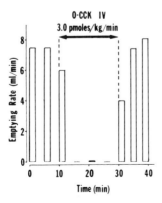

Fig. 8. Gastric emptying of saline with intravenous C-terminal octapeptide O-CCK. [Taken with permission from Moran and McHugh (1982).]

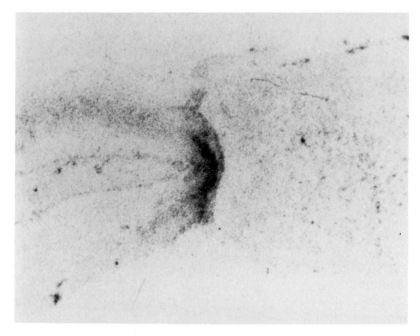

Fig. 9. Receptor autoradiography of ^{125}I-labeled CCK-33.

radiographs in fact resemble a ligature at this site (Fig. 9) (Smith *et al.*, 1984).

Finally, in the doses of CCK that produce the inhibition seen on gastric emptying, there is no effect on food intake. In monkeys prepared with indwelling cannulas, the infusion of this hormone dose for 60 min before a 4-hr feeding period and 10 min into the period had no effect on chow consumption. Only if the CCK infusion was combined with a distending load to the stomach of 150 ml of saline did a clear inhibition on food intake occur which disappeared after the cessation of the hormonal infusion. The total inhibition on food intake did not persist as shown in Fig. 10.

VIII. CONCLUSIONS

As so often with an intervening variable such as satiety, there is a tendency to use it in a blanket fashion, providing a unitary explanation for what are in fact a variety of behavioral aspects—a variety that will depend on different mechanisms and accomplish different functions. The satiety we are demon-

9. The Stomach and Satiety

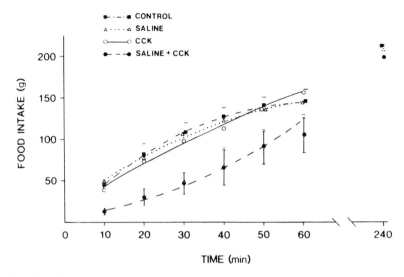

Fig. 10. Cumulative record of chow consumption with and without intravenous CCK and gastric saline loads. [Taken with permission from Moran and McHugh (1982).]

strating here as derived from the state of the stomach is that tied to meals themselves, interrupting them as they reach a certain size. We propose that a physiological basis for this satiety derives from the interactions of the stomach and the intestine.

This physiology makes the distention of the stomach a regulated event that not only will terminate a meal appropriately but also may act to prolong intermeal intervals. This satiety function of the stomach though is likely to function in the short-term, meal-related control of feeding and not in the setting of long-term caloric regulation. It is a clear advance when we can ascribe to a particular set of physiological events a role in feeding and so differentiate that role from other aspects of the same behavior.

ACKNOWLEDGMENT

This work was supported by NIH Grant 2-RO1-AM19302.

REFERENCES

Brener, W., Hendrix, T. R., and McHugh, P. R. (1983). Regulation of the gastric emptying of glucose. *Gastroenterology* **85**, 76–82.

Cannon, W. B., and Washburn, A. L. (1912). An explanation of hunger. *Am. J. Physiol.* **29**, 441–454.
Carlson, A. J. (1916). "The Control of Hunger in Health and Disease." Univ. of Chicago Press, Chicago.
Hamilton, C. L., Ciaccia, P. J., and Lewis, D. O. (1976). Feeding behavior in monkeys with and without lesions to the hypothalamus. *Am. J. Physiol.* **230**, 818–830.
Hunt, J. N., and Spurrell, W. R. (1951). The pattern of emptying of the human stomach. *J. Physiol. (London)* **113**, 157–168.
McHugh, P. R. (1979). Aspects of the control of feeding: Application of quantitation in psychobiology. *Johns Hopkins Med. J.* **144**, 147–155.
McHugh, P. R., and Moran, T. H. (1978). The accuracy of the regulation of caloric ingestion in the rhesus monkey. *Am. J. Physiol.* **235**, R29–R34.
McHugh, P. R., and Moran, T. H. (1979). Calories and gastric emptying. A regulatory capacity with implications for feeding *Am. J. Physiol.* **236**, R254–260.
McHugh, P. R., Gibbs, J., Falasco, J. D., Moran, T. H., and Smith G. P. (1975a). Inhibitions of feeding examined in rhesus monkeys with hypothalamic disconnections. *Brain* **98**, 441–454.
McHugh, P. R., Moran, T. H., and Barton, G. N. (1975b). Satiety: A graded behavioral phenomenon regulating caloric intake. *Science* **190**, 167–169.
McHugh, P. R., Moran, T. H., and Wirth, J. B. (1982). Postpyloric regulation of gastric emptying in rhesus monkeys. *Am. J. Physiol.* **243**, R408–415.
Moran, T. H., and McHugh, P. R. (1982). Cholecystokinin suppresses food intake by inhibiting gastric emptying. *Am. J. Physiol.* **242**, R491–R497.
Shahidullah, M., Kennedy, T. L., and Parks, T. G. (1975). The vagus, the duodenal brake and gastric emptying. *Gut* **16**, 331–336.
Smith, G. T., Moran, T. H., Coyle, J. T., Kuhar, M. J., O'Donahue, T. L., and McHugh, P. R. (1984). Anatomic localization of cholecystokinin receptors to the pyloric sphincter. *Am. J. Physiol.* **246**, R127–R130.
Weisbrodt, N. W., Wiley, J. N., Overholt, B. F., and Bass, P. (1969). A relation between gastroduodenal contractions and gastric emptying. *Gut* **10**, 543–548.
Wirth, J. B., and McHugh, P. R. (1983). Gastric distension and short-term satiety in the rhesus monkey. *Am. J. Physiol.* **245**, R174–R180.

10

The Cephalic Phase of Gastric Secretion

MARK FELDMAN AND CHARLES T. RICHARDSON
Veterans Administration Medical Center and
Department of Internal Medicine
University of Texas Health Science Center
Dallas, Texas

I.   Introduction 181
II. The Cephalic Phase of Gastric Secretion 181
     A. Sham Feeding as a Means of Investigating the Cephalic
         Phase of Acid Secretion 184
     B. Contribution of Different Senses to the Cephalic Phase of
         Gastric Secretion in Humans 189
III. Research Needs 191
     References 191

I. INTRODUCTION

Stimuli arising in the region of the head are capable of stimulating gastric secretion. This is referred to as the cephalic phase of gastric secretion. Although most studies have focused on the cephalic phase of gastric acid secretion, there is also a cephalic phase of gastric pepsinogen, intrinsic factor, bicarbonate, mucus, and nonparietal fluid and electrolyte secretion. Other phases of gastric secretion besides the cephalic phase include the gastric phase, the intestinal phase, and the postabsorptive (circulatory) phase. This chapter will review current concepts of the cephalic phase of gastric secretion, including the role of the chemical senses.

II. THE CEPHALIC PHASE OF GASTRIC SECRETION

The cephalic phase of gastric secretion was identified many years ago by Pavlov (1910). He demonstrated in dogs that swallowing appetizing food stimulates gastric acid secretion even when the food is diverted through an esophagostomy and thus prevented from entering the gastrointestinal tract and being absorbed into the bloodstream. Pavlov also demonstrated that the cephalic phase of gastric acid secretion is abolished by supradiaphragmatic vagotomy, indicating that the vagus nerves mediate the efferent component of this reflex.

The afferent component of the cephalic phase of gastric secretion is activated by the thought, sight, smell, and taste of appetizing food. Afferent signals are transmitted through various pathways within the brain, and ultimately, neurons in the dorsal motor nucleus of the vagus nerve and in the nucleus ambiguus of the vagus are activated (Fig. 1). Both of these vagal nuclei are located in the medulla oblongata. These nuclei contain the cell bodies for long, vagal preganglionic neurons which travel to the stomach. Very little is known about the central neurotransmitters which are involved in the cephalic phase of gastric secretion.

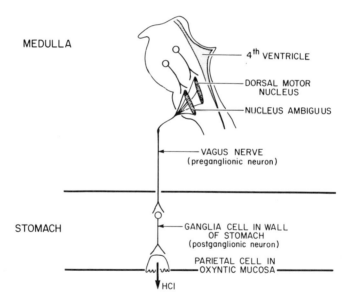

Fig. 1. Model of vagus nerve pathways to the stomach. Cell bodies of the preganglionic neurons of the vagus nerve originate in the dorsal motor nucleus and the nucleus ambiguus. These nuclei in the medulla are innervated by nerves from other parts of the nervous system. Fibers from the vagus travel to ganglia in the wall of the stomach and synapse with postganglionic neurons. The short postganglionic neurons innervate secretory cells. In the figure, a parietal cell in oxyntic gastric mucosa is shown secreting hydrochloric acid (HCl). However, other gastric postganglionic neurons innervate gastric endocrine cells (e.g., gastrin cells) or other secretory cells (e.g., pepsinogen-secreting cells) (see text).

10. The Cephalic Phase of Gastric Secretion

The vagus nerve does not innervate secretory cells of the gastric mucosa directly. Instead, the vagus synapses with groups of neurons (ganglia) within the wall of the stomach (Fig. 1). These short postganglionic neurons contain receptors for acetylcholine which is released at the synapse by vagal, preganglionic fibers. At one time, it was believed that all cholinergic receptors on ganglionic neurons were of the nicotinic type. It is now recognized that some of these ganglionic neurons have muscarinic-type cholinergic receptors rather than nicotinic-type cholinergic receptors (Mutschler and Lambrecht, 1983).

After postganglionic neurons are activated chemically by vagal, preganglionic fibers, they can stimulate gastric secretory cells (Fig. 1). In the case of gastric acid secretion, it is believed that some postganglionic neurons release acetylcholine in the vicinity of muscarinic receptors on parietal cells, which are located in the proximal portion of the stomach. Once activated by acetylcholine, the permeability of parietal cells to calcium ions increases, and as a result of calcium influx into parietal cells, hydrochloric acid (HCl) is secreted into the gastric lumen.

Although not shown in Fig. 1, other postganglionic neurons located in the distal portion of the gastric mucosa stimulate gastrin cells (G cells), releasing the peptide gastrin into the circulation. The neurotransmitter released by postganglionic neurons in the vicinity of G cells to trigger gastrin release has not been identified, but it does not appear to be acetylcholine because atropine enhances, rather than blocks, gastrin release during vagal stimulation (Feldman et al., 1979) (Fig. 2). Also shown in Fig. 2 is the observation that atropine enhances, rather than blocks, the gastrin rise which occurs with gastric distention (Schiller et al., 1980a). The gastrin rise in response to gastric distention is thought to be due to gastric reflexes involving the vagus nerves. Thus, studies with atropine suggest that cholinergic nerves inhibit vagally mediated gastrin release. The leading candidate for the neurotransmitter which releases gastrin during sham feeding or gastric distention is gastrin-releasing peptide, or GRP, a peptide similar to bombesin (Makhlouf, 1983). However, the observation that gastrin release in response to gastric distention is blocked by the β-adrenergic antagonist propranolol suggests that adrenergic nerves may stimulate gastrin release (Peters et al., 1982). After release from G cells, gastrin circulates to parietal cells to elicit acid secretion, probably by acting on specific gastrin receptors. Thus, cephalic vagal stimulation can lead to an increase in acid secretion both via cholinergic muscarinic stimulation of parietal cells and via release of the hormone gastrin.

Histamine plays a major, but still incompletely understood, role in the cephalic phase of gastric acid secretion. When histamine-2 receptors, present on parietal cells, are antagonized by a drug such as cimetidine, acid

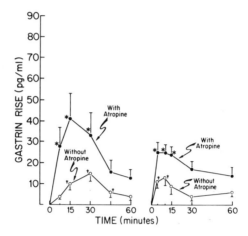

Fig. 2. Effect of intravenous atropine (2.3 µg/kg) on the mean ± SE gastrin rise that occurs in response to sham feeding ($n = 9$) (left) or gastric distention with 700 ml saline at pH 5.0 ($n = 8$) (right) in healthy subjects. Without atropine, mean gastrin rises of 10–15 pg/ml occurred with sham feeding and with gastric distention. Atropine did not block gastrin release; in fact, atropine significantly (*$p < 0.05$) enhanced gastrin release in response to both stimuli, indicating that cholinergic nerves normally inhibit neurally-mediated gastrin release [From Feldman et al. (1979). Reproduced from *The Journal of Clinical Investigation*, 1979, Vol. 63, pp. 294–298, by copyright permission of the American Society for Clinical Investigation; from Schiller et al. (1980a).]

secretion in response to sham feeding is markedly reduced (Schoon and Olbe, 1978). It is unknown whether parietal cells are exposed to a constant amount of histamine released from gastric mast cells or whether gastric mast cells actually release more histamine during cephalic–vagal stimulation than under unstimulated conditions.

A recent study by Feldman and Cowley (1982) has shown that naloxone, a pure opiate receptor antagonist, reduced the cephalic phase of gastric acid secretion in humans. This observation suggests that endogenous opiate peptides (enkephalins, endorphins) also may play a role in the cephalic phase of gastric secretion.

A. Sham Feeding as a Means of Investigating the Cephalic Phase of Acid Secretion

1. Studies in Animals

The cephalic phase of gastric secretion has been studied extensively in dogs and cats subjected to esophagostomy, usually in the neck region (cervical esophagostomy). In animals prepared with an esophagostomy, normal feeding can occur when the esophageal stoma is occluded with a plug. In order to carry out sham feeding, the stoma is unplugged and the esophagus distal to the stoma is occluded by a tourniquet or clamp. Thus, food that is swallowed does not reach the stomach. Gastric secretion can be measured by collecting fluid from a surgically created gastric fistula that still has intact vagal innervation. Alternately, it is possible for the surgeon to construct a proximal gastric pouch that remains vagally innervated (Pavlov pouch). This is in contrast to a proximal gastric pouch that is vagally denervated (Heidenhain pouch). If a surgeon creates a gastric fistula and a Heidenhain pouch in the same animal, it is possible to compare acid secretion from the vagally innervated fistula and from the vagally denervated pouch simultaneously.

The effect of a 10-min period of sham feeding with blenderized dog chow is shown in Fig. 3. Mean acid secretion in 6 dogs is shown, both from the vagally innervated stomach and from the vagally denervated stomach (Heidenhain pouch). It is apparent that sham feeding increased acid secretion in the vagally innervated stomach to about 60% of maximal acid output (MAO). Even after sham feeding ended, acid secretion from the gastric fistula remained much higher than basal for the remainder of the experiment. Sham feeding did not increase acid secretion in the vagally denervated stomach, confirming that intact vagal innervation is necessary for sham feeding to elicit acid secretion.

Recent evidence in dogs suggests that excessive cephalic–vagal stimulation, over a prolonged period of time, can lead to hyperplasia of acid-secreting parietal cells and an increase in MAO (Thirlby and Feldman, 1984). For

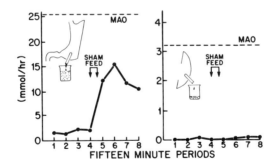

Fig. 3. Effect of a 10-min period of sham feeding with blenderized dog chow on mean gastric acid output (mmol/hr) in 6 dogs with vagally innervated gastric fistulae and vagally denervated gastric pouches. Acid output increased in response to sham feeding in the vagally innervated stomach (left) but not in the vagally denervated stomach (right). Maximal acid output (MAO) in response to a large, intravenous dose of pentagastrin (16 μg/kg/hr) is also shown for comparison. MAO is defined as the sum of the four consecutive 15-min acid outputs after this dose of pentagastrin.

example, after dogs with esophagostomies had been sham fed continuously for 7 hr every day for 6 weeks, MAO in response to pentagastrin increased approximately 30% in the vagally innervated stomach but did not change in the vagally denervated stomach. Since MAO is a reflection of parietal cell mass (Card and Marks, 1960), these studies suggest that cephalic–vagal stimulation is trophic for parietal cell growth. MAO returned toward control values after this 6-week period, indicating that the process was reversible. It is possible that the increased MAO and parietal cell mass seen in some patients with duodenal ulcer disease results from chronically increased endogenous cephalic–vagal stimulation.

2. Studies in Humans

To study the cephalic phase of gastric secretion in humans, most investigators have utilized a "chew-and-spit" sham-feeding technique (modifed sham feeding). In this method, an appetizing meal is presented to the subject or patient, who is permitted to chew and taste the meal but not to swallow it. Instead, the food is expectorated into a basin. Gastric juice is collected through a tube placed into the stomach through the subject's nose or mouth by supplying suction on the tube manually or by using a suction pump. Gastric acid output is calculated by multiplying the volume of gastric juice aspirated by its hydrogen ion concentration. The fluid which has been aspirated from the stomach also can be examined to be certain that no food has been swallowed accidentally, although one study has shown that small amounts of food accidentally swallowed do not stimulate additional gastric acid secretion (Knutson and Olbe, 1974).

The effect of modified sham feeding on gastric acid output in a group of 22 healthy human beings is shown in Fig. 4. After a 60-min basal period, subjects chewed but did not swallow a sirloin steak and french fried potato meal for 30 min. Mean acid output increased from a basal rate before sham feeding of approximately 3 mmol/hr to a rate of approximately 20 mmol/hr during sham feeding. The increased acid output was due to an increase in both gastric juice volume and acid concentration. Acid output remained elevated for at least an hour after sham feeding ended. The peak acid output to a large parenteral dose of pentagastrin in these subjects, determined on another day, averaged 36 mmol/hr (Fig. 4). Thus, sham feeding increases acid secretion to slightly over half of peak or maximal acid output in humans (Fig. 4) and in dogs (Fig. 3).

Why sham feeding does not lead to as much acid output as a maximal parenteral dose of pentagastrin or histamine is uncertain. Part of the explanation may be that sham feeding, unlike gastrin and histamine, also stimulates gastric bicarbonate secretion (Feldman, 1985). Bicarbonate neutralizes some of the acid secreted in response to sham feeding (as much as 6–8

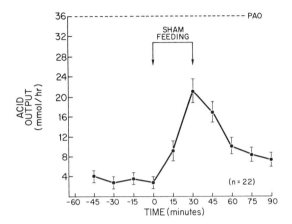

Fig. 4. Effect of a 30-min period of modified sham feeding on mean ± SE gastric acid output in 22 healthy subjects. Peak acid output (PAO) in response to a large, subcutaneous dose of pentagastrin (6 µg/kg/hr) is also shown for comparison. PAO is defined as the sum of the two highest consecutive 15-min acid outputs after this dose of pentagastrin, multiplied by 2 to express results in mmol/hr. [From Feldman and Richardson (1981b). "Physiology of the Gastrointestinal Tract I." Copyright (1981) Raven Press, New York.]

mmol/hr). There is also some evidence that sham feeding may release into the circulation an inhibitor of acid secretion ("vagogastrone") (Sjodin, 1975), although this putative vagal inhibitory hormone has not yet been identified.

The cephalic phase of gastric acid secretion accounts for one-third to one-half of the acid secreted in response to a normally eaten meal in humans (Richardson et al., 1977). If the cephalic phase is bypassed by infusing food directly into the stomach (activating gastric, intestinal, and postabsorptive mechanisms), acid secretion increases to about two-thirds of peak acid output (PAO) (Fig. 5, open circles). If intragastric food infusion is combined with simultaneous sham feeding, all phases of gastric secretion are active, normal eating is stimulated, and gastric secretion approaches PAO (Fig. 5, closed circles). Thus, the cephalic phase acts additively, or perhaps even synergistically, with other phases of gastric secretion, leading to nearly maximal rates of acid secretion after a normal meal. In contrast, cephalic–vagal stimulation induced by sham feeding does not appear to alter the rate at which homogenized food or fluid empties from the human stomach (Schiller et al., 1980b).

As mentioned earlier, the cephalic phase of gastric acid secretion is mediated partly by vagal release of the hormone gastrin. The serum gastrin response to sham feeding is not due to accidentally swallowed food because "partial" sham feeding, which prevents oral contact with food, also increases serum gastrin concentrations and gastric acid output significantly (Feldman

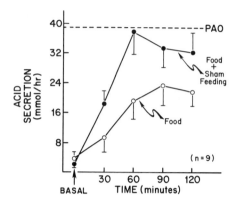

Fig. 5. Effects of 600 ml homogenized food infused directly into the stomach via a nasogastric tube and of 600 ml homogenized food plus sham feeding (from 0 to 30 min) on mean ± SE food-stimulated gastric acid secretion in 9 healthy subjects. Acid secretion was measured by *in vivo* intragastric titration to pH 5.0. Intragastric food alone increased acid secretion in these subjects to approximately two-thirds of peak acid output (PAO), whereas with intragastric food plus sham feeding, acid secretion rates approached PAO. Thus, the cephalic phase acts additively with other phases of gastric acid secretion. Basal acid secretion in these subjects is also shown. [From Richardson *et al.* (1977). Reproduced from *The Journal of Clinical Investigation*, 1977, Vol. 60, pp. 435–441, by copyright permission of The American Society for Clincal Investigation.]

and Richardson, 1981a) (Fig. 6). In our experience in healthy human subjects, serum gastrin concentrations in response to conventional sham feeding usually increase by 10–15 pg/ml (Fig. 2). However, there is considerable intersubject variation, and increases in serum gastrin concentrations cannot be detected during sham feeding in some individuals.

There is evidence to suggest that an average increase in serum gastrin concentration of 10–15 pg/ml during sham feeding is of sufficient magnitude to stimulate gastric acid secretion. First, infusing human gastrin heptadecapeptide in a dose sufficient to increase serum gastrin concentration by 10–15 pg/ml causes a marked increase in gastric acid secretion (Feldman *et al.*, 1978). Second, acidification of the gastric antrum during sham feeding prevents the 10–15 pg/ml increase in serum gastrin concentration and reduces gastric acid secretion in response to sham feeding by approximately 50% (Feldman and Walsh, 1980). Thus, gastrin appears to play an important role in the cephalic phase of gastric acid secretion in many subjects. However, the observation by Feldman and Walsh (1980) that sham feeding can still elicit acid secretion even when gastrin release is blocked by antral acidification indicates that stimulatory mechanisms other than gastrin are also important (e.g., direct stimulation of parietal cells by postganglionic cholinergic neurons).

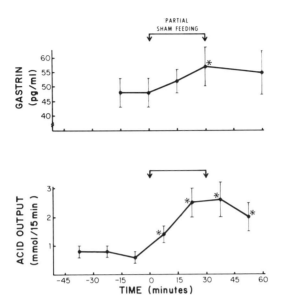

Fig. 6. Effect of "partial" sham feeding, defined as sham feeding which precluded oral contact with food, on mean ± SE serum gastrin concentrations (top) and on mean ± SE gastric acid output (bottom) in 7 healthy subjects. Significant ($p < 0.05$) increases from basal values are shown by asterisks. (From Feldman and Richardson, 1981b.)

B. Contribution of Different Senses to the Cephalic Phase of Gastric Secretion in Humans

The cephalic phase of gastric secretion results from the thought, sight, smell, and taste of appetizing food. All of these are usually activated during conventional sham feeding and during an appetizing meal. Thus, it has been difficult to assess the relative contributions of different senses to the overall cephalic phase of gastric secretion. Some workers have noted that seeing and smelling appetizing food while it is being prepared stimulates gastric acid secretion, even when food is not placed in the mouth and tasted (Moore and Motoki, 1979). Others have shown that simply talking about appetizing food (e.g., under hypnosis) can stimulate gastric acid secretion, indicating that sight and smell may also not be necessary (Feldman and Richardson, 1981a; Klein and Spiegel, 1984). The mechanical act of chewing per se does not affect gastric acid secretion (Richardson et al., 1977).

We have been interested in quantitating the relative importance of different sensory components of the cephalic phase of gastric acid secretion in humans. We have found that patients who were anosmic as a result of Kallmann's syndrome (isolated gonadotropin deficiency) secrete approximately as much acid in response to sham feeding as healthy controls. This finding suggests that smell is a relatively unimportant component of the cephalic phase of acid secretion.

In further studies in healthy volunteers, subjects were permitted to see and smell for 30 min, but not to taste, an appetizing meal being prepared in front of them. Gastric acid secretion during this "see-and-smell" type of partial sham feeding increased from approximately 10% of PAO to almost 30% of PAO. Thus, seeing and smelling food being prepared without tasting it increased acid secretion, although not quite as much as "full" sham feeding, in which acid secretion increased to more than 50% of PAO in healthy subjects. Thus, the taste of food appears to play an important role as well.

We have compared acid secretion rates in healthy subjects who saw and smelled food being cooked with acid secretion rates in subjects who looked at appetizing food without smelling it. The latter condition was accomplished by first cooking the food in a separate room so that the subject could not see or smell it while it was being prepared. Food was then placed in an airtight Plexiglas container so that aromas could not escape. Subjects then looked at the food for 30 min. We have found that seeing food without smelling it is as potent a stimulus of gastric acid secretion as seeing plus smelling, again suggesting that smelling food per se plays a small stimulatory role. This latter conclusion was further supported by another series of experiments, in which subjects were permitted to smell their preselected meal as it was being prepared without seeing it. This was accomplished by cooking the meal on a stove just behind the subject. A small fan was used to allow the food aromas to more easily reach the subject. Smelling food without seeing it or tasting it had a relatively small effect on acid secretion, with acid outputs of only 10% of PAO.

The above results suggest that, whereas smell is relatively unimportant, seeing food and tasting food are important factors in initiating the cephalic phase of gastric acid secretion. Neither sight nor taste is necessary to produce the cephalic phase of gastric secretion, however. This is apparent from additional experiments in which one of us (C.T.R.) talked to subjects about appetizing food for 30 min without any food being present in the room. This food discussion increased acid secretion to almost 30% of PAO. Thus, thinking and talking about food can stimulate gastric secretion in the absence of sight, smell, and taste.

One volunteer agreed to undergo a discussion of appetizing food on two separate days several weeks apart. The food discussion increased acid secretion above basal rates on both study days, and acid secretion remained elevated well beyond the 30-min period of food discussion. It is of interest that the subject secreted more acid during food discussion in Study 2 than in Study 1. Highest acid outputs during food discussion averaged 45 and 24% of

PAO in Study 2 and Study 1, respectively. Although attempts were made to standardize the types of foods discussed on the 2 days, it is difficult to know whether the differences in acid secretion on the 2 days were due to variables on the part of the investigator or on the part of the subject.

III. RESEARCH NEEDS

The cephalic phase of gastric secretion has been studied extensively in the laboratory both in animals and in humans. The afferent neural pathways involved are fairly well understood anatomically, although the relative importance of different chemical and nonchemical senses and of thought processes in the afferent component of the cephalic phase are just beginning to be studied. The efferent component of the cephalic phase of gastric secretion is mediated by the vagus nerves. Very little is known about neurotransmitters involved in the neural pathways which comprise the cephalic phase of gastric secretion, especially neurotransmitters in the central nervous system. Further research is needed to clarify the role of neurotransmitters and of hormones other than gastrin in the cephalic phase of gastric secretion. The role of histamine and of endogenous opiate peptides in the cephalic phase also needs clarification.

ACKNOWLEDGMENT

Supported by Grant AM 16816 from the National Institute of Health and by the Veterans Administration.

REFERENCES

Card, W. I., and Marks, I. N. (1960). The relationship between the acid output of the stomach following "maximal" histamine stimulation and the parietal cell mass. *Clin. Sci.* **19,** 147–163.

Feldman, M. (1985). Gastric H^+ and HCO_3^- secretion in response to sham feeding in humans. *Amer. J. Physiol.* **248,** G188–G191.

Feldman, M., and Cowley, Y. (1982). Effect of an opiate antagonist (naloxone) on the gastric acid secretory response to sham feeding, pentagastrin, and histamine in man. *Dig. Dis. Sci.* **27,** 308–310.

Feldman, M., and Richardson, C. T. (1981a). "Partial" sham feeding releases gastrin in normal human subjects. *Scand. J. Gastroent.* **16,** 13–16.

Feldman, M., and Richardson, C. T. (1981b). Gastric acid secretion in humans. *In* "Physiology of the Gastrointestinal Tract I" (L. R. Johnson, ed.), pp. 693–707. Raven Press, New York.

Feldman, M., and Walsh, J. H. (1980). Acid inhibition of sham feeding-stimulated gastrin release and gastric acid secretion: effect of atropine. *Gastroenterology* **78**, 772–776.
Feldman, M., Walsh, J. H., Wong, H. C., and Richardson, C. T. (1978). Role of gastrin heptadecapeptide in the acid secretory response to amino acids in man. *J. Clin. Invest.* **61**, 308–313.
Feldman, M., Richardson, C. T., Taylor, I. L., and Walsh, J. H. (1979). Effect of atropine on vagal release of gastrin and pancreatic polypeptide. *J. Clin. Invest.* **63**, 294–298.
Klein, K. B., and Spiegel, D. (1984). Hypnosis can both stimulate and inhibit gastric acid secretion. *Gastroenterology* **86**, 1137 (abstract).
Knutson, U., and Olbe, L. (1974). Gastric acid response to sham feeding before and after resection of antrum and duodenal bulb in duodenal ulcer patients. *Scand. J. Gastroenterol.* **9**, 191–201.
Makhlouf, G. M. (1983). Regulation of gastrin and somatostatin secretion by gastric intramural neurons. *In* "Trends in Pharmacological Sciences. Subtypes of Muscarinic Receptors" (B. I. Hirschowitz, R. Hammer, A. Giachetti, J. J. Keirns, and R. R. Levine, eds.), pp. 63–65. Elsevier, New York.
Moore, J. G., and Motoki, D. (1979). Gastric secretory and humoral response to anticipated feeding in five men. *Gastroenterology* **76**, 71–75.
Mutschler, E., and Lambrecht, G. (1983). Selective muscarinic agonists and antagonists in functional tests. *In* "Trends in Pharmacological Sciences. Subtypes of Muscarinic Receptors" (B. I. Hirschowitz, R. Hammer, A. Giachetti, J. J. Keirns, and R. R. Levine, eds.), pp. 39–48. Elsevier, New York.
Pavlov, I. P. (1910). The centrifugal (efferent) nerves to the gastric glands and of the pancreas. Lecture III. *In* "The Work of the Digestive Glands" (W. H. Thompson, trans.), 2nd ed., pp. 48–59. Charles Griffin, Philadelphia.
Peters, M. N., Walsh, J. H., Ferrari, J., and Feldman, M. (1982). Adrenergic regulation of distention-induced gastrin release in humans. *Gastroenterology* **82**, 659–663.
Richardson, C. T., Walsh, J. H., Cooper, K. A., Feldman, M., and Fordtran, J. S. (1977). Studies on the importance of cephalic vagal stimulation in the acid secretory response to eating in normal human subjects. *J. Clin. Invest.* **60**, 435–441.
Schiller, L. R., Walsh, J. H., and Feldman, M. (1980a). Distention-induced gastrin release: Effects of luminal acidification and intravenous atropine. *Gastroenterology* **78**, 912–917.
Schiller, L. R., Feldman, M., and Richardson, C. T. (1980b). The effect of sham feeding on gastric emptying in man. *Gastroenterology* **78**, 1472–1475.
Schoon, I. M., and Olbe, L. (1978). Inhibitory effect of cimetidine on gastric acid secretion vagally activated by physiological means in duodenal ulcer patients. *Gut* **19**, 27–31.
Sjodin, L. (1975). Inhibition of gastrin-stimulated canine acid secretion by sham feeding. *Scand. J. Gastroenterol.* **10**, 73–80.
Thirlby, R. C., and Feldman, M. (1984). Effect of chronic, physiological cephalic-vagal stimulation on maximal gastric acid secretion in the dog. *J. Clin. Invest.* **73**, 566–569.

11

The Gut Brain and the Gut–Brain Axis

WENDY R. EWART AND DAVID L. WINGATE

Gastrointestinal Science Research Unit
The London Hospital Medical College
London, England

I.	Introduction	193
II.	The Gut Brain	195
III.	The Gut–Brain Axis	200
IV.	Peptides	205
V.	Research Goals	206
VI.	Summary	207
	References	208

I. INTRODUCTION

As we respond to the sensations of appetite and satiation and to the demands of ingestion and excretion, the digestive system dominates our conscious existence more than any other system of the body. Long before the dawn of scientific medicine, the digestive system was thought to be well understood by physicians and laity alike, even if the received wisdom and dogma were unsupported by experimental evidence.

Gastrointestinal physiology as a subject for serious study, which began with the studies of the American physician, William Beaumont (1833), on his patient Alexis St. Martin, was thus superimposed on a mass of preconceived ideas, which in turn influenced the course of further research. The view of the gastrointestinal system derived from the work of Beaumont, Pavlov, Bayliss, Starling, and Langley, which might be termed the "classical" view of digestive physiology, provided a scientific basis for ideas that had been in fact widely held for many centuries. Beaumont demonstrated the function of the stomach and the modulation of gastric secretion by the content of the meal. Pavlov showed that exocrine secretion in the digestive tract could be modulated by the central nervous system (CNS). The mechanisms that appeared to mediate these processes were unraveled by Bayliss and Starling (1902) in their discovery of secretin, the first hormone to be described, and by Langley, in his description of the autonomic nervous system. These studies essentially confirmed what was already believed to be true, and portrayed the digestive tract as a series of viscera that respond to the nature of their contents by the release of chemical messengers regulating exocrine secretion, with central modulation through parasympathetic stimulation or adrenergic inhibition.

This view of the digestive tract is familiar to contemporary physicians from their medical school courses in physiology and, indeed, is what is still taught in many medical schools. It is the gut as slot machine. If one considers a coin-operated ticket machine, which accepts a variety of coins (nickels, dimes, quarters, etc.) and which issues appropriate tickets and change, this could be taken to represent the gut in which responses are modulated, by hormone release, according to the input; in this model, the CNS is represented by the ticket purchaser, who dictates the rate and nature of input into the machine. In the slot machine analog, sensors detect the nature of the coins, and determine the appropriate response; in the gut, gastrin, secretin, and cholecystokinin (CCK) are released to determine the exocrine secretory response appropriate to the nutrient load, and sometimes to retard or accelerate transit. This model is consistent with the belief that the digestive tract is silent and relatively inert when void of nutrient, just as the slot machine is inert when not fed with coins.

Over the last two decades, research in three areas of gastrointestinal physiology has challenged established ideas. First, there is a much clearer understanding of the complexity and function of the enteric nervous sytem (ENS)—the intrinsic innervation of the gut composed of the myenteric and submucous plexuses, and an appreciation of the importance of neural regulation that is neither cholinergic nor adrenergic. Second, it is now appreciated that the vagus nerve is largely afferent, and that it provides the opportunity for interaction between the central nervous system and the enteric nervous system; the "classical" view of the vagus nerve is that it is largely efferent and concerned with the activation of exocrine secretion during the cephalic phase, which occurs during the perception of and prior to the ingestion of food. Finally, advances in radioimmunoassay and immunohistochemistry have transformed our views of "gut hormones." It is now clear that what were classically considered to be no more than three "gut hormones" are only a small part of a very large family of regulatory peptides widespread throughout the body, which are unlikely, for the most part or even at all, to be hormones in the generally accepted sense of the word.

It is with the first two of these three areas of developing study that this review is largely concerned; current concepts in these areas lean heavily on recent studies of gastrointestinal motor physiology, and on neurophysiological exploration of the intrinsic and extrinsic innervation of the gut and of the central nervous system. From these studies, new concepts have evolved; in contrast, the complexity of the problem of regulatory peptides, as displayed by Walsh (1981), has destroyed preconceived ideas, such as those of gut hormones, without providing, as yet, a new coherent concept.

II. THE GUT BRAIN

The intrinsic innervation of the digestive tract, comprising the myenteric (Auerbach) plexus, and submucous (Meissner) plexus, was identified by neuroanatomists many years ago, but its function remained a subject for speculation. The concept of digestive tract control by circulating hormones and by the cholinergic and adrenergic extrinsic innervation did not include an obvious function for the intrinsic innervation.

In retrospect, it is now clear that understanding of the role of the intrinsic innervation, now generally known as the enteric nervous system (ENS), has grown in parallel with, and been dependent upon, knowledge of the motor activity of the gut. For most of the twentieth century, the conventional teaching on intestinal motility was based on the peristaltic reflex first described by Bayliss and Starling (1902) and on the radiologic appearances of organized movement conventionally categorized as peristalsis, segmentation, and pendular movement. It was assumed, largely because of the absence of evidence to the contrary and because of the lack of evidence of involvement of the entrinsic innervation, that these movements were organized by the intrinsic plexuses. There was no further progress in this field until the discovery (or rediscovery) of periodic activity allowed the functional dissection of the separate elements of neural control.

The assumption that the digestive tract is inert unless it is fed seems to be, on the face of it, reasonable, and it was Beaumont who first proposed that after the stomach has emptied a meal, it remains quiescent until the next meal. This principle is embodied, even now, in the design of many studies of gastrointestinal function, which comprise a "basal period," assumed to be one of minimal activity, followed by an active response to nutrient challenge. Boldyreff (1905) must take the credit for challenging this assumption. He demonstrated conclusively that, in the dog, the fasting state consists of prolonged motor and secretory quiescence, alternating with brief episodes of active contradictions and copious exocrine secretion. This cycle showed a regular periodicity of approximately 90 min and was abolished by feeding. Although the experimental evidence appeared to be convincing, and it was thought that perhaps this phenomenon might be the explanation of "hunger pains," no convincing functional explanation for the phenomenon was found and, being inconsistent with prevailing concepts, it was ignored and then gradually forgotten (Wingate, 1981).

The development of techniques allowing the prolonged study of gastrointestinal motor activity in conscious animals also allowed the rediscovery of periodic activity; periodic activity is usually abolished by anesthesia and is not seen in *ex vivo* viscera. Szurszewski (1969) showed that periodic activity was seen throughout the stomach and small intestine and that the bands of intense motor activity, which he called the migrating myoelectric complex (MMC), not only recurred at 90-min intervals but migrated distally from the stomach (Fig. 1), taking about 90 min to traverse the entire small bowel. The next development was the study of the effect of feeding on the MMC. Code and Marlett (1975) showed that in the dog, feeding interrupted the MMC pattern at all levels in the small intestine virtually simultaneously, and subsequent studies have shown that the human (Vantrappen *et al.*, 1977) is similar to the dog both in the presence of periodic MMC activity in fasting and the postprandial abolition of MMCs. In contrast, French veterinary scientists (Bueno *et al.*, 1975) showed that in sheep, feeding does not disrupt MMC activity. It seems that the interruption of MMCs by food occurs in species such as carnivores, which take discrete meals that are low in bulk and high in caloric content, whereas grazing animals, which continually ingest bulky material with low nutrient content, exhibit uninterrupted periodic

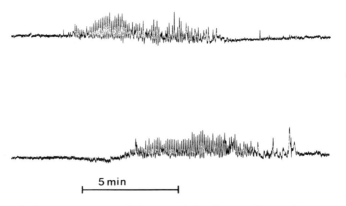

Fig. 1. The human MMC, recorded in an ambulant fasting volunteer from two pressure-sensitive, radiotelemetric capsules stationed (upper trace) 10 cm proximal to the duodenojejunal flexure, and (lower trace) 10 cm distal to the flexure. The "activity front" of the MMC is marked by regular contractions at a rate of 11/min, lasting for about 6 min; the average amplitude of these contractions is 30 cm H_2O. The propagation of the MMC is marked by the later arrival of the activity front at the lower capsule, with a propagation velocity of approximately 7 cm/min.

activity. This was confirmed in the pig, an omnivore, in which MMC interruption can be manipulated by replacing a herbivorous feeding pattern with a carnivorous feeding pattern (Ruckebusch and Bueno, 1976).

With periodic activity established during the 1970s as a subject for serious study, it was inevitable that there would be a search for the likely control mechanisms. This period coincided with the explosive growth in the number of identified gut peptides, which were viewed as "candidate hormones." Perhaps unsurprisingly, it was demonstrated that exogenous gastrin (Weisbrodt et al., 1974) and CCK (Mukhopadhyay et al., 1977) interrupted canine MMC activity, the inference being that the physiological interruption of MMCs on feeding was due to the release of endogenous CCK and gastrin, although systematic study of these effects showed that there were significant differences between the effects of infused peptides and food (Wingate et al., 1978). Subsequently, it has been shown that other exogenous peptides, including glucagon and neurotensin, have the same effect.

Likewise, peptides were implicated in the genesis of periodic activity. Two groups of workers (Itoh et al., 1976; Wingate et al., 1976) showed that exogenous motilin infused into a fasting dog induced a "premature" MMC; this finding seemed consistent with the prevailing view of peptides as hormones.

None of the early studies with peptides required an important role for the intrinsic innervation; the studies that implicated the ENS required different models, which included selective denervation. The traditional model employed at first was the Thiry–Vella loop model. In this model, a segment of small intestine is isolated, with an enterocutaneous fistula at each end, whereas the continuity of the remaining intestine is restored. This provides a segment of intestine isolated from the nutrient stream, with its intrinsic neural connections to the *in situ* intestine severed, but retaining, in the neurovascular pedicle of the segment, intact extrinsic innervation. Carlson et al. (1972) were the first to study the progression of MMCs through a Thiry–Vella segment (TVS) in the dog, and concluded that MMCs progressed through the isolated TVS in sequence as though the segment were still in continuity. This finding was taken by Wingate (1976) to imply some sort of extraenteric control center regulating MMCs through the intrinsic innervation, but shortly afterward it was shown that Carlson et al. had misinterpreted their findings (Bueno et al., 1979; Ormsbee et al., 1981). MMCs in Thiry–Vella loops maintain an independent periodicity that is out of phase with the MMCs in the main bowel (Fig. 2); further, the MMC mechanism in the TVS is not stimulated by motilin (Pinnington and Wingate, 1983). In contrast, in the Thiry–Vella animal, the isolated segment responds to nutrition as though it were in continuity (Pearce and Wingate, 1980). These observations were taken a step further by the work of Sarr and Kelly (1981), who studied autotransplanted isolated canine jejunal segments.

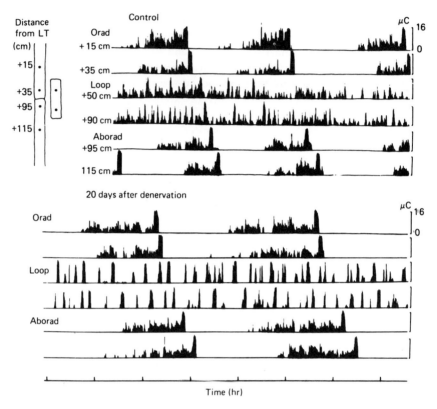

Fig. 2. Autonomous periodicity of a denervated Thiry–Vella loop in a conscious, fasted dog. The illustration shows histograms of spike activity (corresponding to contractile activity) recorded from electrodes implanted on the serosa of the bowel. In the upper (control) panel, regular periodic activity is seen proximal and distal to the Thiry–Vella loop before denervation; a different periodicity is seen in the loop. Twenty days after extrinsic denervation (lower panel), autonomous propagated periodic activity is seen in the Thiry–Vella loop, with a much shorter period than that seen in the remaining anastomosed bowel. (From Bueno et al., 1979.)

These segments resemble Thiry–Vella segments in exhibiting independent MMC activity but, unlike the TVS, their activity does not change on feeding.

From these studies, a number of conclusions can be drawn:

1. Since in the TVS model, intestinal MMC activity is asynchronized, MMC activity does not depend upon periodic fluctuations in plasma peptides (Sarr et al., 1983), although the stomach may differ from the bowel in this respect (Thomas and Kelly, 1979; Pilot et al., 1982).

2. Since in the autotransplanted model, the isolated segment does not

11. The Gut Brain and the Gut–Brain Axis

respond to food, the effect of feeding on intestinal motor activity is mediated by the extrinsic innervation, and not by the release of "digestive" peptides on feeding, although the stomach may differ in this respect (Thomas et al., 1980).

3. The capacity to generate periodic activity, and to organize the appropriate sequence of muscle movements, is a property of the ENS.

The ENS may thus be considered as a neural network that is capable of instituting more than one pattern of organized motor activity in the small intestine; not only is there the MMC, but there is also the postprandial pattern and the adynamic or paralytic pattern, which is induced by, for example, intestinal distension (da Cunha Melo et al., 1981). These patterns of effector activity are changed according to the sensory input (e.g., presence or absence of food, intraluminal pressure, etc.), and are probably modulated by the CNS (Fig. 3). Operationally, these properties are the properties of a brain, hence the term "gut brain." It now seems probable that the organization of gastrointestinal function is carried out by a subsidiary brain, the gut brain, which is in turn modulated by the cephalic brain (McRae et al., 1982); modulation in the opposite direction is also probable (Fig. 4).

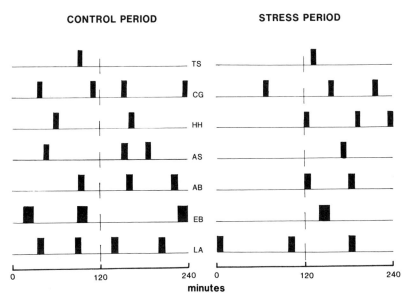

Fig. 3. The effect of psychological stress (the dichotomous listening test) on the incidence of MMCs in human volunteers. Each line represents 8 hr of continuous recording in a single subject; 4 hr of rest was followed by 4 hr of stress. The recording method was similar to that used in Fig. 1; the vertical bars show MMCs. It is seen that there was a marked diminution of MMCs ($p < 0.02$) during the first 2 hr of stress. (From McRae et al., 1982.)

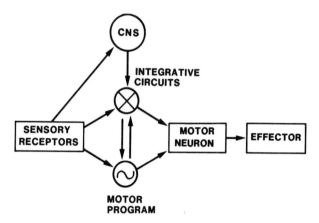

Fig. 4. Heuristic model of intestinal motor control. In this model, the myenteric plexus (or gut brain) contains both the integrative circuits and the motor programs.

How does this brain work? Relatively little is known, but two lines of research have produced some progress. The electrophysiological studies of Wood and his colleagues (Wood, 1981) have characterized the electrical properties of ENS neurons in the guinea pig ileum (Fig. 5), whereas the studies of Costa and Furness (1982) have described not only the morphology of the plexus, but also the distribution of peptides within the plexus. Even so, at the time of writing, it is not possible to relate the functional activity of the ENS to the morphological or electrical properties of components of the plexuses, although speculation is possible (Fig. 5).

III. THE GUT–BRAIN AXIS

The "gut–brain axis" is an oversimplification of the very complex interrelationship between the viscera and the central nervous system (CNS). There is an abundance of anecdotal evidence (Cannon, 1909; Almy, 1978) that supports the existence of a close functional relationship between the gut and the brain, but somewhat surprisingly, experimental documentation of this concept has been a neglected area of investigation. The idea that sensory information from the viscera could influence behavior was shown by Janowitz and Grossman (1949), who demonstrated that food intake in dogs is influenced to a large extent by distension of the stomach. Paintal (1954) subsequently looked for sensory mechanisms at a cellular level that could account for these behavioral observations. These early studies did indeed show a system of stomach stretch receptors presumed to lie in the smooth

11. The Gut Brain and the Gut-Brain Axis

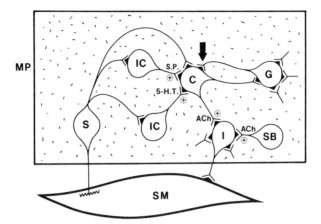

Fig. 5. A simple model of smooth muscle control (SM) by the myenteric plexus (MP) (stippled area). The neural input to the cell is an inhibitory neuron (I), activated either by a command cell (C), or a "steady burst" (SB) cell; this activation is probably cholinergic. The command cell can respond directly to stretch receptors on the muscle cell (S), or through interneurons (IC) that relay to the command cell; equally the command cell respond to other ganglia (G) within the plexus with which it communicates. This model is consistent with the view that the input from the plexus to the smooth muscle is largely inhibitory. (ACh, acetylcholine; 5-H.T., 5-hydroxy-tryptomine; S.P., Substance P.)

muscle wall of the stomach that supplied a detailed description of stomach movements to the CNS by way of the vagus nerve. Subsequent workers (Clarke and Davison, 1978), by studying the patterns of neural information carried in the vagus nerve following mechanical and chemical stimulation of the gastrointestinal tract, have been able to demonstrate that a wealth of different sensory information is relayed to the CNS. It is this "afferent profile" of gut activity that forms the data base on which the CNS elaborates the behavior patterns related to hunger, food intake, and satiation.

As a first step in trying to understand some of these behavioral mechanisms, in recent years a number of investigators have attempted to record neuronal activity in the CNS that is evoked by stimulation of the gut. Morphological studies have traced the central projections of the vagus to the brainstem (Kalia and Sullivan, 1982), identifying the nucleus of the solitary tract (NTS) and dorsal motor nucleus of the vagus nerve (DVN) as primary projection sites for sensory (afferent) information from the gastrointestinal tract. Physiological studies (Harding and Leek, 1971; Mei, 1978; Barber and Burks, 1983; Ewart and Wingate, 1984) have shown that when the peripheral "sensors" are activated, the information that they generate is signaled to the brainstem in recognizable patterns of neuronal activity. which describe, in intricate detail, the onset, duration, strength, and localization of a variety of stimuli. These include mechanical distension of the stomach and duodenum and chemical stimulation of gastric and duodenal mucosa, which may activate osmoreceptors and/or glucoreceptors. Figure 6 schematically demonstrates this type of experimental approach. Figure 7 shows the model that we have used (Ewart and Wingate, 1983a,b, 1984) to further explore the central representation of sensory information from the gastrointestinal tract.

Using microelectrodes, it is possible to record extracellularly the activity of individual neurons within the brainstem of an anesthetized rat; the recording electrode can be positioned accurately in the NTS or DVN by using stereotaxic coordinates determined for the rat brain. Stimulation of gastric mechanoreceptors is achieved by inflation of a water-filled balloon; an alternative stimulus is stimulation of duodenal glucoreceptors by luminal perfusion of the duodenum with isotonic glucose. Using these techniques, it is possible to record the way in which a single neuron in the rat brain responds to specific gastrointestinal tract stimulation, a single neuron's response is, of course, the information with which the rest of the CNS will have to build up its picture of the sensory events occurring in the viscera. The spike trace shown in Fig. 8 is the extracellularly recorded action potentials from a single neuron in the DVN. The horizontal bar indicates the period of time during which isotonic glucose is perfused through the duodenum. It is clearly seen that there is an increase in the firing rate of this neuron following arrival of

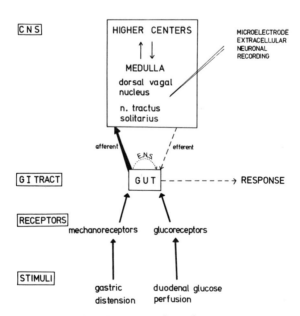

Fig. 6. See text for explanation.

11. The Gut Brain and the Gut–Brain Axis 203

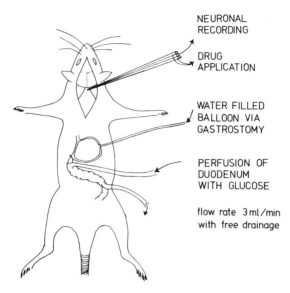

Fig. 7. See text for explanation.

glucose in the duodenum. In this manner, the change in intestinal content is signaled to the brain, and this type of recording demonstrates that changes in individual neuron activity can accurately describe events occurring in the gut. Similarly, the effect of stimulating gastric mechanoreceptors can be recorded in the brainstem; Fig. 9 shows a spike trace of a single neuron in the DVN that is inhibited by gastric distention (GD horizontal bar) but that does not signal the duodenal stimulation (GLU horizontal bar). These differential effects are displayed in a computer-generated spike-frequency histogram shown in the upper panel of this figure. In this model, it is also possible to deliver very small amounts of peptides from the recording electrode to the areas immediately adjacent to the neuron from which the recording is being made. Solutions of peptides such as leu-enkephalin, cholecystokinin-octapeptide (CCK-OP), and bombesin can be placed in the outer barrels of the multibarrel microelectrode; by applying pressure or current to these barrels,

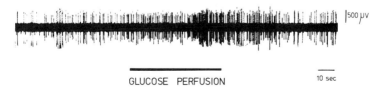

Fig. 8. See text for explanation.

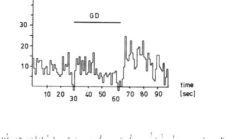

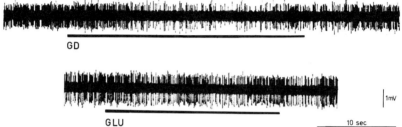

Fig. 9. See text for explanation.

small amounts of these peptides (in the region of 10^{-11} to 10^{-13} fmol/min) can be expressed from the electrode tip, thereby applying these peptides as locally as possible to the neuron being studied. Using these techniques, it has been possible to show that whereas leu-enkephalin is predominantly inhibitory (i.e., when leu-enkephalin is directly applied to these neurons, they show a decrease in firing rate), CCK-OP and bombesin tend to cause an excitation of these neurons, i.e., an increase in firing rate. Figure 10 shows an example of a neuron in the DVN which is excited by the local application of CCK-OP. The peptides leu-enkephalin and CCK-OP have also been shown to be capable of modifying neuronal responses to gastrointestinal tract stimulation (Ewart and Wingate, 1983a,b), an effect that illustrates the power that these substances may have to modulate the pattern of "afferent-like" activity that is being relayed to the brain.

Using this approach, it has been possible to identify packages of information received in the brain that describe neatly and efficiently what is going on in the gastrointestinal tract. The studies of our group and others described so far really represent only the first stage in the central processing of this important sensory information, since the information will be further elaborated and refined in "higher" areas of the brain before it becomes part of any integrative behavioral activity. A clue to this further interpretation of sensory information from the gut has been provided by Jeanningros (1984) in

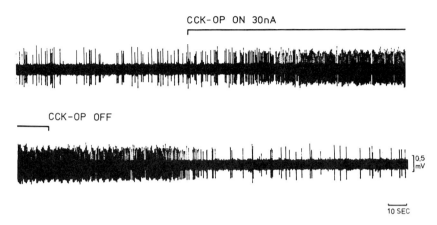

Fig. 10. See text for explanation.

a recent study in which she describes modulation of the activity of single neurons in the lateral hypothalamus elicited by gastric and intestinal distention. Clearly, the framework of a gut–brain axis exists; the question how it actually operates to provide the afferent profile on which behavior is modeled is a subject that will occupy gastrointestinal neuroscience for a long time to come.

IV. PEPTIDES

Although a comprehensive review of the regulatory peptides is beyond the scope of this review, it is pertinent to review the extent to which the new studies in visceral neural function and in ENS–CNS interdependence have altered our perception of the peptides.

Gut peptides were originally viewed primarily as hormones, released from cells scattered in the epithelial lining of the gut, and acting on exocrine secretory systems, and on visceral smooth muscle. Advances in radioimmunoassay, immunohistochemistry, and peptide chemistry have revealed the structure and location of many new peptides, in addition to gastrin, CCK, and secretin. These discoveries, and the increasing evidence of the importance of neural control, permit some interim conclusions on the role of peptides.

1. Many peptides are widely distributed within the nervous system, including the CNS, and also in organs outside the digestive tract.

2. Extrapolation from the results of infusing exogenous peptide to provide physiological conclusions is fraught with error.

3. Although a few gut peptides may have some actions that are true hormonal actions, with physiological effects on distant target tissue that show dose–responsiveness—the action of gastrin on parietal cells is one example—it now seems probable that, in general, peptides interact with the nervous system, and may also be released as a result of neural activity.

4. Peptides may be involved in the regulation of the digestive tract acting peripherally, or within the CNS, as has been shown by the potent effects on the gastrointestinal tract of small concentrations of some peptides injected into the cerebral ventricles.

5. The release of peptides in relation to discrete events such as the MMC, or the presence of specific types of nutrient at various loci in the gastrointestinal tract (such as the release of neurotensin by lipid in the lumen of the distal small intestine) suggests that at any one time, the plasma level of peptides reflects the functional status of the gastrointestinal tract and its contents, and thus provides essential information required for the initiation of appropriate behavior. This suggests that plasma peptides may act as signals.

To illustrate the difficulty of understanding the function of the peptides at the present time, a few examples may suffice. Our studies on the gut–brain axis have shown that the firing rate of neurons in the medulla responding to gastric stimulation may be modified by locally applied peptides that have been isolated in that area of the brain; however, our studies give no indication of the circumstances in which the endogenous peptide in the medulla is involved in the local modulation of neuronal activity. It has been shown that exogenous motilin will induce a premature MMC, but the same effect can be induced by the antibiotic lincomycin. All MMCs, whether spontaneous or induced by exogenous motilin, are accompanied by the release of endogenous motilin (Sarna et al., 1983; Hall et al., 1984). In humans and dogs, pancreatic polypeptide (PP) is a marker of vagally induced pancreatic secretion, but exogenous PP infused in these species has no effect, whereas in fasting ruminants, exogenous PP, but not motilin, induces premature MMCs.

The probable model for the action of most regulatory peptides would seem to be noradrenaline, which is accepted as a classical neurotransmitter of adrenergic nerve transmission, but undoubtedly also has a humoral role when released from the adrenal medulla.

V. RESEARCH GOALS

In less than two decades, our understanding of the motor physiology and neurophysiology of the digestive tract has been radically altered. Although it is unlikely that current concepts will remain intact, recent research has served at least to indicate desirable directions for further study.

The enteric nervous system remains an important field for further exploration, and the approach of Furness and his collaborators, combining morphology, immunohistochemistry, and neurophysiology should prove to be increasingly fruitful. One problem that has to be faced is that of differences between species; most of the work on the myenteric plexus has been carried out on the guinea pig, but this is a species that may differ in important respects from the human, and that has not been much studied *in vivo*. The rat ENS would be preferable, but the human plexus is optimal and appropriate techniques must be developed.

At the functional level, the importance of the gut brain has been demonstrated in relation to motor activity, but it is unlikely that it modulates only motor activity. Techniques are needed for studying the function of the ENS in the modulation of other digestive tract functions, such as exocrine secretion (Vantrappen *et al.*, 1979) and blood flow (Fioramonti and Bueno, 1984).

Pathology provides models such as selective denervation, as in idiopathic degeneration of the myenteric plexus (Schuffler and Jonak, 1982), which cannot be reproduced in the laboratory, but which provide important clues. Wider dissemination among physicians of what is already known is required to stimulate studies of patients, and interdisciplinary collaboration bringing internists and surgeons together with physiologists and morphologists must be encouraged.

It has now been shown that the gut–brain axis, as well as the brain–gut axis is susceptible to direct study; this work needs to be expanded in conjunction with the use of peptides. More needs to be known about sensory receptors in the gastrointestinal tract, and the transmission of afferent impulses peripherally and centrally. The accomplishment of these goals will require trained neurophysiologists to be diverted from their current preoccupation with brain science; the major problem in this field is the shortage of research workers with adequate skills.

All such work will involve the study of regulatory peptides. The revolution in peptides would not have happened without essential chemical and biochemical studies, but further progress demands that those who are skilled in the identification and assay of peptides cast in their lot with scientists concerned with physiological function.

VI. SUMMARY

Recent studies in the control of the digestive tract have emphasized the importance of both extrinsic and intrinsic neural control of the gut, as well as revealing the complexity of the neural control mechanisms. Regulatory peptides, originally viewed as gut hormones or humoral control substances, are likely to act as neuromodulators or even neurotransmitters within the neural networks. This presentation is focused on the autonomy of the intrinsic nerve networks and the importance of the vagal afferent projections within the CNS.

Electromyographic studies on conscious animals have revealed on omnipresent periodic motor activity during fasting, which in carnivores, is replaced by a nonperiodic pattern. It is suggested that the myenteric plexus acts as a "gut brain," which is able to impose different patterns of integrated motor activity in response to different sensory inputs. This activity is not dependent upon the CNS but can be modulated by both the CNS and by exogenous regulatory peptides; in addition, different phases of motor and the associated secretory activity are associated with the humoral release of regulatory peptides.

Although the vagus is predominantly afferent, understanding of the nature and destination of the afferent information requires recording from the brain during controlled physiologic stimulation of the gut. An appropriate model for this has been developed in the rat. Single unit recording in the medulla has shown that about half the units in the dorsal vagal nucleus (DVN) and the nucleus tractus solitarius (NTS) respond to GI stimuli. The spontaneous and stimulus-evoked firing patterns of these neurons can be modulated by the local application of regulatory peptides to the medulla. Further research is required to link functional studies with cellular mechanisms in the enteric nervous system, and to expand neurophysiological knowledge of gut–brain interdependence; all such studies must include consideration of the regulatory peptides.

REFERENCES

Almy, T. P. (1978). The gastrointestinal tract in man under stress. In "Gastrointestinal Diseases" (M. Sleisenger and J. Fordtran, eds.), 2nd ed., Vol. 1, Chap. 1. Saunders, Philadelphia.

Barber, W., and Burks, T. F. (1983). Brainstem response to phasic gastric distension. Am. J. Physiol. 245, G242–G248.

Bayliss, W. M., and Starling, E. H. (1902). The mechanism of pancreatic secretion. J. Physiol. (London) 28, 324–353.

Beaumont, W. (1833). "Experiments and observations on the Gastric Juice and the Physiology of Digestion." F. P. Allen, Plattsburgh.

Boldyreff, W. N. (1905). Le travail périodique de l'appareil digestif en dehors de la digestion. Arch. Sci. Biol. 11, 1–157.

Bueno, L., Fioramonti, J., and Ruckebusch, Y. (1975). Rate of flow of digesta and electrical activity of the small intestine in dogs and sheep. J. Physiol. (London) 249, 69–85.

11. The Gut Brain and the Gut–Brain Axis

Bueno, L., Praddaude, F., and Ruckebusch, Y. (1979). Propagation of electrical spiking activity along the small intestine: Intrinsic versus extrinsic neural influences. *J. Physiol. (London)* **292**, 15–26.

Cannon, W. B. (1909). The influence of emotional states on the functions of the alimentary canal. *Am. J. Med. Sci.* **137**, 480–487.

Carlson, G. M., Bedi, B. S., and Code, C. F. (1972). Mechanism of propagation of intestinal interdigestive myoelectric complex. *Am. J. Physiol.* **222**, 1027–1030.

Clarke, G. D., and Davison, J. S. (1978). Mucosal receptors in the gastric antrum and small intestine of the rat with afferent fibres in the cervical vagus. *J. Physiol. (London)* **284**, 55–67.

Code, C. F., and Marlett, J. A. (1975). The interdigestive myoelectric complex of the stomach and small bowel of dogs. *J. Physiol. (London)* **246**, 298–309.

Costa, M., and Furness, J. B. (1982). Neuronal peptides in the intestine. *Br. Med. Bull.* **38**, 247–252.

Da Cunha Melo, J., Summers, R. W. Thompson, H. H., Wingate, D. L., and Yanda, R. (1981). Effects of intestinal secretagogues and distension on small bowel myoelectric activity in fasted and fed conscious dogs. *J. Physiol. (London)* **321**, 483–494.

Ewart, W. R., and Wingate, D. L. (1983a). Central representation and opioid modulation of gastric mechanoreceptor activity in the rat. *Am. J. Physiol.* **244**, G27–G32.

Ewart, W. R., and Wingate, D. L. (1983b). Cholecystokinin-octapeptide and gastric mechanoreceptor activity in the rat brain. *Am. J. Physiol.* **244**, G613–G617.

Ewart, W. R., and Wingate, D. L. (1984). The central representation of the arrival of nutrient in the duodenum. *Am. J. Physiol.* **246**, G750–G756.

Fioramonti, J., and Bueno, L. (1984). Relation between intestinal motility and mesenteric blood flow in the conscious dog. *Am. J. Physiol.* **246**, G108–G113.

Hall, K. E., Greenberg, G. R. El-Sharkawy, T. Y., and Diamant, N. E. (1984). Relationship between porcine motilin-induced migrating motor complex-like activity, vagal integrity, and endogenous motilin release in dogs. *Gastroenterology* **87**, 76–85.

Harding, R., and Leek, B. F. (1971). The localisation and activities of medullary neurones associated with ruminant forestomach motility. *J. Physiol. (London)* **219**, 587–610.

Itoh, Z., Honda, R., Kiwatashi, K., Takeuchi, S., Aizawa, I. Takayanagi, R., and Couch, E. F. (1976). Motilin-induced mechanical activity in the canine alimentary tract. *Scand. J. Gastroenterol.* **11**(Suppl. 39), 93–110.

Janowitz, H. D., and Grossman, M. I. (1949). Some factors affecting the food intake off normal dogs and dogs with esophagostomy and gastric fistula. *Am. J. Physiol.* **159**, 143–148.

Jeanningros, R. (1984). Modulation of lateral hypothalamic single unit activity by gastric and intestinal distension. *J. Auton. Nerv. Syst.* **11** 1–11.

Kalia, M., and Sullivan, J. M. (1982). Brainstem projections of sensory and motor components of the vagus nerve in the rat. *J. Comp. Neurol.* **211**, 248–264.

McRae, S., Younger, K., Thompson, D. G., and Wingate, D. L. (1982). Sustained mental stress alters human jejunal motor activity. *Gut* **23**, 404–409.

Mei, N. (1978). Vagal glucoreceptors in the small intestine in the cat. *J. Physiol. (London)* **282**, 485–506.

Mukhopadhyay, A. K., Thor, P. J., Copeland, E. M., Johnson, L. R., and Weisbrodt, N. W. (1977). Effect of cholecystokinin on myoelectric activity of small bowel of the dog. *Am. J. Physiol.* **232**, E44–E47.

Ormsbee, H. S., Telford, G. L., Suter, C. M., Wilson, P. D., and Mason, G. R. (1981). Mechanism of propagation of canine migrating motor complex—a reappraisal. *Am. J. Physiol.* **3**, G141–G146.

Paintal, A. S. (1954). A study of gastric stretch receptors: Their role in the peripheral mechanism of satiation of hunger and thirst. *J. Physiol. (London)* **126**, 255–270.

Pearce, E. A., and Wingate, D. L. (1980). Myoelectric and absorptive activity in the transected canine small bowel. *J. Physiol. (London)* **302,** 11P–12P.

Pilot, M-A., Thomas, P. A., Gill, R. E., Knight, P. A., and Ritchie, H. D. (1982). Hormonal control of the gastric motor response to feeding: Cross-perfusion of an isolated stomach using a conscious donor dog. *World J. Surg.* **6,** 422–426.

Pinnington, J., and Wingate, D. L. (1983). The effects of motilin on periodic myoelectric spike activity in intact and transected canine small intestine. *J. Physiol. (London)* **337,** 471–478.

Ruckebusch, Y., and Bueno, L. (1976). The effect of feeding on the motility of the stomach and small intestine in the pig. *Br. J. Nutr.* **35,** 397–405.

Sarna, S., Chey, W. Y., Condon, R. E. Dodds, W. J. Myers, T., and Chang, T. (1983). Cause and effect of relationship between motilin and migrating myoelectric complexes. *Am. J. Physiol.* **245,** G277–G284.

Sarr, M. G., and Kelly, K. A. (1981). Myoelectric activity of the autotransplanted canine jejunoileum. *Gastroenterology* **81,** 303–310.

Sarr, M. G., Kelly, K. A., and Go, V. L. W. (1983). Motilin regulation of canine interdigestive intestinal motility. *Dig. Dis. Sci.* **28,** 249–256.

Schuffler, M. D., and Jonak, Z. (1982). Chronic idiopathic intestinal pseudo-obstruction caused by a degenerative disorder of the myenteric plexus: The use of Smith's method to define the neuropathology. *Gastroenterology* **82,** 476–486.

Szurszewski, J. H. (1969). A migrating electric complex of the canine small intestine. *Am. J. Physiol.* **217,** 1757–1763.

Thomas, P. A., and Kelly, K. A. (1979). Hormonal control of interdigestive motor cycles of canine proximal stomach. *Am. J. Physiol.* **237,** E192–E197.

Thomas, P. A., Schang, J. C., Kelly, K. A., and Go, V. L. (1980). Can endogenous gastrin inhibit canine interdigestive gastric motility? *Gastroenterology* **78,** 716–721.

Vantrappen, G., Janssens, J., Hellemans, J., and Ghoos, Y. (1977). The interdigestive motor complex of normal subjects and patients with bacterial overgrowth of the small intestine. *J. Clin. Invest.* **59,** 1158–1166.

Vantrappen, G., Peeters, T. L., and Janssens, J. (1979). The secretory component of the interdigestive migrating motor complex in man. *Scand. J. Gastroenterol.* **14,** 663–667.

Walsh, J. H. (1981). Gastrointestinal hormones and peptides. *In* "Physiology of the Gastrointestinal Tract" (L. R. Johnson, ed.), pp. 59–64. Raven Press, New York.

Weisbrodt, N. W., Copeland, E. M., Kearley, R. W., Moore, E. P., and Johnson, L. R. (1974). Effects of pentagastrin on electrical activity of small intestine of the dog. *Am. J. Physiol.* **227,** 425–429.

Wingate, D. L. (1976). The eupeptide system: A general theory of gastrointestinal hormones. *Lancet* **1,** 529–532.

Wingate, D. L. (1981). Backwards and forwards with the migrating complex. *Dig. Dis. Sci.* **26,** 641–666.

Wingate, D. L., Ruppin, H., Green, W. E. R., Thompson, H. H. Domschke, W., Wunsch, D., Demling, L., and Ritchie, H. D. (1976). Motilin-induced electrical activity in the canine gastrointestinal tract. *Scand. J. Gastroenterol.* **11**(Suppl. 39), 111–118.

Wingate, D. L., Pearce, E. A., Hutton, M., Dand, A., Thompson, H. H., and Wunsch, E. (1978). Quantitative comparison of the effects of cholecystokinin, secretin, and pentagastrin on gastrointestinal myoelectric activity in the conscious dog. *Gut* **19,** 593–601.

Wood, J. D. (1981). Physiology of the enteric nervous system. *In* "Physiology of the Gastrointestinal Tract" (L. R. Johnson, ed.), pp. 1–37. Raven Press, New York.

12

Cephalic Phase of Digestion: The Effect of Meal Frequency

KATHERINE A. HOUPT AND T. RICHARD HOUPT
Department of Physiology
New York State College of Veterinary Medicine
Cornell University
Ithaca, New York

I. Introduction 211
II. Critical Review and Discussion of Subject Matter 212
   A. A Model Herbivore: The Horse 212
   B. A Model Omnivore: The Pig 224
III. Research Needs 236
    References 237

I. INTRODUCTION

The purpose of this series of studies was to identify a factor or factors that might serve as hunger or satiety signals. Despite decades of research effort, a humoral factor that stimulates intake has not been identified. The search for a satiety factor has been more successful; cholecystokinin does appear to operate as a peripheral satiety factor in rats, monkeys, pigs, and rabbits (Gibbs and Smith, 1977; T. R. Houpt et al., 1978; T. R. Houpt, 1983; Anika et al., 1981), although the evidence for its role as a satiety factor in humans is more tenuous (Kissileff et al., 1981).

Our approach was to measure those hormones and metabolites that were the most likely candidates for hunger and satiety signals. We were most interested in a signal that would operate in the free-feeding animal. Because hunger begins to ebb and satiety to increase with the very first mouthful of food, cephalic factors would be expected to have importance as the initial satiety signal. Therefore, we sampled metabolites, hormones, and the behavior of the animal most intensively just at the beginning of meals.

Most research in ingestive behavior has been performed on rats, but we wished to broaden the scope of our understanding to include ungulates. This comparative approach allows us to study the physiology of feeding in a strict herbivore, the horse, and in the omnivorous pig, the best animal model next to the primates for humans.

II. CRITICAL REVIEW AND DISCUSSION OF SUBJECT MATTER

A. A Model Herbivore: The Horse

1. Social Factors as Stimulants of Intake

Because horses are herd animals living in groups composed of several mares, their offspring, and a stallion, we suspected that social facilitation might be involved in feeding behavior. In order to test this hypothesis, we observed 8 pairs ($n = 9$ ponies) of Shetland-type, pregnant pony mares as a focal dyad (Altmann, 1974) for 2-hr intervals (1000–1200 and 1400–1600) for a total of 117 hr. The ponies were kept in adjacent pipe-rail stalls. Mixed-grass hay was fed *ad libitum*. Fresh hay was placed in the feed troughs 15–20 min before the morning observations.

Each pair of ponies was observed for 2 weeks. During one of the 2 weeks, there was no obstruction between the stalls, enabling the ponies to receive visual cues from each other. During the other week, a 130-cm plywood partition was placed between the two stalls, blocking visual but not olfactory or auditory contact.

We recorded a variety of mutually exclusive behaviors. Of particular interest here are feeding, drinking, and standing.

Figure 1 illustrates the time budget of pony mares in the presence and absence of visual cues (Sweeting *et al.*, 1985), Feeding time decreases and standing time increases in the absence of visual cues. Social facilitation of feeding is apparently dependent, at least in part, on visual contact between the animals because feeding decreased despite the continued presence of olfactory and auditory cues between the animals.

2. Feeding and Drinking Patterns in Horses

Four mature pony geldings were used. A single stall in a room with no other animals was used to house each pony. The room temperature was controlled (21.1 ± 1°C). A 12-hr light–dark cycle was maintained. The

12. Cephalic Phase of Digestion: Meal Frequency Effect

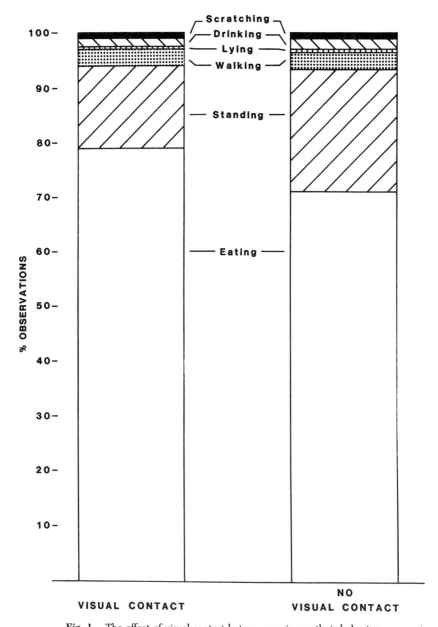

Fig. 1. The effect of visual contact between ponies on their behavior.

ponies were fed *ad libitum* a diet consisting of soybean meal, corn, oats, and bran. The experimentally determined digestibility was 3.4 Mcal/kg feed.

The stall was equipped with two photoelectric cells (Heathkit Model Gd-1021), one positioned above the feed bucket, the other above the water bucket. When the light beam was broken, i.e., when the animal put its head into the feed or water bucket, it was recorded on an event recorder (Esterline Angus). All eating and drinking bouts were measured to the nearest minute, and those less than 30 sec were recorded as 0.

A typical daily record of meals is shown in Fig. 2. It is obvious that the pony initiated a meal nearly every hour, and the length of the meals varied considerably. There were more meals and longer ones during the light

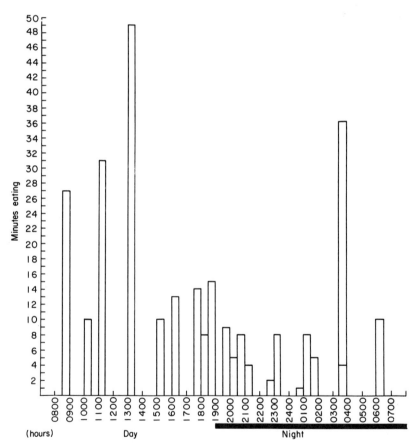

Fig. 2. A typical daily record of meal length and frequency by a pony. [From K. A. Houpt (1982a). Reprinted with permission of Veterinary Practice Publishing Company.]

12. Cephalic Phase of Digestion: Meal Frequency Effect   215

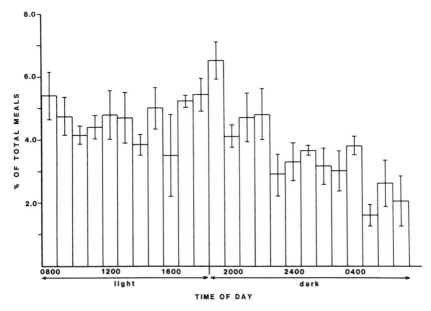

Fig. 3. Percentage of means initiated during every hour by ponies.

portion of the 24 hr than during the night. The percentage of meals begun during each hour of the day is shown in Fig. 3; this figure also indicates the circadian variation of intake. The ponies ate 5 ± 1 kg or 17.3 Mcal/day (K. A. Houpt et al., 1983c). The average meal length was 5 min and the average intermeal interval was 24 min. Approximately 250 min per day were spent eating. Log survivorship of intermeal intervals indicated that breaks less than 10 min in length were within meals, and those longer than 10 min were between meals. When we used that criterion, the ponies ate 24 meals/day. It would appear that even on a high-calorie, low-roughage diet, horses tend to graze, i.e., eat many small meals.

The ponies spent only 27 min/day drinking water, during which time they consumed 8.7 ± 0.9 kg/day at a rate of 552 ml/min. The drinking patterns of the ponies revealed that 89% of their water intake occurred periprandially, that is within 10 min before or 30 min after a bout of feeding (Sufit et al., 1984). This was an unexpected finding because most studies of freely ranging horses have been in arid environments, in which the horses were forced to travel long distances to water and therefore drank only once/day or less (Miller and Denniston, 1979; Berger, 1977). These studies and the practice of watering horses only twice a day, which was common before most stables were furnished with running water, assumed that horses drank infrequently. They can tolerate such scheduled drinking, but obviously do not prefer it.

3. Physiological Responses to Feeding in Horses

These early studies of feeding were mostly descriptive, but did allow us to quantify meal patterns and drinking patterns. Next we wished to investigate the humoral, cardiovascular, and hormonal changes that occur with feeding.

Ten 1- to 3-year-old ponies (9 mares, 1 gelding), weighing 104–198 kg, were tested two at a time. Each pony was kept in a 3.4 × 2.7-m box stall with water *ad libitum* except during the period of blood sampling when no water was available. Feeding is described below. No other animals were present in the barn at any time. Barn temperature ranged from 12 to 32°C depending on weather conditions, time of day, and time of year (July–December). A 60-W light bulb provided light from 0700 to 1900 hr. In good weather, the ponies were allowed 1–4 hr per week of exercise in a small corral.

The ponies were tested in a simple reversal design. For each pair of ponies, one was fed six meals per day, once every 4 hr (6×/day), whereas the other was fed one meal per day, once every 24 hr (1×/day). After 2 weeks on one feeding regime, each pony was placed on the other feeding regime. After each pony of a pair had been on each feeding regime for 2 weeks, a new pair of ponies was obtained and tested.

All ponies were fed a complete, pelleted ration (Agway Choice, 2.76 Mcal/kg; 12.00% crude protein; 20.00% crude fiber). When the ponies were fed one meal per day, they were presented a bucket containing 3000–4000 g of food. They were allowed to eat for 1 hr, after which the food was taken away. To determine food consumption, the leftover food was weighed on a scale accurate to 1 g. While on the 6×/day regime, the ponies received five of the meals from an automatic feeder, which delivered a calibrated amount of food into a bucket by activating a solenoid positioned below the storage container. In order to control for human contact, the sixth meal of the 6×/day regime and the meal on the 1×/day regime were hand delivered into each pony's bucket at 0900. For the ponies on the 6×/day regime, the amount of food fed per day was equal to the percentage body weight its partner ate in its one meal per day. Therefore, daily intake based on percentage body weight did not vary; only the frequency of feeding and meal size varied.

Each week, behavioral observations were collected during three 30-min time periods: prefeeding samples were collected within 2 hr prior to feeding; during-feeding samples were collected during the hand-fed meals; postfeeding samples were collected within 2 hr after the pony finished eating. Observations were made through a one-way mirror, which separated the observer from the stable area. Behaviors were recorded every 5 sec. The categories of behaviors are listed and defined in Table I. They form a set of mutually exclusive behaviors.

TABLE I Categories and Definitions of Behaviors

Behavior	Definition
Eating food	Eating pellets
Eating bedding	Eating sawdust that was used for bedding
Coprophagia	Eating feces
Standing	Standing in one location. Pony could be engaged in a number of behaviors, i.e., grooming, sleeping, nosing inanimate objects, etc.
Walking	Pony moves at least 3 feet
Lying down	Pony lies in either lateral or sternal recumbency
Drinking	Drinking water

Total numbers of observations in each behavior category were calculated for each sample period. Means for each pony and means for each time period were calculated on an Apple II Plus computer. From the latter calculations, percentages of time spent in each behavior during the three time periods were derived. Total observations in each behavior during the three time periods were added together in order to calculate daily time budgets. Differences in behavior between the two feeding regimes and among the three time periods were analyzed using a paired t test.

Electrocardiograms (EKGs) were used to determine heart rate. Standard lead II placement of electrodes on the chest wall near the point of the elbow and on the side just below the withers on the right side was used. The EKG was recorded on a polygraph. The EKG was recorded while the ponies stood quietly (resting) in the stall, while food was shown to them (prefeeding), and while they were eating (feeding). To calculate the rate for the resting condition, three 5-sec blocks of the EKG were averaged just before feed was shown to the horse. Two or more 5-sec blocks were averaged during the prefeeding period, and three 5-sec blocks of the EKG were averaged during the postfeeding period.

Intravenous catheters (PE 60) were placed in the jugular vein on the day of the blood sampling.

The blood samples were taken 30, 15, and 0 min before feeding (0900), and 5, 10, 15, 20, 30, 45, 60, and 90 min after feeding.

The plasma protein concentration for each sample was determined with a hand-held refractometer on the day the sample was taken. To determine osmotic pressure, the samples were thawed and a 3-ml aliquot was analyzed with a freezing point depression osmometer. Plasma glucose was analyzed using the o-toluidine method. Triiodothyronine (T_3) was measured by radioimmunoassay (Reimers et al., 1982).

Each pony was sampled once a week so that two sets of samples were collected on each feeding regime. Mean plasma protein and osmotic pressure values were calculated for each pony at each sampling time for each feeding regime. From these values, mean plasma protein and osmotic pressure values ± SE were calculated.

The results are shown in Figs. 4–7. During and after a large meal, plasma protein rose rapidly to a peak at 30 min and returned to prefeeding levels by 90 min (Fig. 4). In contrast, osmotic pressure rose more slowly, reaching a peak at 60 min and remaining above prefeeding levels for at least 90 min after the meal (90 min after feeding had begun) (Fig. 5). Glucose rose to 150 mg/dl during the absorptive period, as one would expect (Fig. 6). T_3 rose to a peak at 60 min and remained above prefeeding values (Fig. 7). The changes during and after a small meal were similar, but not as marked. There are significant changes between the prefeeding and during-feeding value for all humoral and hormonal elements measured when the ponies were fed only one meal/day. None of the variables was significantly elevated during feeding in the ponies fed six small meals/day. Only T_3 might be considered a putative humoral satiety factor, and then only after a large meal. Glucose may function as a satiety factor, but as a gastrointestinal not a humoral one (see below).

The rise and fall of plasma protein was unexpected. The exact cause of the changes remain unknown. We hypothesize that the secretion of saliva, which occurs only during feeding in the horse (Alexander, 1966) and is approximately 7 liters/day in a pony, results in an isotonic fluid loss from the body, or hypovolemia. Plasma protein drops back to normal when salivation ceases at the end of the meal and when absorption of fluid from the lower tract begins. The changes in osmotic pressure are easier to explain. As food is digested, the number of particles in the tract increases, thus increasing the osmotic pressure of the luminal contents. Water moves osmotically from the interstitial space, from plasma, and eventually even from the cellular compartments in response to this osmotic pressure. As long as digestion continues and absorption is not complete, the osmotic movement of water persists. The changes in fluid balance are probably the stimuli for periprandial drinking in horses.

The increase in plasma glucose is a simple reflection of digestion and absorption of the carbohydrates present in the meal. The trigger for release of T_3 is unknown. An increase in T_3 with feeding has been shown in other species including rats (LeBlanc et al., 1982), cattle (Tveit and Larsen, 1983), and pigs (see below). T_3 may be considered as a possible satiety factor. An increase in metabolic rate stimulated by the increase in T_3 could be the basis of thermostatic control of feeding. The increase in metabolic rate and consequent decrease in efficiency may be one mechanism by which body weight is

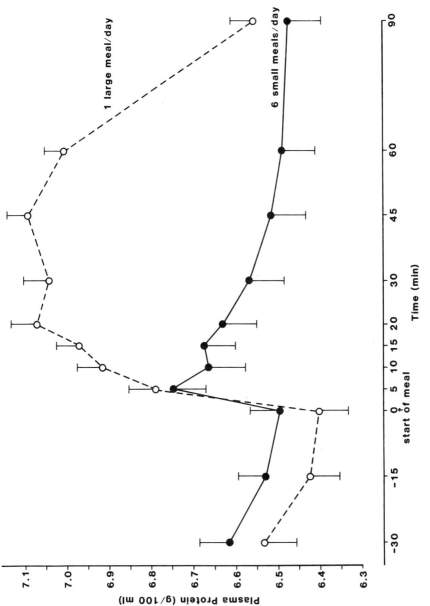

Fig. 4. The effect of meal frequency on plasma protein in ponies.

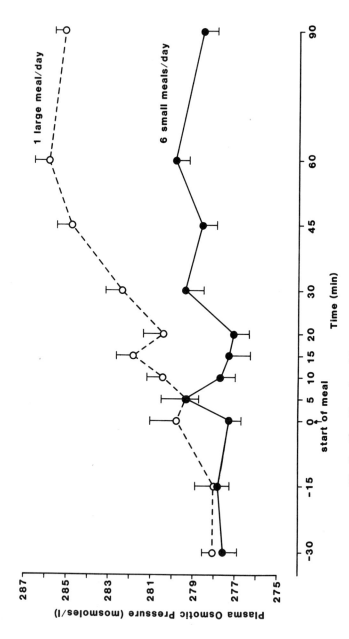

Fig. 5. The effect of meal frequency on plasma osmolality of ponies.

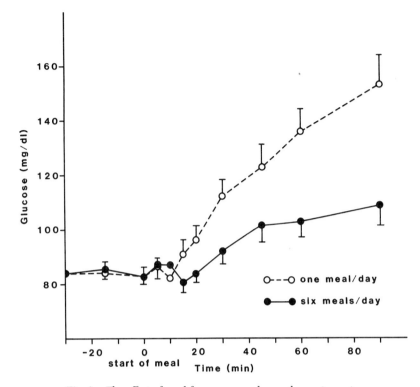

Fig. 6. The effect of meal frequency on plasma glucose in ponies.

regulated, especially in those tissues that lack substantial quantities of brown fat.

Heart rate rose from 42 to 54 beats/min when the ponies could see their meal; it did not rise further when they ate. The changes in heart rate are interesting for two reasons. First, the increase in heart rate occurred when food was visible to the horse, and did not rise further when feeding commenced. The obvious next step is to see the cardiac reaction to the sight of food and its subsequent removal without allowing the horse to feed. Either the anticipation of feed is as potent a sympathetic stimulant as the taste of food, or there is a ceiling effect; the heart rate may go no higher. This is unlikely since equine heart rates as high as 240 beats/min have been recorded (Detweiler, 1984). Second, there was no change with the length of the fast preceding the presentation of food; the heart rate increased as much with the sight of food after a 4-hr fast as after a 23-hr fast.

Most of the ponies learned to obtain a few extra grams of feed by striking the automatic feeder or the wall of the stall that supported it. When the

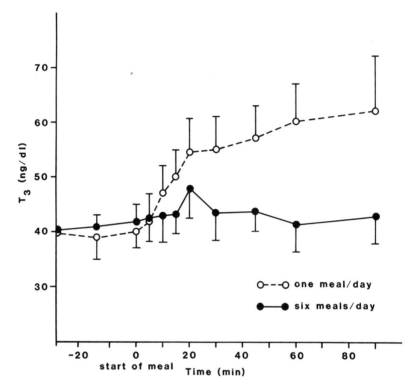

Fig. 7. The effect of meal frequency on plasma triiodothyronine (T_3) in ponies.

supporting structure was modified so that the animals could not reach it, the pony using the feeder at that time began to drink large quantities of water. This reaction is similar to schedule-induced polydipsia in rats (Falk, 1961).

Behavior during the prefeeding, during-feeding, and postfeeding periods was not markedly affected by meal frequency (see Table II). Comparisons during the prefeeding and postfeeding periods indicate that the ponies spent similar amounts of time in each behavior regardless of the feeding regimes. During these two periods, the ponies spent the majority of their time standing and eating bedding. Major differences in behavior patterns occurred during the time period when ponies had access to pellets. Ponies fed one meal per day spent 88% of their time eating, whereas ponies fed six meals spent only 37% of this time eating. These differences were significant ($p < 0.001$). After the ponies on the six-meal schedule consumed their meal, they spent the remaining time standing (37%) and eating bedding (16%). Compared to ponies eating one meal, the ponies eating six meals spent more time standing ($p < 0.005$), walking ($p < 0.01$), drinking ($p < 0.05$), and eating

TABLE II Behavior Patterns of Ponies[a]

Behavior	Before		During[b]		After	
	1×	6×	1×	6×	1×	6×
Walking	5 ± 1	6 ± 1	0.5 ± 0	4 ± 1[c]	4 ± 1	4 ± 1
Standing	39 ± 11	42 ± 3	4 ± 3	22 ± 3[c]	34 ± 5	39 ± 4
Drinking	1 ± 0	1 ± 0	1 ± 0	2 ± 1[c]	2 ± 0	1 ± 0
Eat bedding	13 ± 4	10 ± 3	2 ± 2	10 ± 4[c]	20 ± 4	9 ± 1
Coprophagia	2 ± 1	1 ± 1	0 ± 0	1 ± 1	0 ± 0	2 ± 1
Eating	—	—	53 ± 4	22 ± 3[c]	—	—

[a] Values are in minutes per hour of observation ± SEM.
[b] During indicates the hour when feed was available to the ponies fed 1× a day; a small amount of feed was also available to the ponies fed 6× a day.
[c] 6× significantly different from 1×.

bedding ($p < 0.05$). Eating bedding appears to be a grazing or food-searching behavior, in which the prehensile lips sift through the bedding. The amount of the sawdust bedding actually ingested is unknown and would not be easy to measure. Sawdust might be ingested as a substitute for roughage.

The correlates of feeding in horses have also been measured by Ralston *et al.* (1979) and by Doreau *et al.* (1981). Doreau and co-workers measured a variety of metabolites including glucose, insulin, alanine, and nonesterified fatty acids in mares fed once a day. They found increases in plasma glucose similar to those in our ponies when fed once a day. They also found marked increases in insulin and alanine and a decrease in nonesterified fatty acids. Ralston and associates measured glucose, free fatty acids, and insulin in ponies following a 4-hr fast, and found no significant changes in insulin and glucose or fatty acids taken together. Ralston *et al.* (1979) and Doreau *et al.* (1981) have found trends similar to ours: large changes with feeding after a long fast, but minimal changes after a short one. There was a significant negative correlation between the prefeeding glucose and the size of the subsequent meal (Ralston *et al.*, 1979). The correlation with glucose led to investigations of the effect of glucose on satiety. Intravenous glucose loads did not suppress subsequent intake, but intragastric loads did (Ralston and Baile, 1982a,b, 1983). This indicates that an elevated plasma glucose level per se is not a satiety signal. Intragastric or intestinal receptors are probably involved in satiety.

Cephalic influences are very important in equine satiety. Sham-feeding ponies show little increase in meal size over normally feeding ponies even following 18-hr fasts, although the intermeal interval was shorter in the sham-feeding ponies (Ralston, 1984). These results, which differ so markedly from those in sham-feeding dogs or rats, suggest further study of pregastric stimuli. Oral metering may be important in horses, in which case delivery of prechewed food from one fistulated pony to the stomach of another fistulated pony would not produce satiety. If chewing and tasting food is important, surgical denervation or local nerve blocks might increase intake of horses, in particular sham-feeding ones.

B. A Model Omnivore: The Pig

1. Eating and Drinking Patterns of Pigs

The feeding and drinking patterns of pigs were determined using an operant-conditioning technique. Each pig was trained to push a panel with its snout in order to obtain a food or water reward. The pigs did not need to be taught or "shaped" to make the operant response because rooting with the snout is a very common action pattern in pigs and is used to obtain worms, truffles, and other foods in the wild. The pigs learned to use the panels within 1 day. After initial training in which each response was rewarded, the ratio of responses required for a reward was raised until a fixed ratio (FR) of 10 was reached. That ratio was used for the experiment. Each reinforcement of feed was 5 g of a commercial pig diet (Squealer, Agway). Forty-five milliliters of water was the reinforcement for pushing the water panel. The pigs were housed individually in a 2 × 3-m pen on a concrete floor in a room with no other pigs in the room to avoid social facilitation (Hsia and Wood-Gush, 1982).

Log survivorship curves were used to determine the definition of an eating bout (Fagen and Young, 1978). The interbout interval showed a marked change in slope at 10 min. Pauses in eating of less than 10 min are within meal breaks, and those longer than 10 min are intermeal intervals. The meal patterns of the pigs changed as they grew from 10 to 130 kg. The number of feeding bouts per day decreased from 14 to 7, and the intake per day increased from 924 to 2524 g. The length of time spent eating per day remained relatively constant at 120 min, so that feeding rate increased from 70 to 362 g/bout or from 8 to 20 g/min. This indicates that the speed of prehension is an important factor in the increase of daily feed intake that occurs as the pig becomes older and larger and that neither meal length nor frequency is a factor. To the contrary, the lengthening of the intermeal interval suggests that, when the animal eats larger meals, the satiety factors resulting wane more slowly and delay the onset of the next meal.

A typical daily record is shown in Fig. 8. Pigs eat throughout the 24 hr, but there is a definite diurnal influence. More food is eaten in the light (1281 g) than in the dark (719 g, $p < 0.005$), and there are fewer bouts in the dark

12. Cephalic Phase of Digestion: Meal Frequency Effect 225

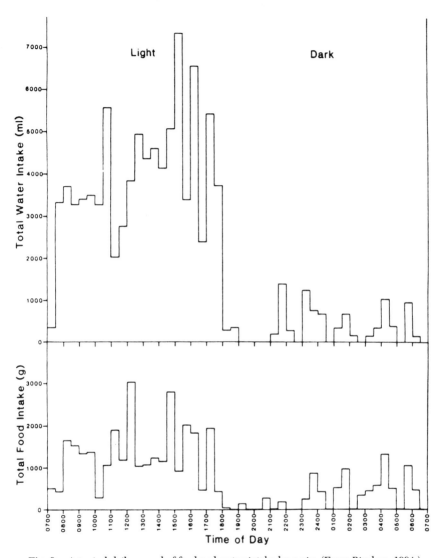

Fig. 8. A typical daily record of food and water intake by a pig. (From Bigelow, 1984.)

(3.3) than in the light (7.5), but bout size is larger in the dark (247 g) than in the light (182 g). Auffray and Marcilloux (1980), Haugse et al. (1965), Friend (1973), Montgomery et al. (1978), and Wangsness et al. (1980) have also studied meal patterns in pigs.

Prandial drinking has been noted in pigs by Yang et al. (1981) and was observed in detail in this study. The pigs drank approximately 2 ml of water per gram of feed ingested. Seventy-seven percent of water intake occurred prandially, that is, within 10 min of a meal. The majority of the drinking occurred within or just before the meal. This study formed the thesis of Bigelow (1984).

2. The Mechanisms Underlying Periprandial Drinking

The striking prandial pattern of drinking in pigs led us to speculate as to the stimulus provoking pre- and intrameal drinking. The most plausible explanation lay in a cephalic phase, in which the anticipation, sight, or smell of food triggered thirst as well as gastric acid secretion. Kraly (1983) has demonstrated in rats the role of histamine in prandial drinking. To determine whether histamine stimulated thirst in pigs, the pigs were operantly conditioned, as described above, to obtain both food and water by pushing panels with their snouts. The role of H_2 blockers, histamine, cholinergic drugs, cholinergic blockers, and a specific gastric secretion stimulant in prandial drinking was investigated.

The pig was deprived of food but not water for an hour before the experiment began. During control experiments, 0.9% NaCl was infused via a surgically implanted jugular catheter for 60 min. Water intake was recorded for 1 hr and during the infusion. The results are shown in Fig. 9. Histamine (1 µg/kg-min) stimulated a significant increase in water intake. An H_2 blocker (cimetidine, 300 mg) was infused before histamine to determine whether H_1 or H_2 receptors were involved. The attenuation of histamine-induced drinking by cimetidine indicated that H_2 receptors are involved.

The primary and best known physiological role of H_2 receptor agonists is to stimulate gastric secretion. Prandial drinking may be a response to the production of gastric acid. If so, it should be stimulated by those things that stimulate release of gastric acid, i.e., parasympathomimetics and gastrin. Gastrin (pentagastrin, 0.05 µg/kg-min) does stimulate thirst in pigs, and the increased intake is attenuated in the presence of the H_2 blocker, cimetidine. The parasympathomimetic drug bethanechol (1 µg/kg-min) also stimulated thirst, but when the bethanechol infusion was preceded by the parasympathetic blocking agent atropine (2 mg/animal), intake remained at control levels. The dose of the drugs used are those that stimulate or inhibit gastric secretion in pigs (Merritt and Brooks, 1970; T. R. Houpt, unpublished results).

Prandial drinking appears to be a response to gastric acid secretion. The pathway from the stomach to the brain has not been identified although vagal afferents are most probable.

3. Humoral Responses to Feeding in Pigs

We have previously reported on taste responses and taste aversion learning in pigs (K. A. Houpt and Houpt, 1977; K. A. Houpt et al., 1979). In this

12. Cephalic Phase of Digestion: Meal Frequency Effect 227

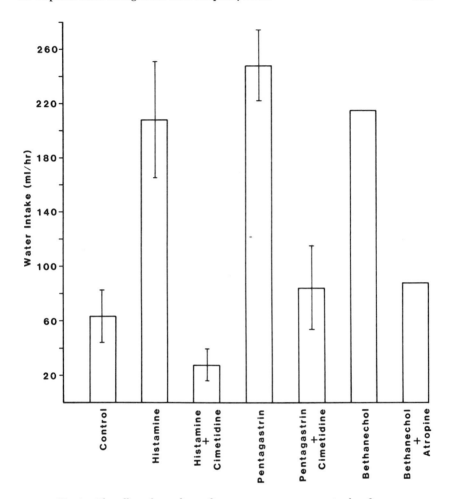

Fig. 9. The effect of stimulants of gastric secretion on water intake of pigs.

discussion, we shall concentrate on events intervening between deglutition and passage of ingested food into the small intestine.

Young pigs (10–40 kg) were surgically implanted with Silastic jugular catheters. Pelleted feed was available either as a 10-g reinforcement for pressing a panel (FR 5) or freely available in a feeder. The pigs were tested when eating a spontaneous meal or after a fast of 5 or 17 hr. When the pigs had been fasted, blood samples were taken 30 min, 15 min, and immediately before feed was made available, and then at 2.5, 5, 7.5, 10, 20, 30, 45, 60, 90, and 120 min after feeding began. In the *ad libitum* or 1-hr fast condition, a blood sample was taken when observations of the pig began and again just as it began to eat. The latter sample was considered the zero time sample.

All sampling was done in the early afternoon to avoid confounding the changes occurring with feeding with those occurring as a result of circadian rhythm.

Glucose was analyzed by the GOD-PERID method on an autoanalyzer (Boehringer Mannheim Gmbh Diagnostica). Insulin was determined by radioimmunoassay (Insulin RIA, The Radio Centre, Amersham, UK). Glucocorticoids were determined by a protein binding assay run automatically through a single G-25 column. T_3 was measured by radioimmunoassay (Reimers et al., 1982) and free fatty acid by chloroform extraction.

A pressure transducer (Gaeltech, Skye, Scotland) was implanted in the carotid artery of three pigs. The end of the catheter exited from the skin at the dorsum of the neck and was covered with an adhesive tape bandage when not in use. Blood pressure was recorded on a polygraph (Beckman Instruments, Schiller Park, IL). The pigs were fasted for 17 hr before each experiment. Water was always available. After 5 or more min of recording had been made, food was presented in a bowl, or the pig pushed a panel for 10 g food rewards. In some of the experiments, milk was presented to the pig in a bowl after the animal had ceased to eat solid food.

Three pigs were surgically prepared with jugular catheters as described above. The pigs were fasted for 17 hr. Blood samples were taken 30 and 15 min and just before (0 min) the pig was allowed to obtain two 10-g food rewards for pushing a panel with his snout. Blood samples were taken 2.5, 5, 7.5, 10, 20, and 30 min after the food had been eaten.

The results are shown in Figs. 10–14. Glucose and insulin rose with feeding, but only if the feeding was preceded by a fast. Even a fast as short as 5 hr was followed by a significant increase in insulin and glucose when food was made available. Corticosteroid levels fell with feeding, but the changes were significant only for the 17-hr fasted pigs. A long fast must be considered a stress for the pig. T_3 rose with feeding, but the rise was significant only following a 17-hr fast. Only free fatty acids change significantly with meals in the free feeding pig.

The increase in plasma glucose might lead one to propose it as a satiety agent. Glucose is a logical choice for a satiety factor since it rises quickly in the blood as carbohydrates are digested in the upper gastrointestinal tract. Unfortunately, infusion of glucose into either the jugular or the portal vein does not affect intake although intraduodenal infusion decreases intake (T. R. Houpt et al., 1979). These results are similar to those of Ralston and Baile (1982a,b) in horses and furnish further proof that gastrointestinal gluco- and osmoreceptors are involved in satiety.

Insulin has also been proposed as a satiety hormone and in fact does suppress intake when given in small physiological doses to pigs (Anika et al., 1980), rats (Vanderweele et al., 1980), and primates (Woods et al., 1979,

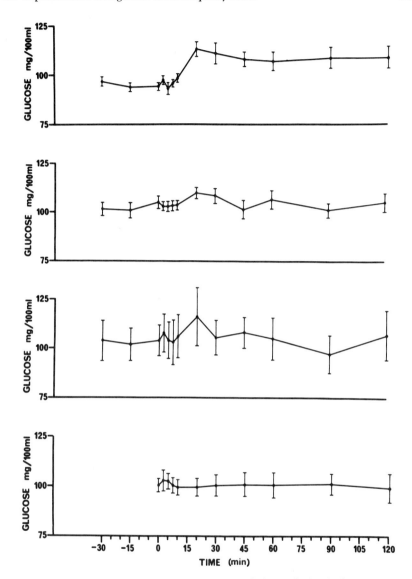

Fig. 10. Changes in plasma glucose concentration before and after feeding. From top to bottom, changes are shown associated with a 17-hr fast and operant feeding, a 5-hr fast and operant feeding, a 5-hr fast and feeding from a bowl, and free feeding. Zero time indicates start of meal. Mean values ± SE. [From K. A. Houpt et al. (1983a). Reprinted with permission of the American Physiological Society.]

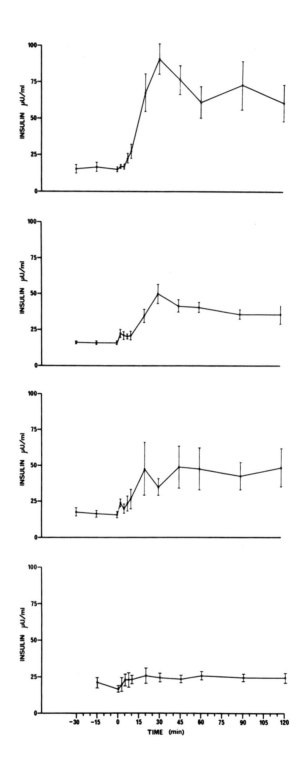

12. Cephalic Phase of Digestion: Meal Frequency Effect

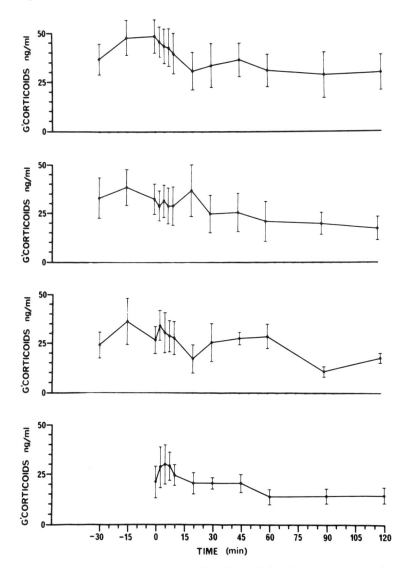

Fig. 12. Changes in plasma glucocorticoids before and after feeding. Experimental conditions are the same as in Fig. 10. [From K. A. Houpt et al. (1983a). Reprinted with permission of the American Physiological Society.]

Fig. 11. Changes in plasma insulin before and after feeding. Experimental conditions are same as in Fig. 10. [From K. A. Houpt et al. (1983a). Reprinted with permission of the American Physiological Society.]

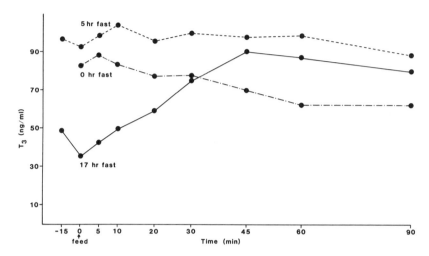

Fig. 13. The effect of feeding on triiodothyronine levels in the pig.

1980). Our results indicate, however, that insulin probably does not play a role in either the initiation or termination of meals in the freely feeding animal. Insulin has also been shown to stimulate intake. In most cases, the levels of exogenous insulin necessary to stimulate intake and the resulting precipitous fall in plasma glucose are large so that the effect is regarded as an emergency condition of glucoprivation. Steffens et al. (1980) and Le Magnen (1980) have detected a rise in insulin just before spontaneous meals. A slight fall was noted in the pigs. It is possible that there is a cephalic phase to insulin secretion. We tested this in three pigs that had been fasted for 17 hr. They were allowed to eat only 20 g of feed. Insulin levels increased from 12 to 21 μU 2.5 min after feeding.

So far, we have not been able to classically condition the pig's cephalic response to feeding. When a buzzer was paired with presentation of food for 10 times, the buzzer alone failed to elicit the increase in insulin seen when even a small amount of food was ingested. Either a greater number of pairings or more close pairing of feed and the conditioned stimulus must be done before we can conclude that classical conditioning of the insulin response does not occur.

Free acids decrease with feeding. The change in free fatty acids is the only change that occurs in freely feeding as well as in fasting pigs. Increases in the levels of free fatty acids in the blood depend on hormone-sensitive lipases to hydrolyze triglycerides. The activity of the lipases increases with decreased availability of carbohydrate. The increase in glucose with feeding provides the carbohydrate—specifically, the α-glycerophosphate—for lipogenesis, so

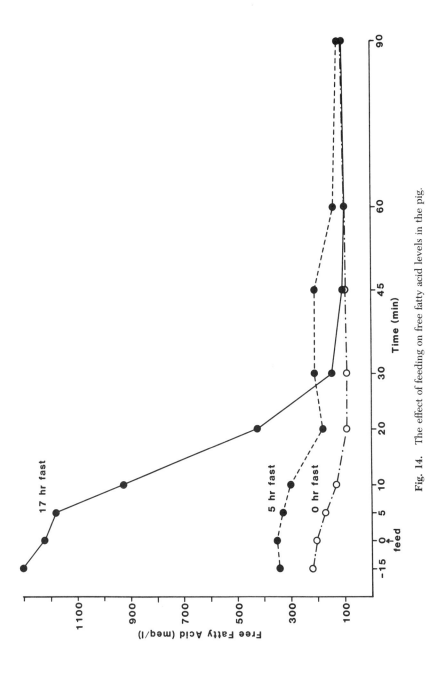

Fig. 14. The effect of feeding on free fatty acid levels in the pig.

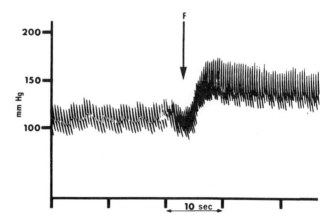

Fig. 15. Rise in carotid arterial blood pressure at initiation of a meal after a 17-hr fast. Feed presented at arrow. [From K. A. Houpt *et al.* (1983a). Reprinted with permission of the American Physiological Society.]

that free fatty acids are no longer released from depot fat, but rather incorporated into triglycerides. Free fatty acids could well serve as a hunger signal, and their disappearance as a satiety signal, but the receptor necessary for transmitting information about body energy status has yet to be identified. Free fatty acids have been implicated in control of feeding, but different conclusions were reached, i.e., that fatty acids suppressed feeding in obese rats (Carpenter and Grossman, 1983).

The changes in T_3 with feeding were similar to those we observed in the ponies. T_3 has been shown to increase with low environmental temperature, with high carbohydrate diet, and with meals (Dauncey and Ingram, 1979; Ingram and Evans, 1980). The meal must be preceded by a long fast for a significant rise in T_3 to occur.

The increase in blood pressure seen in Fig. 15 is a dramatic example of the cardiovascular effect of feeding. Presumably the response is mediated by the sympathetic nervous system. Exposure to a palatable food can cause a rise in blood pressure, even in a pig satiated on a less palatable food.

4. Beyond the Cephalic Phase: Gastric Pressure during Feeding

The role of the stomach itself in feeding was historically considered a major one; Cannon hypothesized it to be the site of hunger sensations. The discovery of central neural influences on feeding behavior led researchers to neglect the gastrointestinal tract until the past decade. Deutsch and Wang

(1977), Wirth and McHugh (1983), and Hunt (1980) have contributed considerably to our knowledge of the influence of upper gastrointestinal factors in hunger. We wished to extend our studies of gastric factors for two reasons: (1) for comparative purposes, to determine whether pigs were responding to the same gastric stimuli, and (2) to directly measure gastric pressure, which has not been thoroughly investigated during meals, in the freely feeding animal.

The procedure was a simple one. An open-tipped silicone rubber catheter was surgically implanted in the stomach through the wall of the body of the stomach with tip extending nearly to the antrum. Pressure was recorded using a pressure transducer and polygraph. The pig was fasted for 4 hr before the experiment.

The experiments are still ongoing, but typical results are shown in Fig. 16. Pressure rose immediately at the beginning of the meal, rose further during the meal until the meal ended, typically with an intragastric pressure of 20–25 cm H_2O. Upon cessation of eating, there was a fall of pressure. These results indicate that intragastric pressures may also serve as satiety signals.

Our previous work on osmotic and glucoreceptive mechanisms involved in intestinal satiety in both suckling and older pigs have been presented elsewhere (K. A. Houpt 1982b; K. A. Houpt et al., 1977, 1983b; T. R. Houpt, 1983; T. R. Houpt et al., 1979, 1983a,b). They tend to confirm our hypothesis that meals occur on a rhythmic basis and that the intermeal interval is determined by waning of satiety signals such as cholecystokinin (CCK) and afferent impulses from intestinal osmo- and/or chemoreceptors.

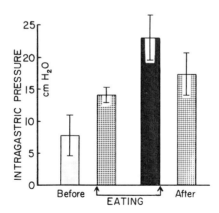

Fig. 16. The effect of meals on intragastric pressure. Mean pressures ± SE for 1- to 2-min periods just before and just after eating began and just before and just after eating stopped (n = 3 pigs; 30 measurements).

III. RESEARCH NEEDS

Specific areas of interest for further research in our laboratories involve periprandial drinking. To determine whether loss of fluid in the saliva is responsible for the increase in plasma protein observed in feeding horses we plan to inhibit salivation. This can be accomplished either permanently by creating salivary duct fistulas or temporarily by adding a local anesthetic to the feed (Alexander, 1966). In the process, we may also determine how important salivation is for deglutition. If the horses reject dry feed, a wet mash can be fed or saliva can be replaced by water introduced through a cheek fistula. In this experiment, plasma protein will be measured and the changes with feeding compared with those in horses that can salivate.

The role of rapid changes in thyroid hormones as part of thermostatic control of feeding should be followed by experiments in which changes in body temperature and metabolic rate during and after feeding are determined. It is now technically possible to measure oxygen consumption in a freely moving horse using a face mask. Unfortunately, that precludes measurements during feeding, but postprandial measurements of specific dynamic action and metabolic rate after meals of different sizes and different composition will strengthen the hypothesis that changes in heat production serve to attenuate increases in body changes in energy content.

The periprandial drinking observed in pigs needs further study. Although it seems clear that stimulants of gastric acid also stimulate drinking, the mechanism is not clear. Vagal afferents may signal the presence of acid in the stomach. The motor portion of the circuit would then stimulate drinking. If that is the case, vagotomy should abolish drinking in response to histamine, cholinergic agents, and gastrin. If, on the other hand, the same stimuli that result in the cephalic phase of gastric secretion, i.e., the anticipation, sight, smell, and taste of food, stimulate drinking behavior directly, then vagotomy would not abolish the response. The latter hypothesis seems unlikely because histamine and gastrin are involved, indicating a gastric origin of stimulation. The role of gastric secretion in feeding also deserves study. There may be a positive-feedback mechanism, whereby a little food stimulates gastric acid secretion, which in turn stimulates more intake. Repeating the experiments described above when food is available to the pig will indicate the importance of gastric secretion in maintaining intake.

Gastric pressure should be measured in various parts of the stomach during and after spontaneous meals and after those following a fast and correlated with feeding and the porcine satiety sequence. A logical next step would be to use a pyloric noose to prevent gastric emptying and carefully measure both feeding behavior and intragastric pressure.

The goal of our research to find a hunger factor or a satiety factor is not entirely realized. It appears that T_3 may serve as either a satiety signal or, like brown fat, serve as a means of reducing the impact of a large meal on body energy stores. The signal from the gut or from the brain to the thyroid gland is unknown. Thyroid stimulating hormone is not involved in the short-term changes in T_3 that follow a meal, but there might be direct neural input. An exciting prospect is that of preventing T_3 release, thereby increasing efficiency. Although an increase in efficiency is not the goal of those prone to obesity, it is to the majority of the people who lack adequate food. Furthermore, increasing efficiency of meat, milk, and fiber production would be beneficial.

More stress should be placed on basic comparative studies. Comparative studies can prove or disprove the general applicability of mechanisms across species. In addition, some hints about the evolutionary advantage of one type of meal pattern or satiety mechanism can be inferred from the habitat of the animal that relies on it.

More emphasis should be placed on multifactoral controls. Study of a single control has led one laboratory to emphasize one control mechanism, whereas another laboratory will defend another. It seems only logical that both hunger and satiety are under multiple controls: hepatic, hormonal, gastrointestinal, mechano-, and chemoreceptor, etc. The additivity or even synergistic effects of two or more manipulations should be investigated.

REFERENCES

Alexander, F. (1966). A study of parotid salivation in the horse. *J. Physiol. (London)* **184**, 646–656.
Altmann, J. (1974). Observational study of behavior: Sampling methods. *Behaviour* **49**, 227–267.
Anika, S. M., Houpt, T. R., and Houpt, K. A. (1980). Insulin as a Satiety Hormone. *Physiol. Behav.* **25**, 21–23.
Anika, S. M., Houpt, T. R., and Houpt, K. A. (1981). Cholecystokinin and satiety in pigs. *Am. J. Physiol.* **240**, R310–R318.
Auffray, P., and Marcilloux, J.-C. (1980). Analyse de la sequence alimentaire du porc, du sevrage à l'état adulte. *Reprod. Nutr. Devel.* **30**, 1625–1632.
Berger, J. (1977). Organizational systems and dominance in feral horses in the Grand Canyon. *Behav. Ecol. Sociobiol.* **2**, 131–146.
Bigelow, J. A. (1984). Feeding and Drinking Patterns and the Effects of Short-term Food or Water Deprivation in Young Pigs. M.S. Thesis, Cornell University, Ithaca, New York.
Carpenter, R. G., and Grossman, S. P. (1983). Plasma fat metabolites and hunger. *Physiol. Behav.* **30**, 57–63.
Dauncey, M. J., and Ingram, D. L. (1979). Effect of dietary composition and cold exposure on nonshivering thermogenesis in young pigs and its alteration by the β-blocker propranolol. *Br. J. Nutr.* **41**, 361–370.
Detweiler, D. K. (1984). Regulation of the heart. *In* "Dukes' Physiology of Domestic Animals" (M. J. Swenson, ed.), pp. 150–162. Cornell University Press, Ithaca, New York.

Deutsch, J. A., and Wang, M.-L. (1977). The stomach as a site for rapid nutrient reinforcement sensors. *Science* **195**, 89–90.

Doreau, M., Martin-Rosset, W., and Barlet, J. P. (1981). Variations au cours de la journée des teneurs en certains constituants plasmatiques chez la jument pouliniere. *Reprod. Nutr. Devel.* **21**, 1.

Fagen, R. M., and Young, D. Y. (1978). Temporal patterns of behaviors: Durations, intervals, latencies, and sequences. In "Quantitative Ethology" (P. W. Colgan, ed.), pp. 79–114. Wiley, New York.

Falk, J. L. (1961). Production of polydipsia in normal rats by an intermittent food schedule. *Science* **133**, 195–196.

Friend, D. W. (1973). Self-selection of feeds and water by unbred gilts. *J. Animal Sci.* **37**, 1137–1141.

Gibbs, J., and Smith, G. P. (1977). Cholecystokinin and satiety in rats and rhesus monkeys. *Am. J. Clin. Nutr.* **30**, 758.

Haugse, C. N., Dinusson, W. E., Erickson, D. O., Johnson, J. N., and Buchanan, M. L. (1965). A day in the life of a pig. *N. Dakota Farm Res.* **23**, 18–23.

Houpt, K. A. (1982a). Feeding problems. *Equine Pract.* **4**, 17–20.

Houpt, K. A. (1982b). Gastrointestinal factors in hunger and satiety. *Neurosci. Biobehav. Rev.* **6**, 145–164.

Houpt, K. A., and Houpt, T. R. (1977). The neonatal pig: A biological model for the development of taste preferences and controls of ingestive behavior. In "Taste and Development: The Genesis of Sweet Preference" (J. M. Weiffenbach, ed.), pp. 86–98. U.S. Department of Health, Education, and Welfare, Bethesda, Maryland.

Houpt, K. A., Houpt, T. R., and Pond, W. G. (1977). Food intake controls in the suckling pig: Glucoprivation and gastrointestinal factors. *Am. J. Physiol.* **232**, E510–E514.

Houpt, K. A., Zahorik, D. M., Anika, S. M., and Houpt, T. R. (1979). Taste aversion learning in suckling and weanling pigs. *Vet. Sci. Commun.* **3**, 165–169.

Houpt, K. A., Baldwin, B. A., Houpt, T. R., and Hills, F. (1983a). Humoral and cardiovascular responses to feeding in pigs. *Am. J. Physiol.* **244**, R279–R284.

Houpt, K. A., Houpt, T. R., and Pond, W. G. (1983b). The effect of gastric loads of sugars and amino acids on milk intake of suckling pigs. *J. Animal Sci.* **57**, 413–417.

Houpt, K. A., Laut, J. E., Kirk, P., and Carter, C. (1983c). Ingestive behavior of ponies on diets varying in caloric density. J. Animal Sci. 57 (Suppl. 1), 138.

Houpt, T. R. (1983). The sites of action of cholecystokinin in decreasing meal size in pigs. *Physiol. Behav.* **31**, 693–698.

Houpt, T. R., Anika, S. M., and Wolff, N. C. (1978). Satiety effects of cholecystokinin and caerulein in rabbits. *Am. J. Physiol.* **235**, R23–R28.

Houpt, T. R., Anika, S. M., and Houpt, K. A. (1979). Preabsorptive intestinal satiety controls of food intake in pigs. *Am. J. Physiol.* **236**, R328–R337.

Houpt, T. R., Baldwin, B. A., and Houpt, K. A. (1983a). Effects of duodenal osmotic loads on spontaneous meals in pigs. *Physiol. Behav.* **30**, 787–795.

Houpt, T. R., Houpt, K. A., and Swan, A. A. (1983b). Duodenal osmoconcentration and food intake in pigs after ingestion of hypertonic nutrients. *Am. J. Physiol.* **245**, R181–R189.

Hsia, L. C., and Wood-Gush, D. G. M. (1982). The relationship between social facilitation and feeding behaviour in pigs. *Appl. Animal Ethol.* **8**, 410.

Hunt, J. N. (1980). A possible relation between the regulation of gastric emptying and food intake. *Am. J. Physiol.* **239**, G1–G4.

Ingram, D. L., and Evans, S. E. (1980). Dependence of thyroxine utilization rate on dietary composition. *Br. J. Nutr.* **43**, 525–531.

Kissileff, H. R., Pi-Sunyer, F. X., Thornton, J., and Smith, G. P. (1981). C-Terminal octapeptide of cholecystokinin decreases food intake in man. *Am. J. Clin. Nutr.* **34**, 154–160.

Kraly, F. S. (1983). Histamine plays a part in induction of drinking by food intake. *Nature (London)* **301** (5903), 65–66.
LeBlanc, J., Dussault, J., Lupien, D., and Richard, D. (1982). Effect of diet and exercise on norepinephrine-induced thermogenesis in male and female rats. *J. Appl. Physiol.* **52**, 556–561.
Le Magnen, J. (1980). The body energy regulation: The role of three brain responses to glucopenia. *Neurosci. Biobehav. Rev.* **4** (Suppl. 1), 65–72.
Merritt, A. M., and Brooks, F. P. (1970). Basal and histamine-induced gastric acid and pepsin secretion in the conscious miniature pig. *Gastroenterology* **58**, 801–814.
Miller, R., and Denniston, R. H., II. (1979). Interband dominance in feral horses. *Z. Tierpsychol.* **51**, 41–47.
Montgomery, G. W., Flux, D. S., and Carr, J. R. (1978). Feeding patterns in pigs: The effects of amino acid deficiency. *Physiol. Behav.* **20**, 693–698.
Ralston, S. L. (1984). Controls of feeding in horses. *J. Animal Sci.* **59**, 1354–1361.
Ralston, S. L., and Baile, C. A. (1982a). Plasma glucose and insulin concentrations and feeding behavior in ponies. *J. Animal Sci.* **54**, 1132.
Ralston, S. L., and Baile, C. A. (1982b). Gastrointestinal stimuli in the control of feed intake in ponies. *J. Animal Sci.* **55**, 243.
Ralston, S. L., and Baile, C. A. (1983). Effects of intragastric loads of xylose, sodium chloride, and corn oil on feeding behavior of ponies. *J. Animal Sci.* **56**, 302.
Ralston, S. L., Van den Broek, G., and Baile, C. A. (1979). Feed intake patterns and associated blood glucose, free fatty acid, and insulin changes in ponies. *J. Animal Sci.* **49**, 838–845.
Reimers, T. J., McCann, J. P., Cowan, R. G., and Concannon, P. W. (1982). Effects of storage, hemolysis, and freezing and thawing on concentrations of thyroxine, cortisol, and insulin in blood samples. *Proc. Soc. Exp. Biol. Med.* **170**, 509.
Steffens, A. B. (1980). The evidence that humoral factors are involved in the regulation of food intake and body weight. *Brain Res. Bull.* **5** (Suppl. 4), 13–16.
Sufit, E., Houpt, K., and Sweeting, M. (1984). Physiological stimuli of thirst and drinking patterns in ponies. *Equine Vet. J.* **17**, 12–16.
Sweeting, M. P., Houpt, C. E., and Houpt, K. A. (1985). Social facilitation of feeding and time budgets in stabled ponies. *J. Animal Sci.* **60**, 369–374.
Tveit, B., and Larsen, F. (1983). Suppression and stimulation of TSH and thyroid hormones in bulls during starvation and refeeding. *Acta Endocrinol.* **103**, 2.
VanderWeele, D. A., Pi-Sunyer, F. X., Novin, D., and Bush, M. J. (1980). Chronic insulin infusion suppresses food ingestion and body weight gain in rats. *Brain Res. Bull.* **5** (Suppl. 4), 7–11.
Wangsness, P. J., Gobble, J. L., and Sherritt, G. W. (1980). Feeding behavior of lean and obese pigs. *Physiol. Behav.* **24**, 407–410.
Wirth, J. B., and McHugh, P. R. (1983). Gastric distension and short-term satiety in the rhesus monkey. *Am. J. Physiol.* **245**, R174–R180.
Woods, S. C., Lotter, E. C., McKay, L. D., and Porte, D., Jr. (1979). Chronic intracerebroventricular infusion of insulin reduces food intake and body weight of baboons. *Nature (London)* **282** (5738), 503–505.
Woods, S. C., McKay, L. D., Stein, L. J., West, D. B., Lotter, E. C., and Porte, D., Jr. (1980). Neuroendocrine regulation of food intake and body weight. *Brain Res. Bull.* **5** (Suppl. 4), 1–5.
Yang, T. S., Howard, B., and Macfarlane, W. V. (1981). Effects of food on drinking behaviour of growing pigs. *Appl. Animal Ethol.* **7**, 259–270.

PART II

Discussion

Naim: Dr. Feldman, have you observed reduced response to repeated trials using modified sham feeding (MSF)? This has been reported by Schwartz using MSF to elicit the cephalic phase response of pancreatic polypeptide. I would just note that we had previously observed reduced response of exocrine pancreatic secretion due to direct stimulation of the oral cavity with taste solutions. But when we mixed the taste stimuli with a cellulose base and fed it to the animals so that they could chew and swallow it (spilling out, of course, through a gastric fistula), response was restored.

Feldman: In healthy human subjects with studies separated by at least 1 week, the gastric acid response to the steak meal under MSF is very reproducible. We have looked at the response from about 10 to 12 individuals.

Naim: Have you measured insulin levels during MSF?

Feldman: Yes we have. We have done two separate studies, one in Dr. Robert Unger's laboratory in Dallas, the other at UCLA. Both measured insulin in humans using an RIA. Results from both laboratories are the same. We have been unable to detect any cephalic phase of insulin release in these normal, healthy human volunteers. We have also infused glucose into the stomach with or without MSF and have been unable to show any potentiation of insulin release by MSF. We have also been unable to demonstrate a cephalic phase to glucagon release in these subjects. In contrast we have confirmed the results from Dr. T. Schwartz's laboratory that there is a cephalic phase to pancreatic polypeptide release in the human using MSF.

Spector: Dr. Feldman, have you systematically varied taste quality of the meals you feed in order to determine the role of hedonics in the cephalic phase of gastric acid release?

Feldman: In the past, we have not varied the quality of the meal primarily because we use this technique in clinical studies with ulcer patients and diabetics. Recently, as we have shown in this presentation, we have begun to separate the various stimuli to the cephalic phase release. It is just a matter of time before we will begin to look at palatability.

Grill: Dr. Ewart, just exactly where are the cells located from which you are recording? I was not sure if they were near the nucleus of the solitary tract or the dorsal nucleus. Second, is there any segregation of one type of cell that responds to a given type of stimulation?

Ewart: We establish the position of the cells using dye marking. Even though we have yet to do a detailed analysis, it would appear that there is no particular topographical organization to these cells with regard to type of stimulation. These conclusions apply to cells within the NTS as well as to those within the dorsal nucleus.

Friedman: I'd like to turn to a discussion of the postabsorptive effects of nutrients. For example, has the experiment been done where the intestine is stimulated without getting a postabsorptive effect to see if that inhibits gastric emptying? As Dr. McHugh has noted, there is a corollary to the very accurate monitoring of gastric acid release, and that is that there is very accurate control of caloric delivery postabsorptively (or at least, postgastrically). Is there some way to determine where the ultimate control is? To say that one does not see an effect of glucose administered systemically in peripheral circulation is not an appropriate experiment. The normal route of glucose movement is through the portal system, and we and others have noted differential effects on food intake with portal vein infusions of different sugars. These also affect gastric acid secretion.

Ewart: There is some evidence on this. In 1983, Hardcastle and Hardcastle reported recordings from mesenteric afferent nerves in the luman. They noted different responses depending on whether they used glucose or nonabsorbable hexoses.

McHugh: I think there are a number of people here who could speak to this issue. When we use mannose in our work we see an immediate effect on gastric emptying and then the effect dissipates. This is in contrast to glucose which, of course, stimulates a longer emptying response.

Friedman: Is the mannose result due to an osmotic effect?

McHugh: It looks like an osmotic effect.

Houpt: If you give a local anesthetic with food intake you can inhibit the functioning of the duodenal receptors, but absorption still takes place.

Fernstrom: Dr. Wingate, I assume from your talk that there is controversy in your field as to the identification of the neurotransmitters involved in the gut–brain relay. I recall you mentioning the possibility that serotonin may be involved, but for that, there is little evidence. Can one administer antagonists or transmitter blockers in your preparation and determine what neurotransmitters may be involved?

Wingate: Whether or not serotonin is involved in our preparation is very uncertain. One of the problems of giving a drug to an enteric nerve is delivery. The pharmacology of the system is such that a wide variety and type of drugs affect motility. I would not take this to mean however, that there are a correlatively wide variety of transmitters directly involved. I think one has to be very certain that the drugs you are using are getting to a site you can define very precisely.

Boyle: Dr. Feldman, in your studies in which you reported gastric emptying, you did not get large differences in emptying, yet you did report large difference is gastric acid secretion. I have a little trouble reconciling that observation with the reports that gastric acid is an inhibitor of gastric emptying.

Feldman: It is true that via receptors in the proximal duodenum, acid does inhibit gastric emptying. Let me clarify the techniques used in the studies to which you refer. In order to measure gastric acid secretion we used a gastric titration technique in which we infuse bicarbonate into the stomach in order to keep the pH constant (Fordtran and Walsh). This means that the amount of acid the duodenum is actually seeing is essentially zero. Even though the amount of acid secretion was increased, during the sham-feeding session, the pH was maintained by this technique so that the $[H^+]$ concentration was very low. If we hadn't controlled the pH, I suspect we would have seen a larger effect on emptying which might have been attributed to sham feeding, but probably would have been due to an indirect effect of acid. In fact there is literature on gastric emptying and sham feeding in which gastric acid is not controlled and it is not possible to determine what the primary cause of the change in emptying is.

Wingate: I'd like to add a small rider to what Dr. Feldman just said. While Dr. Feldman has shown that there is not a cephalic phase effect on emptying, one should not conclude from that that there is no CNS modulation of emptying. It has been very clearly shown by Montaturaie and associates that stress will sharply modulate emptying and acid secretion as well.

Fregly: Dr. Feldman, there are several points I would like clarified. First of all, I have a question on the technique of the Heidenhand pouch. I noticed that the secretion was small compared to the innervated pouch. Have the parietal cells atrophied as a result of the denervation and, therefore, could this be biasing your effects?

Discussion

Feldman: This question brings up a very important point. If you looked at the two scales in the acid secretion figure using the pouch, you noted that secretion was clearly different between the two. First, when we construct the pouch, it is much smaller than the stomach itself, something like 10% of the entire stomach. Thus, there are only $1/10$ of the total stomach parietal cells in the pouch. The maximal acid output from the pouch will be much lower than that from the rest of the stomach. Second, when a vagotomy is performed on a dog, or human, for that matter, the maximal acid output decreases by about 50%. The parietal cells are still there. They have not atrophied. But when they are vagally denervated, they don't work as well. The reason for this is still unknown. It was thought, a number of years ago, that this may be due to loss of cholinergic background activity, but we and others have been unable to replace the acid activity by giving cholinergic drugs.

Fregly: Second, when you gave atropine, there was an increase in gastrin levels in the blood, but yet you decreased acid secretion. You are postulating, according to your figure, two separate receptors, a gastrin receptor and a cholinergic receptor. If your postulate is correct, I don't understand why secretion went down unless gastrin is acting through a cholinergic receptor.

Feldman: That is a very perceptive question. Current thinking is that the receptors interact with each other on the parietal cells. If you infuse gastrin into the peripheral circulation or have a patient with a gastrin-producing tumor (Zolinger–Ellison syndrome), you can give an anticholinergic drug such as atropine and markedly reduce acid secretion. The conclusion is that there is interaction between gastrin and acetylcholine. Apparently, one needs to have all three receptors active in order to get a maximal response—that is, receptors for histamine, gastrin, and cholinergic stimuli. You can reduce the response to gastrin or histamine by giving an anticholinergic drug, or you can block gastrin response by giving an H_2 blocker such as cimetidine, which is how we treat Zollinger–Ellison syndrome.

Kissileff: I noticed on one of your figures, Dr. McHugh, that when you administered saline to the stomach and glucose to the intestine, the animal's intake was inhibited but then at 240 min it was back to normal levels. I was wondering if you knew whether the stomach was emptying faster after the glucose effect had worn off, permitting the animal to eat more and whether the termination of eating occurs when the stomach reaches a critical volume and then starts again when volume decreases? Also, what role might acid secretion play in this?

McHugh: As a matter of fact, Dr. Kissileff, we are in the middle of an experiment now that could answer your question. What I can tell you is that with anything that inhibits gastric emptying, such as a glucose load to the intestine or CCK, you can usually demonstrate that eating starts at slightly smaller gastric loads than if you didn't have these inhibitors present. Gastric distension is one satiety signal, but intestinal factors, hormones, and other metabolic processes also contribute to normal satiety.

Kissileff: Do you think the stomach could be integrating some of these other satiety signals?

McHugh: I think it can be integrating some of them, but there is always a little bit more that comes from other primary sites. The stomach mediates a powerful signal, but not everything is integrated through it.

Feldman: Of course, gastric secretion does affect gastric emptying in a complex way, and if in an experiment it is not controlled, or gastric secretion is not measured, one can make interpretations that are erroneous. As Dr. McHugh pointed out, the rate of emptying of a liquid is partly a function of how much liquid is in the stomach. It is hard to have a perfect preparation, so the best one can do is attempt to control as many variables as possible.

PART III

Consequences of Food Palatability to Nutrition

nondieting subjects (Stunkard, 1981). They were aged between 18 and 35, were not on medication, and knew little about the controls of food intake. In many of the studies, both males and females were studied, but since the differences between the sexes were small, the mean data will be presented. In studies of feeding, it is important to ensure that the subjects like and will consume the test foods. Therefore, usually before the first test day, subjects tasted and rated the test foods. If the potential subject either rated the food as unpleasant (i.e., below neutral) or said he/she would not consume the food, he/she was not tested further.

The experiments were usually run at lunch time after the subjects had eaten a standard breakfast, which helped to ensure that their hunger level was similar on different days. Hunger level was also assessed at the start of every experiment using a visual analog scale. Since the subjects did not eat or drink between breakfast and the test, they were moderately hungry and thirsty at the start. After hunger was assessed, the subjects were given a plate or tray with small portions of 8 or 9 test foods in small clear containers. They were instructed to taste and rate the pleasantness of the taste at that time on a visual analog scale, i.e., a 100-mm line. The foods were always tasted one at a time in the same order, and one of the sample foods was always offered subsequently as a test meal. After this initial rating, subjects were offered as much as they would like to eat of the test meal. The foods were presented in portions that were greater than the subjects were likely to finish, and in an attempt to reduce the probability that subjects would eat according to preconceived notions about portion size, solid foods were presented in small pieces and liquids and semisolids were presented in a large bowl. Two minutes after the subjects finished eating, they were again presented with the tray of foods they had tasted and rated at the start of the meal. We were thus able to analyze the changes in pleasantness of each of the foods from before eating to after eating. By looking at changes in rating rather than absolute ratings and by running subjects as their own controls in the different test conditions, potentially confounding variables such as differences in initial palatability or individual preferences for foods were controlled.

B. Effects of Eating a Food on Palatability

Using these methods, we were able to determine the effect of eating one food to satiety on the subjective pleasantness of that food and of other foods which had not been eaten (Rolls et al., 1981a). Twenty-four subjects rated the taste of eight foods (cheese on cracker, sausage, chicken, walnuts, bread, raisins, banana, cookies). After this initial rating, they were given a plate of either cheese on crackers or sausages (these foods were of similar energy density and palatability) and instructed to eat as much as they liked. Two minutes after the end of the meal they rerated the taste of the eight foods sampled before the meal. It was found that the liking for the taste of the foods eaten in the meal decreased significantly more than for the foods not eaten. Thus satiety appeared to be specific to the food that had been consumed, and it was of interest to know whether such changes in the hedonic response to foods would be related to the amount eaten subsequently. To test this, the subjects were given an unexpected second course just after they completed the second series of ratings. If the subjects were given the same food in the second course that they had consumed in the first course, their intake fell to about half that eaten in the first course, whereas if they were given the different food, intake was the same in the second course as it was in the first course. The changes in liking over the first course were found to be significantly correlated with the amount eaten in the second course.

Thus as a food is consumed, its taste is liked less, and this change in hedonic response is related to the amount of food that will be consumed during the rest of the meal. Such changes in palatability, which we have called "sensory-specific satiety," will serve to promote the selection of a varied diet.

C. Pleasantness Changes in a Varied, Four-Course Meal

In a recent experiment (Rolls et al., 1984), we have looked at the changes in the hedonic response to foods over a four-course meal which consisted of two savory courses (i.e., foods that were not sweet) followed by dessert and fruit. In this experiment, we determined whether in a meal with very varied foods the specific decreases in pleasantness of foods already eaten gave way to a more general satiety after several courses, so that all foods became unpleasant. We also determined whether food intake in successive courses in a meal was related to changes in pleasantness which had already occurred in the meal.

Forty-eight male and female subjects of normal weight were tested twice at lunch time. On one occasion they were given a varied meal consisting of four successive courses, which were sausages, bread and butter, chocolate whipped dessert, and bananas. On the other occasion they were given a plain meal which consisted of just one of the four foods offered repeatedly in the four courses. Subjects rated the pleasantness of the taste of eight foods at the start of each course. The way in which the hedonic response to the taste of food changed during this four-course meal is shown in Fig. 1. To avoid the effect of individual preferences on the results, data are shown only for the same subjects eating the same food in a particular course. For example, the pleasantness ratings for course three are just for the 12 subjects that had

13. Changing Hedonic Responses to Foods   251

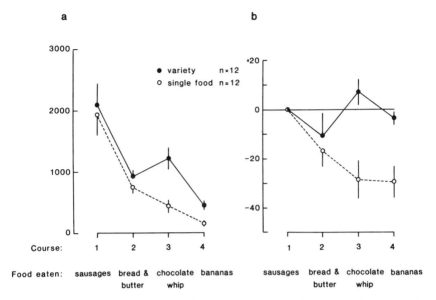

Fig. 1. Data for course 1 are from subjects given sausages in the plain meal; for course 2, from subjects given bread and butter; for course 3, from subjects given chocolate dessert; and for course 4, from subjects given bananas. (a) A paired comparison of the mean (±SEM) energy intake in each course by subjects when given plain and varied meals. (b) A paired comparison of the mean (±SEM) change in pleasantness of the taste of the foods from the start of the meal to the start of the course in which that food was eaten. Changes in pleasantness before a food was eaten were correlated with the intake of that food in the following course. [From Rolls et al. (1984). With permission from *Appetite* 5, 337–348. Copyright 1984, by Academic Press Inc. (London) Ltd.]

chocolate dessert in the plain meal. Thus changes in the plain meal, when subjects had already had two courses of chocolate dessert can be compared with the varied meal, when they had already eaten sausages and bread and butter. In the plain meal, the pleasantness of the taste of the foods showed a consistent decline with time, whereas there was little change in the pleasantness of the uneaten foods in the varied meal. It was found that these pleasantness changes correlated significantly with the amount of a particular food that was eaten in the subsequent course (see Fig. 1). The relative lack of change in the pleasantness of foods that had not yet been eaten in the varied meal may explain why variety in the diet stimulates food intake (see below).

The way in which eating a particular food can affect the pleasantness of its taste, and the taste of other noneaten foods was shown most clearly for the plain meals when the same food had been offered in the four courses. The changes in pleasantness from the start of the meal to the end of the fourth course are shown in Fig. 2 for each type of plain meal. It can be clearly seen

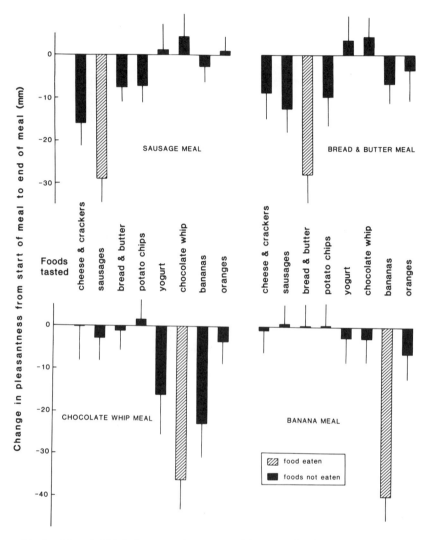

Fig. 2. Mean (±SEM) change in pleasantness of the taste of eight sample foods from the start to the end of the plain meal, showing data for the four different plain meals separately. The eaten foods (hatched bars) declined the most in pleasantness. [From Rolls et al. (1984). With permission from *Appetite* **5**, 337–348. Copyright 1984, by Academic Press Inc. (London) Ltd.]

that the pleasantness of the eaten food showed the largest decline. There were, however, some interactions between foods so that some of the uneaten foods decreased in pleasantness more than other uneaten foods. The basis of this interaction appeared to be that consumption of sweet foods caused some decline in the pleasantness of other sweet foods but had little effect on savory

13. Changing Hedonic Responses to Foods   253

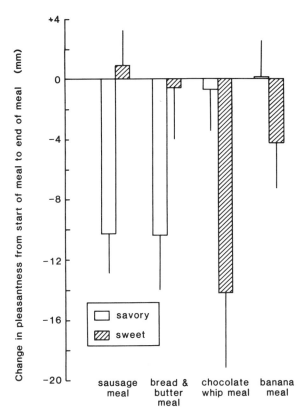

Fig. 3. Mean (±SEM) changes in pleasantness of the taste of the uneaten savory and sweet foods (see Fig. 2) from the start to the end of the plain meal, showing data for the four different plain meals separately. Foods similar to the eaten foods in savoriness or sweetness declined more in pleasantness than dissimilar foods. [From Rolls et al. (1984). With permission from *Appetite* **5**, 337–348. Copyright 1984, by Academic Press Inc. (London) Ltd.]

foods, whereas the consumption of savory foods decreased the pleasantness of other savory foods and not sweet foods (see Fig. 3). The savory foods were higher in fat content than the sweet foods, so it is not clear whether such interactions take place just on the basis of differences in flavor, or whether energy or even specific macronutrients may also be involved.

III. NUTRIENT-SPECIFIC SATIETY

There has been little work on the role of different macronutrients in the specificity of satiety or the changing hedonic response to foods. Ingestion of peanut oil was reported not to modify the perception of other alimentary odors and tastes (Rabe and Cabanac, 1974), whereas proteins did so, moderately (Guy-Grand and Sitt, 1976; Cabanac, 1979). Carbohydrates were reported to produce the biggest changes in food-related tastes and smells (Cabanac, 1979). We saw previously that after consumption of a particular food, the biggest decrease in pleasantness was for the food consumed, but that pleasantness for some other, uneaten foods also declined. We suggested that such interactions between foods could take place on the basis of similarities of tastes such as sweetness or saltiness. It is possible that such interactions could occur between foods with similar macronutrient composition. This type of interaction would be likely if the physiological usefulness of foods is the main factor affecting hedonic responses, as has been suggested by Cabanac (1971).

To determine the role of different macronutrients in the changing hedonic responses to foods, we offered subjects equicaloric amounts of foods high in one macronutrient and low in the other major nutrients. The experimental design was similar to that already described in that subjects tasted and rated the pleasantness of a series of foods before and after eating the test meal. The design differed in that this was the only study in which the subjects were required to eat a fixed amount rather than the amount they desired. The test meals consisted of a high-protein food, a high-fat food, a high-carbohydrate food, a high-sucrose food, and a food of mixed composition. In the tastings, each food high in one major nutrient was paired with a food that was tasted but not eaten which was similar in nutrient composition, but which had different sensory properties. Thus to test for interactions between foods on the basis of macronutrient composition, the change in pleasantness for an eaten food could be compared with that for an uneaten food with similar composition. The test foods were as shown in the tabulation below.

	Eaten food	Uneaten food
High protein	Chicken	Cottage cheese
High fat	Cream cheese (on celery)	Brazil nut
High carbohydrate	Pasta shells (in tomato sauce)	Pretzel
High sucrose	Turkish Delight (confectionery)	Raspberry yogurt
Mixed composition	Chocolate bar	

Ten normal-weight, unrestrained (Stunkard, 1981) female subjects were tested five times in a counterbalanced design. Having fasted overnight, subjects arrived in the laboratory at 11:00 AM. After tasting and rating the pleasantness of the nine foods listed above, they were given the test meal, which was 300 kcal (made up to 390 g with water) of the five eaten foods. After finishing the entire meal, they tasted the nine foods 2, 20, 40, 60, and 120 min later.

The changes in the pleasantness at 2 and 120 min after the meal are shown in Fig. 4. Clearly the largest change in pleasantness was for the food which had been consumed. To examine the possibility of interactions between foods on the basis of nutrient composition, the uneaten food with a similar composition to the eaten food is shown separately from the other uneaten foods. There are no significant differences in the changes in pleasantness of the taste of the uneaten foods with the same composition or a different composition to that of the eaten food. Thus this data does not support the hypothesis that hedonic changes are due to the nutrient composition of the foods. Furthermore, there are no significant differences in the magnitude of the changes following the different nutrients, and therefore these results do not support the previous suggestion (Cabanac, 1979) that some nutrients are more effective than others in modifying the hedonic response to foods.

IV. THE ROLE OF SENSORY PROPERTIES OF FOODS IN SATIETY

Shortly after Cabanac (1971) suggested that the pleasantness or palatability of food depends upon its physiological usefulness, his suggestion was challenged. Wooley *et al.* (1972) found that ingestion of a noncaloric sweet solution of cyclamate was just as effective in decreasing the pleasantness of 20% sucrose as was glucose. We (B. Rolls and E. Rolls, 1982 and unpublished) and Drewnowski *et al.* (1982) have also found that the pleasantness of solutions declines when they are tasted but not ingested. These findings indicate that the sensory properties of foods could be important for the changing hedonic response to foods during and after ingestion. We have conducted a series of experiments investigating the role of the sensory properties of foods in satiety.

A. Time Course of Changes in Pleasantness

If the hedonic changes that occur with eating are due to sensory stimulation by the foods, these changes should occur rapidly and then decline with time after eating. On the other hand if the changes are primarily due to postabsorptive effects of foods, they should increase in magnitude over the hour after eating. To distinguish between these possibilities, we have determined the time course of the changes in the pleasantness of a variety of sensory properties (taste, smell, appearance, and texture) of an eaten food and a selection of uneaten foods. Thirty-two women tasted and rated the pleasantness of the appearance, smell, texture, and taste of nine foods (cheese on cracker, orange jello, raspberry jello, tomato soup, consomme, orange drink, tomato segment, orange segment, and chocolate), then ate as

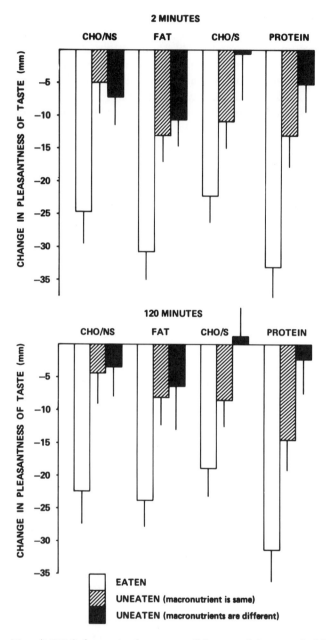

Fig. 4. Mean (±SEM) changes in pleasantness of the taste of the eaten food for all macronutrient groups, the uneaten food of the same macronutrient group, and the uneaten foods of different macronutrient groups at 2 and 120 min after the preload. There is no significant difference between the uneaten foods regardless of macronutrient content. Thus, no evidence of nutrient-specific satiety was found.

13. Changing Hedonic Responses to Foods

much cheese on cracker as they liked and rerated the pleasantness of the nine foods at 2, 20, 40, and 60 min after the meal. The results are shown in Fig. 5. For all of the sensory properties, there was a bigger decrease in the pleasantness of the eaten food than of the uneaten foods. The changes tended to be greatest at 2 min after the meal, with a gradual recovery in pleas-

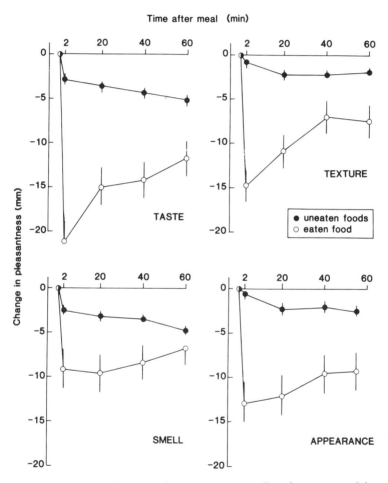

Fig. 5. Mean (±SEM) changes in the taste, texture, smell, and appearance of the eaten food (open symbols) and the uneaten foods (closed symbols) following a meal of cheese on crackers from 2, 20, 40, and 60 min after the meal. For all sensory variables except smell, for which the magnitude of response was smaller, the most significant decline in pleasantness occurred at 2 min after the meal for the eaten food. In addition, there was a significant increase in the pleasantness of the eaten food with time for the taste and texture variables, indicating a reversal in the decline in pleasantness observed at 2 min.

antness over the hour after eating. It is clear from this data that the largest changes in the hedonic response to foods occur before most of the meal will have been absorbed.

Although Cabanac (1971) suggested that the physiological usefulness of foods affected their palatability, he did not think it was the postabsorptive effects of foods which were critical because injections of glucose into the superior mesenteric artery did not affect the pleasantness of sweet solutions (Pruvost et al., 1973). Changes in the gastrointestinal tract were thought to be the more critical influence on palatability because intubation of mannitol, a nonabsorbed sugar, into the stomach reduced the pleasantness of sweet solutions. Receptors in the duodenum appear to be particularly important since glucose was more effective when tubed directly into the gut than when delivered to the stomach (Cabanac and Fantino, 1977).

At present, we do not know whether the presence of food in the stomach and duodenum contributed to the decrease in pleasantness seen after eating. To establish this, the time course of the passage of different foods through the gastrointestinal tract could be determined, and this could be related to the time course of the hedonic changes for different foods. It is unlikely, however, that changes in the stomach or gut can account for the specificity of satiety, which is more likely to be due to the sensory stimulation accompanying eating and/or the cognitive awareness that a particular food has just been eaten.

B. Influence of Low-Calorie Foods on Pleasantness Changes

In order to further clarify the position regarding the relative importance of sensory stimulation and caloric value, we conducted an experiment using foods which had similar sensory qualities (appearance, smell, texture, and taste) but which differed in energy density. We tested 24 normal-weight, unrestrained females (Stunkard, 1981). Twelve subjects were assigned to a soup condition and 12 to a jello condition depending upon which food they found more palatable. Half of the subjects in each group received the "high-calorie" version of the test meal on the first test day (tomato soup, 0.49 kcal/g; orange jello, 0.54 kcal/g), and half received the low-calorie version on the first day (soup, 0.07 kcal/g; jello, 0.09 kcal/g). Before the meal, subjects were asked to rate hunger, stomach fullness, and the pleasantness of the nine sample foods on visual analog scales and then to eat as much of the test meal as they liked. At 2, 20, 40, and 60 min after consumption, subjects were asked to rerate hunger, stomach fullness, and the pleasantness of the foods. The changes were studied over an hour to allow time for absorption to take place. After the 60-min rating, subjects were offered a second course of cheese on crackers to test whether there would be compensation for differences in intake in the first course. Intake data were recorded for both courses, and subjects served as their own controls.

Despite the large differences in energy density between the two versions of the test meals, there were no significant differences in the weight of food eaten in either the first or second course for either the soups or jellos. Hunger ratings and stomach fullness did not differ following the high- and low-energy foods (see Fig. 6 and 7). For both versions of the soup and jello test meals, the greatest decline in pleasantness occurred at 2 min after consumption for the eaten food, and there were no significant differences in the pattern of response over the subsequent hour. Thus, the differences in caloric value of the test meals had no differential effect upon the development, time course, or magnitude of sensory-specific satiety (Figs. 6 and 7). Subjects were unable to detect differences in energy density, as shown by their subjective responses to the foods. Further evidence for this, however, is provided by the observation that no compensation occurred in the second course following the low-calorie test meals.

In conclusion, the sensory properties of the test meals appeared to have a major role in the development of satiety. However, in our study, we did not compare the physiological effects of the high- and low-energy-density foods. Even foods with near zero energy, such as those used here, cause gastric and gut distension, gastric acid secretion, and release of hormones such as insulin. It seems unlikely that the time course of these changes would be the same for the high- and low-energy foods. Nevertheless, further work is required to clarify the role of gastric, duodenal, and postabsorptive changes following low-calorie foods in the specificity of satiety.

V. CHANGES IN THE PALATABILITY OF UNEATEN FOODS

Although the largest changes in palatability that occur after eating are for the food which has been consumed, there are some interactions between foods so that some uneaten foods show a decrease in palatability. Such interactions could occur because the foods have similar sensory properties, or because cognitively the foods are considered to be of the same type, or perhaps because the foods have the same macronutrient content. Our data help to distinguish between these possibilities.

In the experiment already described, in which we examined the effects of high- and low-energy soups and jellos on satiety, the uneaten foods were chosen so that the basis of the interactions between foods could be clarified. Some of the uneaten foods were selected to be of the same flavor as the test meal (i.e., for the tomato soup a piece of fresh tomato, and for the orange jello a piece of orange segment and an orange drink were tasted) or the same

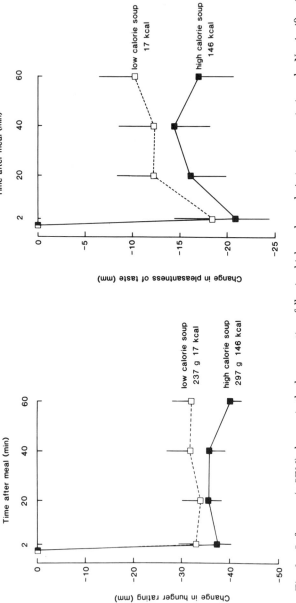

Fig. 6. Left: mean (±SEM) changes in the hunger ratings following high- or low-calorie tomato soup test meals. No significant difference between the two conditions was observed. Right: mean (±SEM) changes in the pleasantness of the taste of the high- and low-calorie tomato soup. No significant difference between the two conditions was observed.

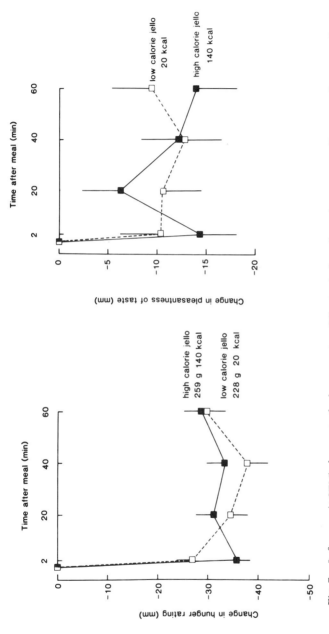

Fig. 7. Left: mean (±SEM) changes in the hunger ratings following high- or low-calorie orange jello test meals. No significant difference between the two conditions was observed. Right: mean (±SEM) changes in the pleasantness of the taste of the high- and low-calorie orange jello. No significant difference between the two conditions was observed.

general food type as the test meal (i.e., for the tomato soup, consomme and for the orange jello, raspberry jello were tasted).

After the meal of tomato soup (Fig. 8), the sweet/savory divide was not so obvious as noted previously. For example, the cheese on cracker remained virtually unchanged after the soup. This could perhaps be the large textural differences between the foods overriding an interaction on the basis of taste or flavor. Consumption of the tomato soup did, however, decrease the pleasantness of another soup, consomme, indicating that interactions between foods can occur if the foods are of the same type, even when they are different in flavor, smell, and appearance. Over time the orange drink also decreased in pleasantness, indicating that since all of the liquids declined, there could be interactions based on texture or mouth feel. The satiety did not appear to be flavor specific in that the pleasantness of the taste of fresh tomato was not significantly depressed by the consumption of tomato-flavored soup.

After the meal of orange jello (Fig. 9), there was some interaction between foods on the basis of flavor in that all of the orange-flavored foods declined in pleasantness after the orange jello, although the decline in the taste of the orange segment did not persist. The strongest interaction was on the basis of food type in that the raspberry jello showed a strong decrease in pleasantness following the orange jello despite the difference in color and flavor.

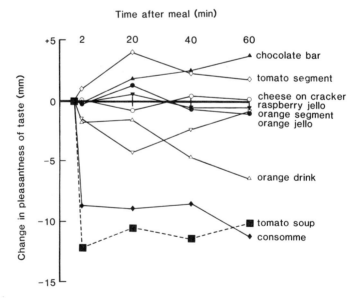

Fig. 8. Mean changes in the pleasantness of the taste of all nine foods following a meal of tomato soup.

13. Changing Hedonic Responses to Foods 263

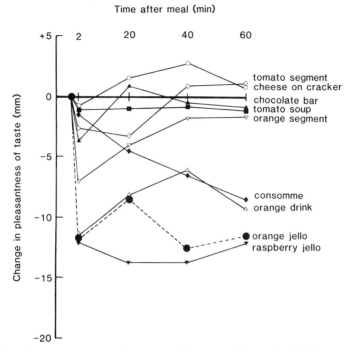

Fig. 9. Mean changes in the pleasantness of the taste of all nine foods following a meal of orange jello.

These studies present a basis from which an understanding can develop of how consumption of particular foods can affect the hedonic response to other foods. For example, we find that consumption of sweet foods can decrease the pleasantness of other sweet foods and the consumption of savory foods can affect other savory foods, but such interactions may not occur if the foods are very different in sensory properties other than taste, e.g., texture. There may be interactions on the basis of flavor although this type of interaction is not as strong as that between foods of the same basic type. Within a short time period (2 hr) after the meal, there was little indication that macronutrient composition is a major factor in food interactions.

VI. EFFECTS OF VARIATION IN THE SENSORY PROPERTIES OF FOODS ON FOOD INTAKE

If satiety is specific to particular foods or to particular properties of foods, then during a meal more should be consumed if a variety of foods are available than if just one food is presented. We have found in a series of experiments that variety in a meal can increase energy intake, and that although varying just one sensory property of the foods can enhance intake, the more different are the foods the greater the enhancement will be.

Variations in the flavor of food can enhance intake. We found (Fig. 10) that successive courses of three different flavors of cream cheese sandwiches (salt, curry, and combined lemon and saccharin) enhanced intake by 15% ($p < 0.05$) compared to the successive presentation of the favorite food (Rolls et al., 1982b,c). On the other hand, we found no enhancement of intake with three different flavors of pink yogurt (raspberry, strawberry, and cherry) or with three different flavors of chocolates which looked the same (orange, coffee, and mint Matchmakers, Rowntree Mackintosh Ltd.). The lack of effect could be because all the foods in the experiments were sweet and of the same type (Rolls et al., 1981b, 1982a).

Variations in the shape of food, which affect both the appearance and mouth feel, can also affect intake. Thus we found that successive courses of

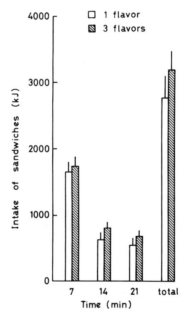

Fig. 10. Mean (±SEM) amount (kilojoules) of sandwiches eaten in three successive 7-min courses when subjects ($n=24$) were given either just the favorite flavor, or three flavors successively. The mean total intakes for the two conditions are also shown. [From Rolls et al. (1982c). Reprinted with permission from *Physiology and Behavior* **29**, Rolls, B. J., Rowe, E. A., and Rolls, E. T. How sensory properties of foods affect human feeding behavior. Copyright (1982), Pergamon Press.]

13. Changing Hedonic Responses to Foods 265

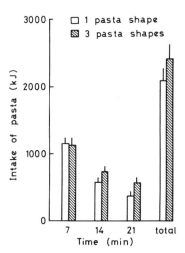

Fig. 11. Mean (±SEM) amount of pasta eaten in three successive 7-min courses when subjects (n=24) were given just their favorite shape, or three shapes successively. The mean total intakes for the two conditions are also shown. [From Rolls et al. (1982c.) Reprinted with permission from *Physiology and Behavior* 29, Rolls, B. J., Rowe, E. A., and Rolls, E. T. How sensory properties of foods affect human feeding behavior. Copyright (1982), Pergamon Press.]

three different shapes of pasta enhanced intake by 14% ($p < 0.025$) compared with the successive presentation of the favorite shape (Fig. 11). In the condition in which just one shape was presented, it was found that the pleasantness of the food eaten decreased more than that of the foods not eaten, and this could explain why the variety of shapes increased energy intake (Rolls et al., 1982b,c).

In the pasta study, it was not clear whether it was the variation in appearance or mouth feel which was the more important influence on intake or whether changes in just one of these properties would affect intake. We have looked at the effect of varying appearance by offering candy-coated chocolates (Smarties, Rowntree Mackintosh, Ltd.) of different colors both successively and simultaneously, and comparing intake to that of the favorite color presented alone. We found no effect of the variety of colors on intake, although we did find that when just one color was available, the pleasantness of that color decreased significantly more after eating ($p < 0.01$) than that of uneaten colors (Rolls et al., 1982b,c). We propose that this hedonic change would affect subsequent food selection so that it would be more likely that foods of a different color would be chosen, but this hypothesis has not been tested.

Thus when just the flavor or shape of foods was varied, the enhancement of intake was around 15% over three successive courses. When more properties of the foods differ, the enhancement is greater. When we offered four successive courses of sandwiches with very different fillings (cheese, tomato, egg, or ham), intake was a third more ($p < 0.002$) than when the same filling was offered throughout (Rolls et al., 1981b, 1982a).

In that study, although the sandwiches differed in appearance, texture, smell, flavor, and nutritional composition, they were still the same food type, i.e., sandwiches. We find an even greater enhancement of intake when four successive courses of very different foods are presented (Fig. 12). In a four-course meal of sausages, bread and butter, chocolate whipped dessert, and bananas, intake was 60% more than the mean of the intakes when just one of the foods was presented. These foods differed in basic type, nutrient composition, appearance, smell, texture, and taste (Rolls et al., 1984). Thus, it is clear that the greater the differences between foods, the greater the enhancement of intake by variety.

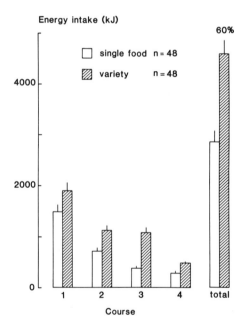

Fig. 12. Mean (±SEM) energy intake (kilojoules) of foods eaten by subjects in each course and the total energy intake by subjects given the plain meal, in which the four courses consisted of the same food, and the varied meal, in which each successive course consisted of a different food. [From Rolls et al. (1984). With permission from *Appetite* **5**, 337–348. Copyright 1984, by Academic Press Inc. (London) Ltd.]

13. Changing Hedonic Responses to Foods   267

VII. CONCLUSIONS

The sensory properties of foods have a major influence on the hedonic changes occurring during and after eating a meal. The reasons for this conclusion are as follows.

1. Foods with similar sensory properties but differing energy densities are equally effective in producing decreases in the pleasantness of the eaten food.
2. The decreases in the pleasantness of the eaten food occur very rapidly after eating (i.e., within 2 min) and decline in magnitude over the following hour.
3. Not only the taste, but also the appearance, smell, and texture of the eaten food decrease more than for uneaten foods.
4. Consumption of a food (i.e., sweet or savory) can decrease the pleasantness of other foods with similar sensory properties, but there do not appear to be interactions between foods on the basis of macronutrient content.
5. Changing the sensory properties of foods, i.e., flavor or shape, without affecting energy or nutrient content, can increase energy intake in a meal.

ACKNOWLEDGMENTS

This research was supported by the Medical Research Council of Great Britain.

REFERENCES

Aristotle. (1973). "De Sensa," (G. R. T. Ross, trans.) Arno Press, New York.
Cabanac, M. (1971). The physiological role of pleasure. *Science* 173, 1103–1107.
Cabanac, M. (1979). Sensory pleasure. *Q. Rev. Biol.* 54, 1–29.
Cabanac, M., and Duclaux, R. (1970). Specificity of internal signals in producing satiety for taste stimuli. *Nature (London)* 227, 966–967.
Cabanac, M., and Fantino, M. (1977). Origin of olfacto-gustatory alliesthesia: Intestinal sensitivity to carbohydrate concentration? *Physiol. Behav.* 18, 1039–1045.
Cole, T. J., and James, W. P. T. (1978). The slimdicator. A slide rule device for assessing obesity. *Practitioner* 220, 628–629.
Drewnowski, A., Grinker, J. A., and Hirsch, J. (1982). Obesity and flavor perception: Multidimensional scaling of soft drinks. *Appetite* 3, 361–368.
Duclaux, R., Feisthauer, J., and Cabanac, M. (1973). Effect of eating a meal on the pleasantness of food and nonfood odors in man. *Physiol. Behav.* 10, 1029–1033.
Guy-Grand, B., and Sitt, Y. (1976). Origine de l'alliesthésie gustative: Effets comparés de

charges orales glucosées ou protido-lipidiques. *C. R. Séances Acad. Sci. [D]* **282,** 755–757.

Metropolitan Life Insurance Company. (1959). New weight standards for men and women. *Stat. Bull. Metropol. Life Insur. Co.* **40,** 1–4.

Pruvost, M., Duquesnel, J., and Cabanac, M. (1973). Injection of glucose into the superior mesenteric artery in man: Absence of negative alliesthesia in response to sweet stimuli. *Physiol. Behav.* **11,** 355–358.

Rabe, E., and Cabanac, M. (1974). Origine de l'alliesthésie olfacto-gustative: Effets comparés d'une huile végétale et du glucose intragastrique. *C. R. Séances Acad. Sci. [D]* **278,** 765–768.

Rolls, B. J., Rolls, E. T., Rowe, E. A., and Sweeney, K. (1981a). Sensory specific satiety in man. *Physiol. Behav.* **27,** 137–142.

Rolls, B. J., Rowe, E. A., Rolls, E. T., Kingston, B., Megson, A., and Gunary, R. (1981b). Variety in a meal enhances food intake in man. *Physiol. Behav.* **26,** 215–221.

Rolls, B. J., Rolls, E. T., and Rowe, E. A. (1982a). The influence of variety on human food selection and intake. *In* "Psychobiology of Human Food Selection" (L. M. Barker, ed.), pp. 101–122. A.V.I. Publishing Co., Westport, Conn.

Rolls, B. J., Rowe, E. A., and Rolls, E. T. (1982b). How flavour and appearance affect human feeding. *Proc. Nutr. Soc.* **41,** 109–117

Rolls, B. J., Rowe, E. A., and Rolls, E. T. (1982c). How sensory properties of foods affect human feeding behavior. *Physiol. Behav.* **29,** 409–417.

Rolls, B. J., van Duijvenvoorde, P. M., and Rolls, E. T. (1984). Pleasantness changes and food intake in a varied four course meal. *Appetite* **5,** 337–348.

Rolls, E. T., and Rolls, B. J. (1982). Brain mechanisms involved in feeding. *In* "Psychobiology of Human Food Selection" (L. M. Barker, ed.), pp. 33–62. A.V.I. Publishing Co., Westport, Conn.

Stunkard, A. J. (1981). "Restrained eating": What it is and a new scale to measure it. *In* "The Body Weight Regulatory System: Normal and Disturbed Mechanisms" (L. A. Cioffi, W. P. T. James, and T. B. Van Itallie, eds.), pp. 243–251. Raven Press, New York.

Wooley, O. W., Wooley, S. C., and Dunham, R. B. (1972). Calories and sweet taste: Effects of sucrose preference in the obese and nonobese. *Physiol. Behav.* **9,** 765–768.

14

Role of Variety of Food Flavor in Fat Deposition Produced by a "Cafeteria" Feeding of Nutritionally Controlled Diets

MICHAEL NAIM,* JOSEPH G. BRAND,[†]
AND MORLEY R. KARE[†]
*Faculty of Agriculture
The Hebrew University of Jerusalem
Rehovot, Israel
[†]Monell Chemical Senses Center, and
University of Pennsylvania
Philadelphia, Pennsylvania

I.	Introduction	269
II.	"Cafeteria" Feeding as a Model for Dietary Obesity	271
III.	Preference Tests for Food Flavors and Texture	273
IV.	"Cafeteria" Feeding Experiments with Nutritionally Controlled Diets	281
V.	Conclusions	289
	References	290

I. INTRODUCTION

Experimental animals are often able to regulate the selection and consumption of nutrients according to their physiological needs. Under choice situations, animals may discriminate and identify each food item by cuing on the unique sensory character of each food. When a nutrient deficiency occurs, they may develop a specific hunger for this particular nutrient, using a combination of innate and learned processes that allow them to associate the sensory properties of the food with its nutritive quality (e.g., Harris *et al.*,

1933; Richter, 1936; Rodgers, 1967; Rogers and Leung, 1977). Food consumption can be stimulated in the absence of physiological need by altering the sensory quality or the composition of the food. Rats eat more food if the flavor character of the food is changed during a meal, compared with that eaten during a meal composed of a single-flavored food (LeMagnen, 1956; Treit *et al.*, 1983), and in humans the pleasantness rating of a specific flavor decreases during ingestion whereas the pleasantness rating for uneaten foods remains unchanged (Rolls *et al.*, 1981a). Thus, the tendency for selecting a variety of food items rather than selecting only a single item seems to be operative in rats and humans and may help to ensure a surfeit of essential nutrients.

Although the above mechanisms of consumatory behavior may ensure that intake of energy and nutrients does not fall below a certain level, they do not seem to protect the organism against overconsumption. It was proposed that caloric intake was regulated by postingestional effects produced after food was eaten, and that this regulation occurred even without oropharyngeal sensation (Janowitz and Grossman, 1949; Epstein and Teitelbaum, 1962; Jorden, 1969). Yet, this postingestional caloric regulation can be overcome. The overconsumption induced by a meal containing a variety of flavored diets was suggested as a possible contributing factor to the development of dietary obesity (LeMagnen, 1956; Treit *et al.*, 1983; Louis-Sylvestre *et al.*, 1984). Indeed, offering rats a variety of snack-type foods normally consumed by humans in a multichoice ("cafeteria") procedure produces hyperphagia and obesity in rats (Sclafani and Springer, 1976; Rothwell and Stock 1979; Rolls *et al.*, 1983). This paradigm is believed to be a reliable model for dietary obesity in humans, since humans also tend to overeat when offered a variety of palatable foods (Rolls *et al.*, 1981b; Porikos *et al.*, 1982). Given this result, one can therefore appreciate the survival value that the preference for selecting a variety of foods may have had for animals in deficiency or emergency situations. However, its benefit is questionable during circumstances in which nutrients are available in excess of survival needs. It is probable that an excess of nutrients was not normally available during the evolution of mammalian species, and therefore there was no selective advantage for the development of a mechanism against overconsumption.

The physiological mechanisms of dietary obesity are not well understood (Bray, 1982). Although the "cafeteria" feeding of rats is an appropriate model to study dietary obesity, the hypothesis that the variety of food flavors in the "cafeteria" model is the stimulating factor for the increased energy intake which leads to obesity is not proven. In addition to a variety of sensory properties, the cafeteria diet is rich in fat and sucrose. Evidently, the macronutrient composition of the diet itself contributes to obesity (Schemmel *et al.*, 1970; Herman *et al.*, 1970). Thus, the high palatability that characterizes the cafeteria diet could be due to sensory and/or postabsorptive signals (e.g., Mook, 1963; Cabanac, 1971; Naim et al., 1977). In the current popular model using the cafeteria design, both of these variables are present, and therefore the interpretation of the results is confounded. Indeed, the sensory signals and composition cannot easily be separated in the popular approach that uses snack food items or commercially available, but compositionally variable, diets (Rolls et al., 1983; Louis-Sylvestre et al., 1984).

In this chapter, we shall present the results of a study designed to quantify the manner in which the sensory properties of the food and the macronutrient dietary components induce hyperphagia and obesity in rats under "cafeteria" feeding.

II. "CAFETERIA" FEEDING AS A MODEL FOR DIETARY OBESITY

Dietary obesity can be induced in rats by feeding a high fat diet (Mickelsen et al., 1955; Schemmel et al., 1970), or one containing granular sucrose (Kratz and Levitsky, 1979), by offering sucrose in solution along with a stock diet (Kanarek and Hirsch, 1977), and by offering a variety of snack items high in fat and sucrose in a "cafeteria" paradigm (Sclafani and Springer, 1976). An increase in energy intake is believed to be the major cause of dietary obesity in the above models. Genetic factors may also be important since the degree to which animals respond to this manipulation is strain dependent (Schemmel et al., 1970). Individual variation within a single strain has also been reported (Drewnowski et al., 1984). In genetic models of obesity, hyperphagia may not be a necessary prerequisite for the abnormally high rate of fat deposition (Zucker, 1967; Cleary et al., 1980).

Investigators classify the cafeteria diet as "highly palatable" or "mixed and varied" (e.g., Sclafani and Springer, 1976; Rothwell and Stock, 1979). Because of this classification, the current practical conclusion among many nutritionists and public health professionals is that obesity induced by "cafeteria" feeding is primarily due to the variety in the sensory properties of the cafeteria foods. By definition, the snack foods offered are palatable since rats preferred these components over the stock diet (Sclafani and Springer, 1976). However, diet preference is determined not only by taste, smell, and texture, but also by postabsorptive feedback (Mook, 1963; Cabanac, 1971; Naim et al., 1977). Thus, the determinants of palatability include factors other than those that stimulate peripheral sensory receptors. Determining the contribution of each of these factors to the development of hyperphagia and obesity produced by the cafeteria feeding has only recently been attempted.

Using the classical model of cafeteria feeding, it is impossible to determine which factors in the foods are responsible for producing the hyperphagia. Moreover, because each animal displays individual preferences for individual foods and since each of the "cafeteria" foods contains a different macronutrient profile, composition of the ingested food cannot be controlled from one animal to the next. In order to quantify the contribution of the sensory stimuli to the overeating response during cafeteria feeding, the nutritional composition of the diet must be controlled. Furthermore, the consumatory behavior of rats should be determined in long-term experiments in which a variety of sensory stimuli are administered into the animal's diet, but where the presence of obesity-inducing factors, such as high fat and sucrose, is eliminated. Surprisingly, very limited attention has been given to this question. In one study (Stock and Rothwell, 1979), adding flavors such as vanilla and saccharin to the stock diet of rats was insufficient to induce any marked changes in food intake. Rolls et al. (1983) conducted an experiment in which rats were offered three lipogenic foods of high palatability either simultaneously or in succession, or as single components. All rats offered the palatable foods were hyperphagic compared to chow-fed controls. Rats exposed to foods presented simultaneously were most hyperphagic and showed greatest body weight and fat accretion. Those exposed to foods presented successively were less hyperphagic. The effect of variety on inducing hyperphagia was evident. However, although the various foods contained a similar energy content, they were not identical in their nutritional composition, and therefore the contribution of the sensory factors of these foods to the hyperphagia was not separated from other dietary constituents.

Very recently, Louis-Sylvestre et al. (1984) attempted to separate the sensory properties of food from other dietary components. They fed rats a multichoice ("isocafeteria") diet of commercial powder or pelleted chows of either rat, hamster, or rabbit, in which some of these chow diets were modified with the addition of aspartame, vaselin, corn oil, butter, and chocolate flavor. During a 10-day experiment, the rats exposed to this isocafeteria diet ate more and gained more weight than controls fed rat chow. The other two group of rats, which were fed either a high-fat diet or a conventional cafeteria diet, also ate and gained more weight than controls. These investigators concluded that variety and high palatability per se were sufficient factors to overcome regulatory mechanisms. However, based on the experimental design of this study, it is not certain whether the variety in sensory properties was the only factor of variety in the isocafeteria diet. For example, the nutritional composition of the isocafeteria foods varied from one food to another. Protein and carbohydrate levels varied and lipid content among the isocafeteria foods ranged from 3.8 to 15.7%. Further, the use of aspartame as

14. Flavor Variety in Fat Deposition 273

a taste stimulus for rats is questionable. Rats find aspartame tasteless or aversive (Nowlis and Frank, 1977; Naim et al., 1982). The aim of the following experiments was to investigate the effect of offering a variety of preferred flavors and textures in nutritionally controlled semipurified diets upon the induction of hyperphagia and fat deposition in rats during a multichoice feeding experiment. Three research stages were conducted: (1) Preference for food flavors derived from foods commonly used in the current model of "cafeteria" experiments was determined with rats using nutritionally balanced diets. In addition, various textural forms of the diet were also evaluated. (2) Once a catalog of preferred flavored diets was developed, the flavored, nutritionally balanced diets were offered to rats in a multichoice feeding design in order to maximize flavor and texture variety (Naim et al., 1985). (3) Diets high in fat and sucrose but also containing a variety of preferred flavors were offered to rats, also in a multichoice design (Naim et al., 1985). Control groups received diets without added flavors. Results of these experiments should permit one to evaluate the relative importance of flavor variety and diet composition on hyperphagia and the development of obesity.

III. PREFERENCE TESTS FOR FOOD FLAVORS AND TEXTURE

A catalog for preferred flavors and textures was developed for the Sprague–Dawley rats using nutritionally balanced diets (Naim et al., 1986). The composition of these diets was as follows (g/kg diet): casein, 200; DL-methionine, 3; corn oil, 50; cellulose, 20; cornstarch, 673; salt mixture (AIN-76), 50; vitamin mixture (equivalent to levels in AIN-76 but containing no added sucrose), 2; choline chloride, 2. Food flavors (IFF, New York) were added to the diets at the expense of cornstarch. To determine the preference–aversion curve for each flavor, two-choice preference tests were conducted for both brief (1 hr) and long-term (up to 5 days) periods (for experimental details, see Naim et al., 1977).

One hundred Sprague–Dawley male rats weighing 150 g were divided randomly into 10 groups of 10 rats each. Rats were housed in individual cages throughout all experiments described in this chapter. During an experiment of 5 days, each group was subjected to a preference test to determine intake of a diet containing a specific concentration of a single flavor versus intake of an unadulterated diet. Five concentrations of each flavor were tested at levels indicated in Figs. 1 and 2. Thus, the preference for two flavors could be determined in one experiment. When one experiment of 5 days was completed, an interval of 9 days was allowed before the next

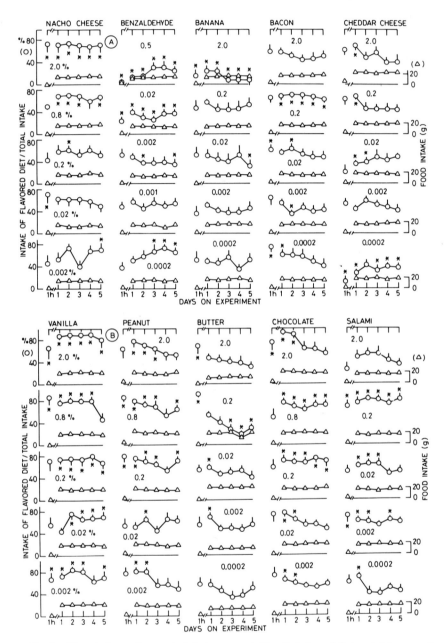

Fig. 1. Two-choice preference tests between a nutritionally balanced diet containing one of five concentrations of a single flavor versus an unadulterated, nutritionally balanced diet. Data are expressed as a percentage intake of the flavored diet divided by total intake of both diets (○)

14. Flavor Variety in Fat Deposition

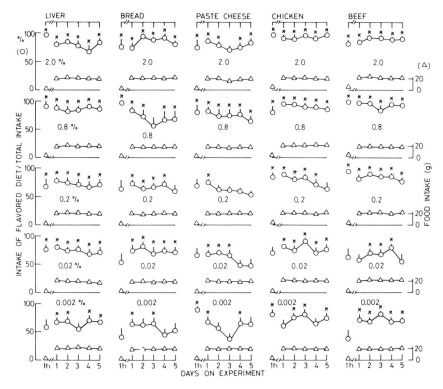

Fig. 2. Two-choice preference tests between a nutritionally balanced diet containing one of the five concentrations of a single flavor versus an unadulterated, nutritionally balanced diet. Data are expressed as a percentage preference for the flavored diet (○) and as total food intake (△) for both 1-hr tests and 5-day tests. Flavors tested were liver, bread, natural cheese paste, chicken, and beef. Asterisks indicate significant preferences or aversions at $p < 0.05$. See legend to Fig. 1. [From Naim et al. (1986). Reprinted with permission from Physiol. Behav. (in press). Preference of rats for food flavors and texture in nutritionally controlled semi-purified diets. Copyright 1986, Pergamon Press.]

---

and as total food intake (△) for both a 1-hr test (values at the extreme left of each test) and a 5-day test. Numbers within each test segment indicate the percentage concentration of each designated flavor in the diet. (A) Preference tests for various concentrations of nacho cheese, benzaldehyde, banana, bacon, and cheddar cheese flavors. (B) Preference tests for various concentrations of vanilla, peanut, butter, chocolate, and salami flavor. Values are the mean and SEM of 10 rats per each concentration of each flavor. Values above the 50% preference level with asterisks indicate a significant preference ($p < 0.05$) for the flavored diet over the unadulterated diets. Values below the 50% preference level with asterisks indicate that the intake of the flavored diet was significantly lower than that of the unadulterated diet. [From Naim et al. (1986). Reprinted with permission from Physiol. Behav. (in press). Preference of rats for food flavors and texture in nutritionally controlled semi-purified diets. Copyright 1986, Pergamon Press.]

preference experiment was begun using the same trained animals. During the 9-day interval, animals were fed the control, unadulterated diet. Rats were always naive to the flavor being tested at the beginning of each experiment and were assigned randomly into new groups at the end of each experiment. After four experiments, all rats were replaced with new animals (150 g each) of the same strain.

The results of the preference tests for the flavors are shown in Figs. 1 and 2. Some conclusions are clear from these experiments. First, one can see that minor amounts of these intensive food flavors could be easily detected by rats when incorporated in the semipurified diets. Second, the preference for some flavors such as liver, bread, cheese paste, and beef was well above the 50% level. This indicated that these flavors were highly preferred. Flavors such as banana, benzaldehyde, and butter, on the other hand, were not appealing to the rats, especially at high concentrations. Third, preference or rejection of flavored diets was concentration dependent. More importantly, whether a specific diet was either preferred or not, total food intake was not affected during the length of these two-choice preference tests. Thus, as we and others have previously reported (Kenney and Collier, 1976; Naim and Kare, 1977), the presence of a single appealing or single aversive taste stimulus does not affect total food intake. Interestingly, levels of 0.2% benzaldehyde or 0.4% citral in the diet (data for citral are not shown), which were used previously (LeMagnen, 1956) to induced a flavor variety and hyperphagia in rats, were found to be unpreferred levels in our experiments.

These results led us to select the following flavors at the concentration indicated (g/100 g diet) for further experiments on texture preference and feeding: peanut, 0.8; bread, 0.04; beef, 0.8; chocolate, 0.4; nacho cheese, 0.8; cheese paste, 0.8; chicken, 0.2; cheddar cheese, 0.4; bacon, 0.3; salami, 0.3; vanilla, 0.4; liver, 0.8.

One might also assume that texture variety is an important sensory factor in the cafeteria feeding model. During a short-term study (Rolls et al., 1981a), human subjects consumed more of three yogurts that differed in flavor, texture, and color when presented successively than if only one flavored yogurt was offered. However, when texture and appearance of the three yogurts were kept constant, successive changes in flavor did not increase consumption compared with the intake of one flavor. Offering school children a successive presentation of four different flavored chocolates also resulted in no increase in intake if appearance and texture were controlled (Rolls et al., 1982). Thus, altering flavor alone may not produce a maximal variety effect. Furthermore, various sensory factors in the food may interact in a complex manner. Rats, for example, usually prefer the texture of dietary pellets rather than the powder form (Corbit and Stellar, 1964). However, the

manipulation of flavor and texture can reverse this preference (Naim et al., 1977).
The objective of the experiment described in Fig. 3 was to determine for each flavored diet what type of pellet size the rats prefer. One hundred twenty naive rats were divided into 12 groups of 10 animals each. Each group was subjected to a three-choice preference test, first for 1 hr then for 24 hr. All three food cups contained the same nutritionally balanced diet, adulterated with the same preferred level of a specific flavor. The types of flavors used and their concentrations are indicated above. The difference among the three cups was in texture. One cup contained a powder form, the second contained small pellets (tablets) with a diameter of 0.4 cm, and the third food cup contained large pellets of 1 cm diameter. Pellets of both sizes were prepared by P. J. Noyse Co., New Hampshire, from diets prepared in our laboratory. To permit the compression of our diets into pellets, all diets were prepared with cornstarch type 78-1551 purchased from National Starch Co., New Jersey. The location of the three food cups was randomized within each cage. One can see (Fig. 3) that rats generally preferred the flavored diets in small and large pellet form over the powder form during the brief and the 24-hr preference tests. These results are in line with those of Corbit and Stellar (1964). In the 24-hr tests, all flavored diets were significantly preferred in the form of small pellets.

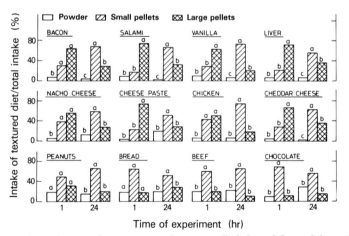

Fig. 3. Three-choice preference tests with nutritionally balanced flavored diets offered in three texture forms: powder, small, or large pellets. Tests were conducted for periods of 1 and 24 hr. Values are the mean of 10 rats per each flavored diet. Values not sharing the same letter for each test are significantly different at $p < 0.05$. [From Naim et al. (1986). Reprinted with permission from *Physiol. Behav.* (in press). Preference of rats for food flavors and texture in nutritionally controlled semi-purified diets. Copyright 1986, Pergamon Press.]

The taste of some of the conventional foods of the "cafeteria" diet such as cheese and salami is accompanied by the fatty or greasiness texture. Greasiness has indeed been reported to be appealing to rats (Hamilton, 1964). The experiment whose results are described in Fig. 4 was designed to determine the preference for diets containing fat or oil. One hundred Sprague–Dawley male rats (200 g) were divided randomly into 10 groups of 10. Rats were then subjected to two-choice preference tests between a diet containing a single concentration of either corn oil, lard, or hydrogenated vegetable oil (Crisco) and an unadulterated, nutritionally balanced diet. Tests were conduced for short (1 hr) and long (up to 5 or 6 days) time periods. Oil and fat were administered into nutritionally balanced diets at the expense of corn starch at levels specified in Fig. 4. The specified doses of oil, as indicated in Fig. 4, include the basic 5% level of corn oil incorporated in all diets. The preference for diets containing either oil or fat was found to

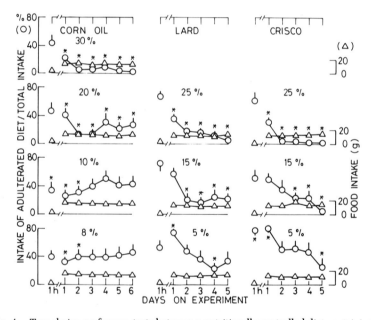

Fig. 4. Two-choice preference tests between a nutritionally controlled diet containing one concentration of either corn oil, lard, or saturated vegetable fat ("Crisco") versus a nutritionally balanced diet. Data are expressed as gram percentage intake of the oil- or fat-containing diet normalized to total intake (O) and as total food intake from both diets (Δ). Values are the mean and SEM of 10 rats per each test. Values above 50% preference level with asterisks indicate a significance preference ($p < 0.05$) for the fat containing diet over the unadulterated diet. Values below the 50% preference level indicate that oil- or fat-containing diets were less preferred than the unadultered diet.

be, in most cases, below the 50% level, which indicates that these diets were less preferred than the unadulterated ones. Only diets containing the low level of 5% of either lard or Crisco were preferred over the control diet. This preference for the 5% fat disappeared by day 2, and the animals preferred the control diet by day 4 or 5. This change in diet selection could be explained by assuming that either the taste of the 5% fat was not appealing to the animals in the long run or, alternatively, that a physiological feedback might have induced the change.

The lack of preference for diets containing fats at concentrations higher than 5% is an important observation since a major goal in our study was to mimic as much as possible the appealing sensory properties of the "cafeteria" foods with a minimum effect on macronutrient composition. Adding a noncaloric fat substitute at a level which might be "isotextural" with 5% fat would result in a minimal effect on caloric dilution. Sucrose polyester (SPE), a nonabsorbable food additive, has the desirable textural properties of fat (Fallat et al., 1976). Thus, we determined the preference for a diet containing 4% SPE (Procter and Gamble Co.). The SPE was administered into nutritionally balanced diets at the expense of cornstarch. Since the manufacturer recommended that SPE be added together with a 1% hydrogenated palm oil, all diets were prepared with 1% palm oil at the expense of cornstarch. Two-choice preference tests were then performed between a flavored diet containing 4% SPE and the same flavored diet with no SPE added (Table I). Flavors were added to the levels found to be preferred in the initial preference tests. The results, shown in Table I, indicate that SPE did not markedly affect diet preferences. In six preference tests, the presence of SPE led to a significant preference, suggesting that the level of 4% SPE could be detected by rats. There were no tests in which SPE was found to induce an aversion to the diet.

In summary, the above preference tests with nutritionally balanced diets provided a catalog of 12 preferred food flavors for the Sprague–Dawley rats. The preference level for four texture forms was also determined. One should not conclude, however, that the same preferred level of flavors and the same preference for texture will always be appealing to rats of other strains, nor should it be concluded that every individual of a given strain will demonstrate the same preference. The above preference tests allow flavor and textural variety to be introduced into semisynthetic diets at levels in which the animals can certainly detect them. This large catalog of preferred flavors and textures was required since, as hypothesized previously (LeMagnen, 1956; Rolls et al., 1981b; Treit et al., 1983; Rolls et al., 1983), it is the variety in sensory properties of the food that stimulates overeating rather than the palatability or the preference for each individual food item.

TABLE I Two-Choice Preference Tests between a Flavored Diet Containing 4% SPE versus a Flavored Diet with No SPE Added[a]

Time of experiment (hr)	Peanuts	Bread	Beef	Chocolate	Nacho cheese	Cheese paste	Chicken	Cheddar	Bacon	Salami	Vanilla	Liver
1	62	46	61	52	86[b]	66	76[b]	63	54	48	62	35
24	55	48	54	45	48	71[b]	56	58	80[c]	69[b]	67[b]	52

[a] Values are intake of SPE diet as percentage of total intake.
[b] Significant preference for SPE diet, $p < 0.05$.
[c] Significant preference for SPE diet, $p < 0.01$.

IV. "CAFETERIA" FEEDING EXPERIMENTS WITH NUTRITIONALLY CONTROLLED DIETS

The following experiments were performed to determine the effect that food flavor variety has upon the consumatory behavior of rats fed two types of nutritionally controlled diets (Naim et al., 1985). The first diet (NB) was formulated to contain nutrients at a recommended physiological level and has been described (see beginning of Section III). The second diet (HFHS) was a high-saturated fat (25%), high-sucrose (25%) diet, which contained the same micronutrients and protein concentration as the first, but with a fat and sucrose content similar to that found in the current "cafeteria" diet of conventional foods. Fat (saturated vegetable fat, Crisco) and sucrose were added to the diet at the expense of cornstarch. The saturated fat that was added was in addition to the 5% corn oil and 1% palm oil already in the NB formula.

The first question to be addressed is whether variety in sensory stimuli administered into nutritionally balanced diets can produce an increase in energy intake. In the studies of LeMagnen (1956) and Treit et al. (1983), a successive change in the flavor of a chow diet produced a hyperphagia. However, the design of these studies does not permit one to extrapolate that the observed hyperphagia would be operable in long-term studies of dietary obesity. Both studies were conducted for a short duration (a few days). Furthermore, in these studies rats were meal fed rather than fed ad libitum, their normal pattern of eating. Louis-Sylvestre et al. (1984) suggested recently that the variety in sensory properties alone was sufficient to induce hyperphagia and overeating in low-fat diets. However, as discussed before, in their work dietary components varied from one food item to the next. Moreover, a 10-day dietary treatment period is too short to allow one to extrapolate their results to the problem of the development of obesity by dietary means. The suggestion that a variety in sensory stimuli will produce hyperphagia and obesity in rats when the foods contain high energy (fat) and sucrose seems to be more justified (Rolls et al., 1983).

The following experiments used semipurified diets to investigate what effect flavor variety has on energy intake in rats, when macronutrient content is completely controlled. Sprague–Dawley male rats weighing 280–300 g were acclimated to a reversed 12-hr, light–dark (dark starts at 11:30 hr) cycle for a 10-day period during which they were fed ad libitum a nutritionally balanced, unflavored diet. Because 4% SPE was administered into some diets in the following experiment, a 1% hydrogenated palm oil was added to all diets, including the one used during acclimation. Animals were then divided randomly into 6 groups of 13–15 rats each. Two ad libitum feeding sessions were conducted daily during a 23-day experiment. The first feeding session started at 11:30 hr (first part of dark period) and the second session at 17:30 hr. Rats were fed the following diets: Group 1 was offered in a no-choice situation an unadulterated nutritionally balanced diet (designated as "NB"). Group 2 was offered a choice ("cafeteria") of 3 food cups during each feeding session. All 3 food cups contained the same nutritionally balanced diet, but each was adulterated with a different flavor. The type of flavors used and their levels were the same as those found to be preferred in intial preference tests. In addition to the 12 flavored diets, an unadulterated diet (control) was also used as one of the choices. The order of presentation of the 13 diets during each feeding session was randomized with the restriction that no identical flavored diets were offered in the same session. During each feeding session, a diet with one texture form of either powder, small pellets, large pellets, or 4% SPE was supplied. Texture was changed in a randomized order from one session to another. Group 2 was designated "NB-CAF." Rats of Group 3 were offered during each session, in a no-choice manner, a high-fat (25% saturated vegetable fat, 5% corn oil, 1% palm oil), high-sucrose (25%) diet (designated as "HFHS"). Group 4 was offered, during each feeding session, a choice of three flavored diets formulated with flavor levels identical to those in NB-CAF but using the HFHS diet mix. The sequence of flavor presentations was identical to that presented to the NB-CAF rats. No texture modifications were possible in this diet. Group 4 was designated "HFHS-CAF." Rats of Group 5 were offered, in a no-choice manner, a high-fat diet (25% saturated vegetable fat, 5% corn oil, 1% palm oil) (designated "HF") with no sucrose added. Rats of Group 6 were offered, in a no-choice manner, a high-sucrose diet (25%) with no fat added above the level in the NB diet and designated "HS."

Food intake was measured for each session (two sessions per 24 hr), and all groups received fresh food in the cage at the beginning of each session. Body weight of each animal was measured every 48 hr. At the end of the 23-day experiment, rats were sacrificed by decapitation without prior fasting. Retroperitoneal and epididymal fat pads were removed and weighed from all rats. Liver and brown adipose tissue (BAT) were removed and weighed, and blood serum was collected from the NB, NB-CAF, HFHS, and HFHS-CAF groups. Serum triiodothyronine (T_3) and thyroxine (T_4) were obtained by radioimmunoassay (RIA) using commercially available RIA kits (Amersham), and serum insulin (IRI) was determined by radioimmunoassay using the method of Herbert et al. (1965). Lipid content of fat pads was determined according to Folch et al. (1957).

The cumulative energy intake and body weight gain data are presented in Table II. The most interesting phenomenon in this experiment was the consumatory behavior of the NB-CAF group. Although these animals were subjected to a flavor and texture variety, their energy intake as well as body weight gain was almost identical to that of the NB group of rats fed the

TABLE II Cumulative Energy Intake and Body Weight Gain of Animals Fed Nutritionally Controlled Diets with and without Flavor and Textural Variety[a]

Dietary treatment group[b]	Days on experiment								
	Energy intake (kcal)					Body weight gain (g)			
	1	5	15	23		3	6	15	23
NB	100 ± 3[b] (100)	497 ± 17[c] (100)	1464 ± 41[b] (100)	2210 ± 56[b] (100)		15 ± 2[b] (100)	31 ± 3[b] (100)	68 ± 5[b] (100)	96 ± 6[c] (100)
NB-CAF	100 ± 3[b] (100)	505 ± 12[c] (102)	1436 ± 42[b] (98)	2196 ± 61[b] (99)		20 ± 1[ab] (133)	35 ± 3[b] (113)	74 ± 4[b] (109)	107 ± 5[bc] (111)
HFHS	116 ± 7[ab] (116)	544 ± 22[bc] (109)	1495 ± 43[b] (102)	2228 ± 54[b] (101)		19 ± 2[ab] (127)	36 ± 2[b] (116)	75 ± 4[b] (110)	101 ± 5[bc] (105)
HFHS-CAF	127 ± 4[a] (127)	624 ± 12[a] (126)	1725 ± 36[a] (118)	2554 ± 54[a] (116)		25 ± 1[a] (167)	47 ± 2[a] (152)	97 ± 4[a] (143)	128 ± 5[a] (133)
HF	110 ± 6[b] (110)	571 ± 23[b] (115)	1641 ± 56[a] (112)	2417 ± 79[ab] (109)		20 ± 2[ab] (133)	39 ± 3[b] (126)	93 ± 6[a] (137)	119 ± 7[ab] (124)
HS	107 ± 4[b] (107)	520 ± 15[bc] (105)	1484 ± 53[b] (101)	2236 ± 87[b] (101)		19 ± 1[ab] (127)	37 ± 2[b] (119)	75 ± 4[b] (110)	103 ± 7[bc] (107)

[a] Data from Naim et al. (1985). Values are the mean ± SEM of 13–15 rats per each treatment. Values not designated by the same superscript letter are different at $p < 0.05$ or less. Numbers in parentheses indicate the percentage ratio between the indicated value and that obtained from the NB group.

[b] NB, Group offered the nutritionally balanced diet; NB-CAF, group offered the flavored, nutritionally balanced diets in a "cafeteria" design; HFHS, group offered the high-fat, high-sucrose diet; HFHS-CAF, group offered the flavored, high-fat, high-sucrose diets in a "cafeteria" design; HF, group offered the high-fat diet; HS, group offered the high-sucrose diet.

unadulterated diet. Rats of the NB-CAF group indeed detected the variety of sensory stimuli during our "cafeteria" feeding sessions since their relative preference varied from one flavored diet to another (Table III). The beef-flavored diet was found to be preferred the most whereas the vanilla-flavored diet was preferred the least. Further, for the NB-CAF animals, a significant difference in diet selection among the three flavor choices was found during one-third of the feeding sessions. In Table IV, one can see that at least during the second session of the dark period, the intake of flavored diets by the NB-CAF rats was higher in sessions having a texture of either small or large pellets than that found in sessions where animals were offered the powder or SPE forms. Thus, the modification in texture from one session to another also contributed to the variety in sensory properties of the NB-CAF diet. In contrast with the recently reported results of Louis-Sylvestre et al. (1984), our data indicate that variety in flavor and texture, when offered in a nutritionally balanced semisynthetic diet, does not stimulate eating above that seen in a nonflavored powder form diet of identical composition (Table II).

Data from animals of the NB-CAF group demonstrate that individual preferences were expressed using the current multichoice feeding design. In many cases, individuals selected a diet of a particular flavor (e.g., chicken) over any other flavors present during all sessions, while other individuals displayed little or no preference for that flavor at any time. This was a consistent observation over all flavors and all individuals, and thus confirms the assumption that animals were able to discriminate a preferred flavor from among the three choices. The selection of a particular flavored food over all others on an individual basis is a problem in the conventional cafeteria model, since macro- and micronutrient content of each food differ widely. In the model presented here, however, that expression of individual preference is not a hindrance, since nutrient composition of all flavored diets within a particular group is identical.

In contrast to the animals of the NB-CAF group, rats of the HFHS-CAF group, which were offered a choice of flavored diets using the high-fat, high-sucrose diet base, consumed 26% more calories and gained 52% more weight than the NB rats during the first 5–6 days of the experiment. At this time, the HFHS-CAF group had a higher energy intake and weight gain level than all other groups. By the end of the twenty-third day of the experiment, the cumulative energy intake of the HFHS-CAF rats was 16% above baseline, and weight gain was increased above baseline by 33%. The increased weight gain of the HFHS-CAF rats compared with that of the NB group is similar to that reported previously by some investigators (e.g., Rothwell and Stock, 1979) for rats fed the "cafeteria" of conventional foods relative to those fed rat chow. Furthermore, the increase in serum T_3 level

TABLE III Relative Preference for Each Flavored Diet and Variety Effect observed during "Cafeteria" Experiments of Nutritionally Controlled Diets[a]

Dietary treatment group[b]	Flavors											Variety effect[c]
	Peanuts	Bread	Beef	Chocolate	Nacho cheese	Cheese paste	Chicken	Cheddar cheese	Bacon	Salami	Vanilla	Liver	Control	
NB-CAF	37[bcd]	31[bcde]	49[a]	28[cde]	31[bcde]	35[bcde]	33[bcde]	31[bcde]	29[cde]	24[de]	23[e]	41[ab]	40[abc]	33
HFHS-CAF (Exp. 1)	34[abc]	35[ab]	38[a]	31[abc]	35[ab]	35[ab]	31[abc]	36[ab]	27[c]	29[bc]	32[abc]	37[a]	31[abc]	22
HFHS-CAF (Exp. 2)	28[abc]	34[abc]	26[bc]	34[abc]	34[abc]	39[ab]	31[abc]	35[abc]	24[c]	27[bc]	36[abc]	42[a]	38[ab]	46

[a] Data from Naim et al. (1985). Data are expressed as the relative intake of a specific flavored diet as a percentage of total intake observed in each feeding session. Each value is the mean of 10–12 feeding sessions for each flavored diet. Values not designated by the same superscript letter are different at $p < 0.05$.

[b] NB-CAF, Group offered the flavored, nutritionally balanced diets in a "cafeteria" design; HFHS-CAF, group offered the flavored, high-fat, high-sucrose diets in a "cafeteria" design.

[c] The percentage of feeding sessions in which a significant difference in diet selection among the three flavor choices was found.

TABLE IV Food Intake (Grams) Levels in Rats for Each Textural Form during "Cafeteria" Feeding Sessions of Nutritionally Balanced Diet[a]

Feeding sessions during the dark period	Powder	SPE[b]	Large pellets	Small pellets
First	8.6 ± 0.3[a]	8.6 ± 0.3[a]	9.7 ± 0.1[a]	9.5 ± 0.5[a]
Second	9.5 ± 0.3[a]	9.8 ± 0.4[a]	12.4 ± 0.2[b]	12.5 ± 0.3[b]

[a] Data from Naim et al. (1985). Values are the mean ± SEM of 5–6 feeding sessions for each form of texture performed either in the first or second part of the dark period. Values not sharing the same superscript letter are different at $p < 0.05$ or less.

[b] Four percent sucrose polyester.

found in rats of the HFHS-CAF group (Table V) is also in line with the T_3 response reported for rats fed the conventional "cafeteria" foods (Rothwell and Stock, 1981; Tulp et al., 1982). Unexpectedly, energy intake and body weight gain of the HFHS group were not increased above the NB baseline. As described below, feeding of the HFHS diet was repeated in an additional experiment in order to verify this result. Feeding a diet containing high fat was reported to induce overeating and obesity (Schemmel et al., 1970). Indeed, the intake and the body weight gain of the HF group were increased significantly after day 6 compared to values of the NB baseline. However, rats of the HFHS-CAF group did not consume more calories nor did they gain more weight than the HF rats by the end of the 23-day experimental period.

TABLE V Liver and Brown Adipose Tissue (BAT) Weights and Serum Insulin, T_3, and T_4 Levels of Animals Fed Nutritionally Controlled Diets with and without Flavor and Textural Variety[a]

Dietary treatment group[b]	Liver wt. (g/100 g body wt.)	BAT wt. (g/100 g body wt.)	T_3 (ng/ml)	T_4 (μg/dl)	IRI (ng/ml)
NB	3.01 ± 0.06[a]	0.13 ± 0.006[a]	0.40 ± 0.02[b]	4.93 ± 0.20[a]	3.34 ± 0.28[a]
NB-CAF	3.11 ± 0.07[a]	0.15 ± 0.008[a]	0.39 ± 0.03[b]	5.48 ± 0.34[a]	3.27 ± 0.35[a]
HFHS	3.10 ± 0.11[a]	0.12 ± 0.006[a]	0.37 ± 0.02[b]	5.23 ± 0.17[a]	2.40 ± 0.17[a]
HFHS-CAF	3.17 ± 0.07[a]	0.14 ± 0.008[a]	0.51 ± 0.05[a]	5.96 ± 0.30[a]	3.11 ± 0.45[a]

[a] Data from Naim et al. (1985). Values are the mean ± SEM of 13–15 rats per group. Values not sharing the same superscript letter are different at $p < 0.05$ or less.

[b] NB, Group offered the nutritionally balanced diet; NB-CAF, group offered the flavored, nutritionally balanced diets in a "cafeteria" design; HFHS, group offered the high-fat, high-sucrose diet; HFHS-CAF, group offered the flavored high-fat, high-sucrose diets in a "cafeteria" design.

14. Flavor Variety in Fat Deposition

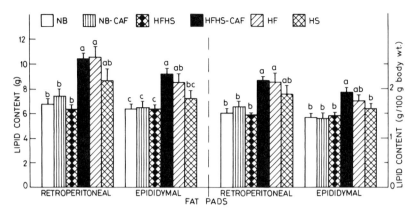

Fig. 5. Lipid content in fat pads of rats fed nutritionally controlled diets containing a variety of food flavors during "cafeteria" experiments. Values are the mean and SEM of 13–15 rats per group. Values not sharing the same letter are different at $p < 0.05$. NB, Animals offered the nutritionally balanced diet; NB-CAF, animals offered the flavored, nutritionally balanced diet; NB-CAF animals offered the flavored, nutritionally balanced diet in a "cafeteria" design; HFHS, animals offered the high-fat, high-sucrose diet; HFHS-CAF, animals offered the flavored, high-fat, high-sucrose diet in a "cafeteria" design; HF, animals offered the high-fat diet; HS, animals offered the high-sucrose diet. [From Naim et al., 1985. © J. Nutr. 115, 1447–1458, American Institute of Nutrition.]

The lipid content of fat pads of the six groups is shown in Fig. 5. Changes in fat content reflected the energy intake and body weight gain data. The HFHS-CAF and HF groups accumulated more fat in their fat pads than the NB, NB-CAF, and HFHS groups. A trend toward an increase in content of fat pads was also seen in rats of the HS group. The lack of difference in energy intake, weight gain, and fat pad content between the HF and HFHS-CAF rat groups suggests that the variety of flavors per se in the HFHS-CAF diet did not induce, in the long run, more hyperphagia and fat deposition than did the HF diet. Since a significant difference in diet selection among the three flavor choices of the HFHS-CAF diet was found to occur in only 22% of the feeding sessions (Table III, Experiment 1), one may assume that the variety effect of these flavored diets was not sufficient to further stimulate energy intake during the 23-day experiment. Part of the explanation for this low effect of variety may be explained by the fact that manipulations in the texture of the HFHS-CAF diet were not possible.

The objective of an additional feeding experiment was twofold. First, it was necessary to test again the effect of feeding the HFHS diet upon energy intake and body weight gain of different rats because the HFHS diet did not induce increased energy intake in the first experiment as did the HF diet, even though both diets had the same fat content. Second, it was necessary to replicate the results of feeding the HFHS-CAF diet. Forty-five Sprague–Dawley male rats weighing 250–300 g were acclimated to a reversed light–dark cycle as described before. Animals were then divided randomly into 3 groups of 15 rats each. The experimental design was identical to that of the first experiment, but the experiment was conducted for 12 days. Group 1 was fed the NB diet. Group 2 was offered the HFHS diet, and Group 3 was exposed to the HFHS-CAF treatment. The results of this experiment are shown in Fig. 6. The HFHS-CAF group demonstrated increases in energy intake and body weight gain above the NB baseline level similar to those found in the first experiment (Table II). Rats fed the HFHS diet also showed

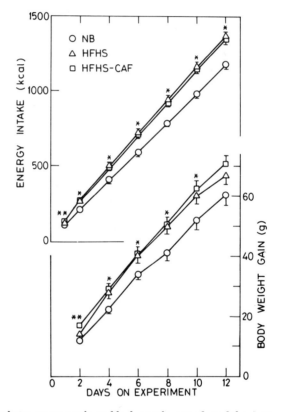

Fig. 6. Cumulative energy intake and body weight gain of rats fed nutritionally controlled diets in a "cafeteria" regimen for 12 days (Experiment 2). Values are the mean and SEM of 15 rats per group. NB, Group fed the nutritionally balanced diet; HFHS, group fed the high-fat, high-sucrose diet; HFHS-CAF, group fed the flavored, high-fat, high-sucrose diet offered in a "cafeteria" design. *Significantly ($p < 0.05$) different value for both the HFHS-CAF and HFHS groups compared with the NB group. **Significant difference ($p < 0.05$) between the HFHS-CAF and the NB group only. [From Naim et al., 1985. © J. Nutr. 115, 1447–1458, American Institute of Nutrition.]

an increase in energy intake and body weight gain. Rats fed the HFHS-CAF diet did not consume more calories nor did they gain more weight than the HFHS rats, even though a significant variety effect of 46% (Table III) was observed during the "cafeteria" feeding of this diet. Therefore, in our experiments, the effect of variety in food flavors on hyperphagia in rats seems to be of minor importance compared to the stimulating effect of the fat in the diet.

Should one conclude that diet taste and flavor do not contribute to hyperphagia and obesity in rats? At present, the answer to this important question is still not available, even though the above experiments were designed to separate the sensory properties from other dietary components. Offering rats a sucrose solution in addition to water along with rat chow produces obesity (Kanarek and Hirsch, 1977), whereas a saccharin solution is ineffective (Kenney and Collier, 1976). Further, rats given a fructose solution along with their rat chow had a higher serum triglyceride level than those offered a glucose or sucrose solution (Kanarek and Orthen-Gambill, 1982). Perhaps, a taste stimulus in itself does not stimulate an overconsumption of calories, but when coupled with a physiological feedback (e.g., the appealing taste of sucrose and its postabsorptive effect), the overall consumatory behavior is then altered.

V. CONCLUSIONS

1. A catalog of preferred food flavors and texture for nutritionally balanced diets was developed for the Sprague–Dawley rats.

2. The presence of preferred flavors and texture varieties in nutritionally balanced diets did not induce hyperphagia in rats during *ad libitum* "cafeteria" feeding.

3. Feeding rats a high-fat, high-sucrose diet containing a variety of flavors in a "cafeteria" paradigm produced an increase in energy intake, body weight gain, lipid content in fat pads, and serum T_3 levels.

4. The high-fat diet in itself induced increased energy intake, body weight gain, and lipid content in fat pads to a level which was either equal or close to that produced by the "cafeteria" feeding of a flavored, high-fat, high-sucrose diet.

5. The stimulating factors for the hyperphagia and obesity produced during "cafeteria" feeding of conventional foods should be further characterized.

ACKNOWLEDGMENTS

The authors acknowledge and are grateful for the advice and aid of the following individuals: Dr. L. Scharph of International Flavors and Fragrances for providing the flavor additives and for helpful discussions; Dr. E. Hunter of Procter and Gamble for providing sucrose polyester and for helpful discussions; Dr. R. Carpenter, Dr. R. Threatte, Dr. Y. Katz, Ms. Susan Van Buren, Mr. Douglas Bayley, Ms. Melissa Katz, Mr. Stanley Lewis, Ms. Martha Levinson, and Ms. Kathy Gniecko for excellent technical assistance; and Ms. Janice Blescia for typing the manuscript.

REFERENCES

Bray, G. A. (1982). Regulation of energy balance: studies on genetic, hypothalamic and dietary obesity. *Proc. Nutr. Soc.* **41**, 95–108.
Cabanac, M. (1971). Physiological role of pleasure. *Science* **173**, 1103–1107.
Cleary, M. P., Vasselli, J. R., and Greenwood, M. R. C. (1980). Development of obesity in zucker obese (fafa) rat in absence of hyperphagia. *Am. J. Physiol.* **238**, E284–E292.
Corbit, J. D., and Stellar, E. (1964). Palatability, food intake and obesity in normal and hyperphagic rats. *J. Comp. Physiol. Psychol.* **58**, 63–67.
Drewnowski, A., Cohen, A. E. Faust, I. M., and Grinker, J. A. (1984). Meal-taking behavior is related to predisposition to dietary obesity in the rat. *Physiol. Behav.* **32**, 61–67.
Epstein, A., and Teitelbaum, P. (1962). Regulation of food intake in the absence of taste, smell, and other oropharyngeal sensation. *J. Comp. Physiol. Psychol.* **55**, 753–759.
Fallat, R. W., Glueck, C. J., Lutmer, R. F., and Mattson, F. H. (1976). Short-term study of sucrose polyester, a non-absorable fat-like material as dietary agent for lowering plasma cholesterol. *Am. J. Clin. Nutr.* **29**, 1204–1215.
Folch, J., Lees, M., and Sloanestanley, G. H. (1957). A simple method for the isolation and purification of total lipids from animal tissues. *J. Biol. Chem.* **226**, 497–509.
Hamilton, C. L. (1964). Rats' preference for high fat diets. *J. Comp. Physiol. Psychol.* **58**, 459–460.
Harris, L. J., Clay, J., Hargreaves, F., and Ward, A. (1933). Appetite and choice of diet. The ability of the vitamin B deficient rat to discriminate between diets containing the lacking vitamin. *Proc. R. Soc. Lond. (Biol.)* **113**, 161–190.
Herbert, V., Lau, K.-S., Gottlieb, C. W., and Bleicher, S. J. (1965). Coated charcoal immunoassay of insulin. *J. Clin. Endocrinol.* **25**, 1375–1384.
Herman, R. H., Zakim, D., and Stifel, F. B. (1970). Effect of diet on lipid metabolism in experimental animals and man. *Fed. Proc.* **29**, 1302–1307.
Janowitz, H. D., and Grossman, M. (1949). Some factors affecting food intake of normal dogs and dogs with esophagostomy and gastric-fistulae. *Am. J. Physiol.* **159**, 143–148.
Jorden, H. A. (1969). Voluntary ingragastric feeding: Oral and gastric contributions to food intake and hunger in man. *J. Comp. Physiol. Psychol.* **68**, 498–506.
Kanarek, R. B., and Hirsch, E. (1977). Dietary-induced overeating in experimental animals. *Fed. Proc.* **36**, 154–158.
Kanarek, R. B., and Orthen-Gambill, N. (1982). Differential effects of sucrose, fructose and glucose on carbohydrate-induced obesity in rats. *J. Nutr.* **112**, 1546–1554.
Kenney, J. J., and Collier, R. (1976). The effects of saccharin on the sucrose consumption of rats. *J. Nutr.* **106**, 388–391.
Kratz, C. M., and Levitsky, D. A. (1979). Dietary obesity: differential effects with self selection and composite feeding techniques. *Physiol. Behav.* **22**, 245–249.
LeMagnen, J. (1956). Hyperphagie provoquée chez le rat blanc par alteration du mechanisme de satieté peripherique. *C. R. Soc. Biol. (Paris)* **150**, 32–34.

Louis-Sylvestre, J., Giachetti, I., and LeMagnen, J. (1984). Sensory versus dietary factors in cafeteria-induced overweight. *Physiol. Behav.* **32**, 901–905.

Mickelson, O., Takahashi, S., and Craig, C. (1955). Experimental obesity. I. Production of obesity in rats by feeding high-fat diets. *J. Nutr.* **57**, 541–554.

Mook, D. G. (1963). Oral and postingestional determinants of the intake of various solutions in rats with esophageal fistulas. *J. Comp. Physiol. Psychol.* **56**, 645–659.

Naim, M., and Kare, M. R. (1977). Taste stimuli and pancreatic function. *In* "The Chemical Senses and Nutrition" (M. R. Kare and O. Maller, eds.), pp. 145–162. Academic Press, New York.

Naim, M., Kare, M. R., and Ingle, D. E. (1977). Sensory factors which affect the acceptance of raw and heated defatted soybeans by rats. *J. Nutr.* **107**, 1653–1658.

Naim, M., Rogatka. H., Yamamoto, T., and Zehavi, U. (1982). Taste responses to neohesperidin dihydrochalcone in rats and baboon monkeys. *Physiol. Behav.* **28**, 979–986.

Naim, M., Brand, J. G., Kare, M. R., and Carpenter, R. G. (1985). Energy intake, weight gain, and fat deposition in rats fed flavored, nutritionally controlled diets in a multichoice ("cafeteria") design. *J. Nutri.* **115**, 1447–1458.

Naim, M., Brand, J. G., Christensen, C. M., Kare, M. R., and van Buren, S. (1986). Preference of rats for food flavors and texture in nutritionally controlled semi-purified diets. *Physiol. Behav.* (in press).

Nowlis, G. H., and Frank, M. (1977). Qualities in hamster taste: Behavioral and neural evidence. *In* "Sixth International Symposium on Olfaction and Taste" (J. LeMagnen and P. MacLeod, eds.), pp. 241–248. Information Retrieval, Ltd., London.

Porikos, K. P., Hesser, M. F., and van Itallie, T. B. (1982). Caloric regulation in normal weight men maintained on a palatable diet of conventional foods. *Physiol. Behav.* **29**, 293–300.

Richter, C. P. (1936). Increased salt appetite in adrenalectomized rats. *Am. J. Physiol.* **115**, 155–161.

Rodgers, W. L. (1967). Specificity of specific hungers. *J. Comp. Physiol. Psychol.* **64**, 49–58.

Rogers, Q. R., and Leung, P. M. B. (1977). The control of food intake: When and how are amino acids involved? *In* "The Chemical Senses and Nutrition" (M. R. Kare and O. Maller, eds.), pp. 213–249. Academic Press, New York.

Rolls, B. J., Rolls, E. T., Rowe, E. A., and Sweeney, K. (1981a). Sensory specific satiety in man. *Physiol. Behav.* **27**, 137–142.

Rolls, B. J., Rowe, E. A., Rolls, E. T., Kingston, B., Megson, A., and Gunary, R. (1981b). Variety in a meal enhances food intake in man. *Physiol. Behav.* **26**, 215–221.

Rolls, B. J., Rowe, E. A., and Rolls, E. T. (1982). How flavour and appearance affect human feeding. *Proc. Nutr. Soc.* **41**, 109–117.

Rolls, B. J., van Duijvenvoorde, P. M., and Rowe, E. A. (1983). Variety in the diet enhances intake in a meal and contributes to the development of obesity in the rat. *Physiol. Behav.* **31**, 21–27.

Rothwell, N. J., and Stock, M. J. (1979). A role for brown adipose tissue in diet-induced thermogenesis. *Nature (London)* **281**, 31–35.

Rothwell, N. J., and Stock, M. J. (1981). A role for insulin in the diet-induced thermogenesis of cafeteria-fed rats. *Metabolism* **30**, 673–678.

Sclafani, A., and Springer, D. (1976). Dietary obesity in adult rats: Similarities to hypothalamic and human obesity syndrome. *Physiol. Behav.* **17**, 461–471.

Schemmel, R., Mickelsen, O., and Gill, J. L. (1970). Dietary obesity in rats: Body weight and body fat accretion in seven strains of rats. *J. Nutr.* **100**, 1041–1048.

Stock, M. J., and Rothwell, N. J. (1979). Energy balance in reversible obesity. *In* "Animal Models of Obesity" (M. F. W. Festing, ed.), p. 141–151. Oxford University Press, New York.

Treit, D., Spetch, M. L., and Deutsch, J. A. (1983). Variety in the flavor of food enhances eating in the rat: A controlled demonstration. *Physiol. Behav.* **30,** 207–211.

Tulp, O. L., Frink, R., and Danforth, E., Jr. (1982). Effect of cafeteria feeding on brown and white adipose tissue cellularity, thermogenesis, and body composition in rats. *J. Nutr.* **112,** 2250–2260.

Zucker, L. M. (1967). Some effects of caloric restriction and deprivation on the obese hyperlipemic rat. *J. Nutr.* **91,** 247–254.

15

Quantitative Relationship between Palatability and Food Intake in Man

HARRY R. KISSILEFF
Obesity Research Center
St. Luke's-Roosevelt Hospital and
Department of Psychiatry
Columbia University College of Physicians and Surgeons
New York, New York

I.	Introduction	293
	A. Objectives	293
	B. Definition	295
	C. Previous Studies	296
II.	A Brief-Exposure Taste Test for Measuring Intrinsic Palatability	298
	A. Procedure	298
	B. Subjects	299
	C. Intrinsic Palatability Ratings	299
III.	Relationship of Intrinsic Palatability to Food Consumption	307
	A. Objective	307
	B. Methods	308
	C. Results	309
IV.	Needs for Future Research	313
V.	Summary and Conclusions	315
	References	315

I. INTRODUCTION

A. Objectives

In order to compare controls of food intake in subjects who differ in variables such as age, sex, and body composition, it is important to use a food that is as similar as possible in acceptability to all groups. In fact, even when controls are studied within the same type of individuals, uncontrolled variability could conceivably be reduced by minimizing differences in acceptability level of food among subjects. Furthermore, it also conceivable that in certain types of studies, such as the effects of food preloads, the independent variable (e.g., the preload type or amount) could have different effects depending upon the test meal type that followed it, either within or across subjects, and that the subject's hedonic reaction to the food could influence intake in different ways under different conditions. It would seem, therefore, that before undertaking systematic studies of physiological controls of food intake in any species, it would be important to have a method for evaluating the possible effect of hedonic reactions to the test foods and to select foods that will be acceptable to the largest possible portion of the population to be studied.

When I began studying food intake in man in 1976, there was no universally acceptable food analogous to laboratory chow in animal-feeding studies, and no procedure for determining which of a variety foods normally consumed by man could be considered as "the food" for studying human feeding. I was interested in determining whether intakes would be different for meals consisting of liquefied as opposed to semisolid, chewable foods. Gary Klingsberg accepted the challenge of developing an equipalatable diet that could be served as either a solid or a liquid. He devised a yogurt and fruit diet (Kissileff et al., 1980), which contained chunks of apples and bananas to make it chewable, but which could be liquefied simply by putting it into a blender. However, we had no way of knowing how this diet compared in acceptability with other foods people had used in laboratory feeding studies, such as liquid formula diets (e.g., Jordan et al., 1966). I surmised that future progress would probably be facilitated if we could develop procedures for rapidly assessing food from the standpoint of acceptability, and if we could, using these procedures, arrive at a food that would have a uniformly high level of acceptability across a variety of subject characteristics. Such foods could be developed either by trial and error or by considering features that make foods acceptable. In addition, it would also be important to know whether, in the absence of a uniformly acceptable food, intakes could be corrected for differences in palatability, assessed independently of intake. As Pfaffmann (Pfaffmann et al., 1977), and no doubt others, have remarked, definitions of palatability in terms of intake tend to be circular. In man however, this circularity can be broken by questioning the subject rather than by relying solely on food consumption. It should then be possible, at least in theory, to scale the relationship between intake and palatability, since we would have independent measures of each.

15. Palatability and Food Intake in Man

The foregoing considerations led to four objectives: (1) development or refinement of a quick and reliable procedure for assessing palatability independently of food intake; (2) validation of the procedure by comparing its results with results using the same substances in classical psychophysical procedures; (3) comparison of intrinsic palatabilities of several foods that have been or could be used in laboratory food intake tests in man; and (4) determination of the quantitative relationship between palatability and food intake in man, both in terms of the amount of variance in intake that could be explained by palatability, and the magnitude of a unit change in intake per unit change in palatability.

B. Definition

What I mean by "palatability" is the rating a subject gives to an item when asked "how much do you like or dislike" the item in question, under standardized reference conditions. Perhaps a better designation would be to call this rating the "intrinsic palatability," meaning that it is a property of the food, to distinguish it from "learned palatability," which is a response based on prior associations between a food and its postingestive consequences. The "reported palatability" would then be the sum of intrinsic palatability, measured under standardized conditions, and other factors, such as prior experiences or what the subject has recently eaten. The preceding definition is not meant to ignore the large literature showing that there are both learning and direct physiological effects on reports of food likes and dislikes (e.g., Booth, 1978; Le Magnen, 1971; Rolls et al., 1981; Cabanac and Duclaux, 1970). I am simply trying to compare foods under standardized conditions and to use this information to predict intake under standardized conditions. Adding other variables, such as manipulations of physiological state or experiences, in no way invalidates or detracts from this procedure, nor do I claim that these procedures will give identical results in other conditions of testing. However, I do believe that if the conditions are duplicated, the results will also be duplicated. I hope that the addition of other variables will not destroy the relationships among foods I shall show here, but that instead they will be additive with what I would consider a basal condition. In order not to unduly encumber this chapter with verbiage, unless otherwise specified, "palatability" will refer to "intrinsic palatability," as described above. Although it could be argued that intrinsic palatability, as defined, could be measured accurately only in the newborn, prior to its first feeding, we shall assume that intrinsic palatability can be approximated by the procedures described below. We shall also show that these procedures provide a stable response measure and that, in the absence of imposed physiological challenges or associations, these measures can be considered to reflect intrinsic palatability.

C. Previous Studies

Although many studies have been done on food intake and sensory evaluation of foods in man, very few studies have been done in which both intake and hedonic response have been jointly evaluated. For example, Nisbett (1968) showed that when ice cream was adulterated with quinine to make it bitter, men ate more of the ice cream they rated more palatable. This procedure is not the same as determining the effect of palatability on a nutritionally balanced meal. Hill and McCutcheon (1975) showed a similar result with a multicourse meal, but they manipulated palatability by assigning subjects to meals based on palatability ratings of the items in the meal on a food preference inventory given prior to testing. This procedure confounds palatability with food composition. That is, the differences in amounts consumed between diets of different palatability and composition could be due as much to differential satiating effects of the foods (e.g., see Kissileff, 1984) as to differences in their palatabilities. Price and Grinker (1973) showed that when subjects were given 5 varieties of crackers to rate after a preload, they ate more of the crackers as a function of their rating on a 5-point scale. In this study, palatability was probably confounded with composition of the crackers (although the composition was not specified), and the quantities consumed (2–12 crackers) were not large enough to be considered a meal. Bellisle and Le Magnen (1980, 1981) have carried out similar studies using bread covered with a variety of spreads. Although they focused primarily on chewing responses and not intake, they reported that within half the subjects there was a significant correlation between palatability ratings and amount consumed. However, these data do not provide the kind of quantitative prediction one would want in developing new products or theories about intake control. Rolls (1986, this volume) has reported repeatedly that palatability declines for foods that are eaten in a meal but not for foods that have not been eaten, or in some cases more for eaten than for uneaten foods, but this information tells us little about the relationship between intake and palatability. In an attempt to manipulate palatability by more careful control of sensory effects, Witherly et al. (1980) examined the hedonic responses and rates of consumption of fixed quantities of foods whose palatability was manipulated by sodium or sucrose content. Both of these variables could influence ingestive response independently of taste alone, either by osmotic postingestion effects for sodium (Houpt et al., 1979) or possibly by insulin release, as suggested by Louis-Sylvestre and Le Magnen (1980). It is doubtful, however, that insulin alone plays a role in controlling short-term food intake, since Woo et al. (1984) have shown that intake of a test meal is neither increased nor decreased by a fourfold increase in insulin when blood glucose is held constant. Finally Booth et al. (1982) have investigated conditioning effects on palatability. They showed that the postingestive consequences of eating foods with distinctive cues can modify palatability ratings of those foods, but they did not quantify the relationship between change in palatability and change in intake.

None of the above studies has directly explored the following questions that concern investigators interested in the relationship between palatability and food intake: (1) What is the magnitude and shape of the relationship, e.g., is it linear, and if so, what is the slope? (2) What percentage of the variability in intake can be accounted for by palatability? (3) What is the physiological mechanism by which palatability affects intake?

Besides the work cited above on the relationship between palatability and food intake, several important findings have emerged from the hedonic testing of solutions, as well as foods and beverages. In planning the present work, it was deemed important to incorporate their replication as a validation procedure. Thompson et al. (1976) showed that, as the concentration of a sucrose solution is increased from threshold to 1 M, some individuals exhibited an increased liking, whereas others showed decreased liking. This phenomenon apparently could not be attributed to either body weight or sex variables, and the sample was probably too small to be sure, in any case. A similar dichotomy was also noted by Enns et al. (1979), which was clearly attributable to a sex difference. Men's preferences increased whereas women's decreased, as the sucrose concentration rose. Rodin (1975) also observed a dichotomy, but of a different sort. When sucrose concentration in milk was increased, obese individuals' preferences rose more than did those of lean individuals. Thus sucrose added to water gives different palatability/concentration relations than sucrose in milk (i.e., nutritive vs nonnutritive base). This difference in sucrose palatability appears to have been confirmed by Drewnoski and colleagues (Drewnoski and Greenwood, 1983; Drewnoski et al., 1982, 1983), in that obese/nonobese differences in hedonic ratings of stimuli occurred when sucrose was added to liquid dairy products but not when it was added to a water-based drink.

Of specific importance for the data reported here are the sex differences in hedonic ratings of sucrose and the two response types. These phenomena should be demonstrable with a single, highly concentrated sucrose solution. It was predicted that the frequency of sucrose ratings on a 9-point scale would be bimodally distributed if both sexes were included in the same distribution, or, that peaks in the distribution would be on opposite sides of neutral if each sex's distribution were plotted separately. We, therefore, included a concentrated sucrose solution in all of our taste tests.

II. A BRIEF-EXPOSURE TASTE TEST FOR MEASURING INTRINSIC PALATABILITY

A. Procedure

Intrinsic palatability of selected foods, fluids, solutions, and water was measured by presenting 8–15 cups of food (in aliquots of 5 ml or 3–7 g) to subjects under standardized conditions. The cups were embedded in a styrofoam block to maintain temperature (45–55 or 120–130°F). Subjects reported to the laboratory after an overnight fast, and were positioned in front of a row or two of cups with a sink at the side. A cup of distilled water and an additional container were provided for rinsing. A 9-point category scale (Peryam and Pilgrim, 1957) was placed in the front of the cups. The left side of the scale was labeled "like" (9 to 6); the right side was labeled "dislike" (4 to 1); and the middle was labeled "neither" (5). It was further qualified on each side by the additional descriptors, extremely (1 and 9), very much (2 and 8), moderately (3 and 7), and slightly (4 and 6). The subject was instructed by a timed tape recording exactly when to sample, when to report, and when to rinse. It was timed so that a new sample was taken every 60 sec. Reports were requested 15 sec after sampling, and rinsing was requested 5 sec after reporting. The subject was told to swish the food around in the mouth for a few seconds and swallow it. The subject was also told to rinse until all aftertaste was gone and to rinse again 40 sec after sampling if all the aftertaste was not gone by then. Items were presented in random order except that 6.1% sucrose was always first and distilled water was second to familiarize the subjects with the procedures. The selection of items for the tests was guided empirically and intuitively. We began by using items that we were studying on intake tests, and we were interested initially in comparing our yogurt and fruit diet with formula diets previously used in laboratory feeding studies by others. When the yogurt and fruit diet proved unusable in the obese population, we switched to other foods we thought might be more appealing but would still be nutritionally balanced.

The test takes about 1 min per sample to run and 2 min for the instructions. It is therefore a very quick test and is much better than magnitude estimation procedures for screening large numbers of subjects, because it does not require the time-consuming process of training subjects. It differs in several respects from more traditional published procedures. First, only a few items (8 to 15) and small amounts (5 ml) were used in each session. Other tests have frequently used much larger quantities (e.g., 2–3 oz in Moskowitz and Sidel, 1971, p. 680) or more items at a time (e.g., 18 in Enns et al., 1979 and 64 in Drewnowski, et al., 1982). Second, the subject swallows the item so that the back of the mouth and esophagus are stimulated, in contrast to tests by others in which the items are typically expectorated. Third, subjects were instructed by tape recording; the interval between stimuli was carefully controlled and was longer than intervals typically reported so that possible interactions among stimuli would be minimized. The only drawback of the present procedure is that it may not be as precise in "refining the measurement of degree of acceptability" (Moskowitz and Sidel, 1971) as magnitude-estimation procedures are. In particular, hedonic ratings tended to be invariant when log-magnitude estimates of the same foods varied from 1 to 2, whereas when log-magnitude-estimated hedonic ratings were below 1, the two were linearly related.

Two variants of the basic procedure were also employed to assess its reliability. In the basic procedure, each participant was tested with each item once with each of eight items, and the test was only done once. In the first variant, subjects were given an item three consecutive minutes in a row before rinsing. The purpose of this variant was to determine whether repeated exposure without rinsing would result in a change of rating. This test was employed because in one of our studies we had found initially that ratings of palatability made after the meal did not correlate as well as we had hoped with either the taste test ratings or with intake. We hoped that a longer exposure might bring out a stronger relationship between amount consumed and rating. The other variant was to test seven women on four separate occasions, two different days and two different sessions on each day. In this case, the second session occurred 3.5 hr after the first and 3 hr after a standardized breakfast.

B. Subjects

The data presented here are based on subsamples from a population of 362 subjects divided into 10 groups by sex, age (under or over 30 years), weight [within 10% of desirable weight for height (Bray, 1975) or greater than 120% of desirable weight for height], and whether tested with a rinse after each item (single) or after three items (multiple). The groups are not completely balanced for two reasons. First, the obese patients came from a clinic which included few young males. Second, the nonobese individuals had been recruited for experimental studies in which the criteria for invitation were primarily ages 18–35 but in which the predominant group was under 30. Demographic data are summarized in Table I.

C. Intrinsic Palatability Ratings

1. Water and Solutions

The results of the palatability rating test will be presented in two formats: first, as frequency distributions showing the number or percentage of indi-

TABLE I Demographic Information[a]

Age category	Rinse frequency	N	Age	Ht	Wt	%[b]	N	Age	Ht	Wt	%[b]
			Men					**Women**			
					Lean						
Young	Single	65	21.55	179.88	73.29	98.67	72	21.22	163.56	57.96	98.0
± SD			2.35	6.57	9.31	10.45		2.74	6.10	9.87	8.6
Young	Multiple	22	19.82	177.05	70.41	101.73	46	19.91	163.91	55.32	98.3
± SD			2.52	5.50	6.21	5.71		1.77	6.25	5.50	6.6
Old	Single	2	40.50	175.15	86.65	120.00	9	36.56	164.99	66.79	104.4
± SD			6.36	3.75	3.18	0.00		3.28	4.92	6.54	22.4
					Obese						
Young	Single	9	23.11	180.62	112.17	157.78	27	23.04	163.17	89.13	156.3
± SD			4.14	5.67	21.61	31.93		4.45	6.42	15.12	24.9
Old	Single	18	45.56	177.71	126.35	177.78	92	45.28	163.67	106.82	186.4
± SD			9.81	8.35	29.81	42.64		10.51	8.18	25.06	42.4

[a] Subjects are classed by age (young < 30, old > 30), weight (lean within 10% of desirable weight, obese greater than 120% of desirable weight), and number of items before rinse (one = single, three = multiple).

[b] The percentage of desirable weight (Bray, 1975).

viduals reporting a given rating, and second, as mean ratings for those items in which distributions were unimodal and similar across age, sex, and weight categories. The frequency histogram of liking for distilled water (Fig. 1) exhibited remarkable indifference (mode at 5) and uniformity across groups, just as one would expect if the procedure worked properly. Approximately 60% of each group gave it a 5, and 90% rated it between 4 and 6. Under the conditions of this test, water appears to be hedonically neutral. If we had deprived or overhydrated the individuals, it is likely that the rating of water would be different. The frequency distribution of 6.1% sucrose was similar to that of distilled water, except that it was spread out more (see Fig. 2).

The frequency distributions of 34% sucrose were bimodal in all groups except lean young males (see Figs. 3 and 4). Although most of the individuals disliked this concentration, there were a few who liked it. The strength of the bimodal relation is particularly striking when one considers that of 183 subjects, representing all groups tested with a single sample, except for lean young men and obese young men (the latter excluded for paucity of observations), only 5 (4 obese old females, and 1 obese old male) gave 34% sucrose an indifferent rating. There were no significant differences in the shapes of these curves among the four groups shown in Fig. 3 (females: young, lean and obese, and old obese; males: young obese). However, there was a significant difference in the shapes of the curves for the lean males and females

15. Palatability and Food Intake in Man

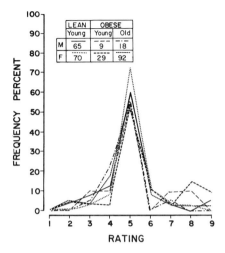

Fig. 1. Frequency distribution of intrinsic palatability ratings of distilled water. Numbers (Ns) in the key to lines representing different groups indicate numbers of subjects in each group. M indicates males; F indicates females in this and succeeding figures. Frequency percent is the percentage of the total in each group, not percentage of all subjects. Raw frequency counts can be obtained by multiplying the ordinate value by N shown in the key.

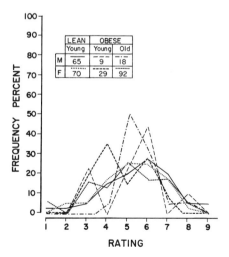

Fig. 2. Frequency distribution of intrinsic palatability ratings of 6.1% sucrose solution. Key as in Fig. 1.

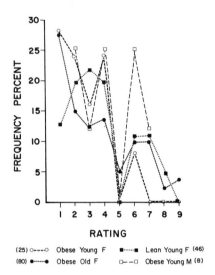

Fig. 3. Frequency distribution of intrinsic palatability ratings of 34% sucrose. Numbers in parentheses in key to lines at bottom indicate numbers of subjects in each group.

shown in Fig. 4 [$\chi^2(2)=25.8$]. It is clear that the bimodality seen in the lean young females did not extend to the lean young males. In addition, the two peaks in the females were different in magnitude. The peak in the dislike range was higher than in the like range. For the men, the distribution was skewed toward the dislike direction but the mode was definitely in the like range (8). Results with rinses after each item (single) were similar to results with the same item sampled three times in a row (multiple) for 34% sucrose. However, the peaks appeared sharper. This result is probably a consequence of averaging the three ratings and rounding to the nearest whole number. The important point is that the shapes of the distributions were similar under both single and multiple conditions. Therefore a single sample followed by a rinse is as good an indicator of the palatability as multiple samples. We have, therefore, replicated the findings of both Enns *et al.* (1979) of a difference in sucrose liking between men and women, and of Thompson *et al.* (1976) of the bimodal distribution in sucrose liking. We have also shown an apparently new finding that the bimodal distribution of sucrose liking was not found in lean young men, although it was seen in all other groups.

2. Repeated Testing before Rinse

When subjects were given repeated samples without rinsing, the reports changed minutely (mean changes ranged from 0.07 to 0.68 units) from one

15. Palatability and Food Intake in Man   303

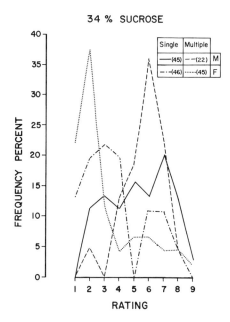

Fig. 4. Frequency distribution of intrinsic palatability ratings of 34% sucrose in subjects tested with a rinse after each item (single) or with one rinse after every three items (multiple).

report to the next. However, these changes were significant for four of the five items tested this way. Ratings for distilled water and 34% sucrose dropped by 0.3 to 0.4 units, whereas ratings of banana colada drink and another yogurt and fruit drink rose by 0.2 to 0.6 units. When the banana colada drink was adulterated with cumin, there were no changes across reports. The mean rating of the three reports was at most 1.48 units less and at least 0.01 units more than ratings of the same items on single sample tests. In only a third of the comparisons of single- vs multiple-sample tests was the difference between the two greater than half a unit, although the average ratings ranged over different foods from 3.72 to 6.83, in the single-sample tests. Thus the single-sample test gives the same basic pattern of response as tests in which the same substance is sampled three times before a rinse. In the seven women tested repeatedly for four sessions, there were no differences across sessions in any of the items sampled. The items included both liked and disliked foods. This result does not preclude the possibility that certain items could change in hedonic tone across time of day, as Birch et al. (1984) have shown. However, with the exception of macaroni and beef, which might be considered more of a lunch than a breakfast, all the other items in our test (distilled water, 34% sucrose, apple juice, English muffin) could be eaten any time, whereas the items used by Birch et al. (1984) deliberately included foods that would be classified as primarily eaten at certain meals. We are therefore confident that our basal intrinsic palatability test gives reproducible results both within a session and across sessions.

3. Real Foods and Drinks

Having established the reproducibility and validity of this particular test, we shall now examine the reports on various foods and drinks we have tested. The first is the yogurt and fruit diet in liquid form, originally developed to compare intakes on solid and liquid foods of the same composition. We tested only three of the eight original (single test) groups on this diet because, although it turned out to be quite palatable for lean young individuals, it was not palatable in the older obese subjects (see Fig. 5). We therefore abandoned use of this diet, although we did use it in our studies on cholecystokinin (CCK) (Kissileff et al., 1981; Pi-Sunyer et al., 1982) but selected only subjects who liked it. Note that this diet, which probably contained at least 10% sucrose (mainly from the Dannon vanilla yogurt product), did not exhibit a bimodal frequency distribution, but was clearly liked by nonobese and disliked by obese. It seemed therefore that our "ideal" food was not quite ideal for all populations. But how did it compare

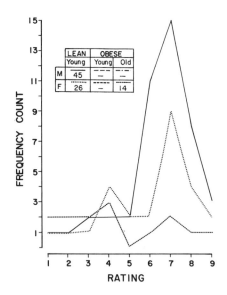

Fig. 5. Frequency distribution of intrinsic palatability ratings of yogurt and fruit diet (see Kissileff et al., 1980, for its composition). Only three groups were tested on it. Numbers, in the key to lines, representing different groups, indicate numbers of subjects in each group.

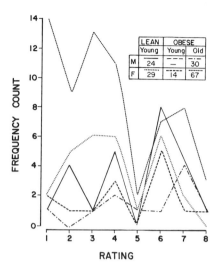

Fig. 6. Frequency distribution of intrinsic palatability ratings of vanilla Ensure. Explanation to key as in Fig. 1.

with other foods that have been, or could be, used for feeding studies in man? One such food that we have tested extensively is the formula diet, Ensure (Ross Laboratories). The frequency distribution of this diet (see Fig. 6) was typical of several others that we tested (vanilla Sustacal, vanilla Nutriment, vanilla Nutri-1000, and vanilla Instant Breakfast). We show data with Ensure because we tested more people on it. Like 34% sucrose, and unlike our yogurt and fruit diet, the frequency distribution of liking for Ensure was bimodal, with the majority of the population disliking it. Because the bimodal distribution extended to all groups, it is probably not attributable to sweetness, typified by 34% sucrose, in which a bimodal distribution was not seen in young lean men. It would be important to explore possible flavor and taste combinations that could lead to such distributions, but at present we have no idea what could produce this distribution. The strong vanilla flavor is a possibility, but vanilla was also present in our yogurt and fruit diet, though associated with the flavor of fruit.

Lest one get the impression that all sweet fluids may have dichotomous distributions, one that clearly does not is apple juice (Fig. 7). Apple juice was highly palatable to all groups. This finding probably comes as no surprise, since apple flavors appear ubiquitously in almost all types of ingested items from baby food to liquor. Perhaps its early association with nutrition in infancy is responsible for this effect. There was almost no dislike for apple juice.

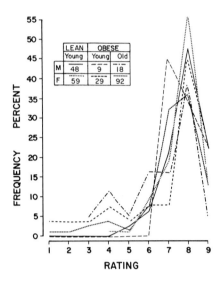

Fig. 7. Frequency distribution of intrinsic palatability ratings of apple juice (Red Cheek, Natural). Key as in Fig. 1.

The last food for which a distribution will be presented is banana colada frozen yogurt drink, prepared by mixing 1 part of banana colada drink mix (Lakewood Co., Miami, Florida) with 3 parts Dannon vanilla frozen yogurt. This diet was studied because it was reported by one of our graduate students to be highly palatable, and we were looking for a highly palatable, potable food that could be adulterated to reduce its palatability without changing its nutrient content, as we shall describe shortly. The frequency distribution of this relatively sweet diet, unlike that of Ensure, was unimodally distributed (see Fig. 8) and highly palatable. There was no significant difference in the distribution between those tested with a single rinse after each item and those tested with the items three times in a row before rinsing.

In order to enable easy comparison among many of the items we tested, the means, standard deviations (STD), and numbers tested (N) are shown in Fig. 9. Items were included only if at least 10 subjects were tested, and only if the item's distribution was unimodal. The yogurt and fruit diet we previously developed (Kissileff et al., 1980) was slightly palatable (5.9 on 9-point scale), not unpalatable as some critics had claimed. It was not, however, as palatable as other foods that have been, or could be, used in laboratory eating experiments. Surprisingly, tunafish salad (made with white tuna in water, mayonnaise, minced onion, and fresh celery) was the most palatable of all. The banana colada frozen yogurt drink was also very palatable, but

15. Palatability and Food Intake in Man   307

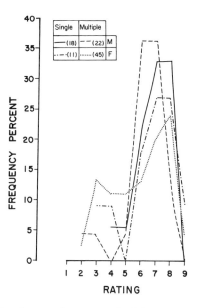

Fig. 8. Frequency distribution of intrinsic palatability ratings of banana colada frozen yogurt drink (see Bobroff and Kissileff, 1985, for its composition). See Fig. 4 for explanation of key.

addition of 1 part in 250 of powdered cumin reduced its palatability to a low level (3.3). Another interesting point is that adding 1/500 cumin reduced the palatability of the banana colada diet by 2.4 units, but adding 1/500 cinnamon to a vanilla yogurt and fruit diet only raised its palatability by 0.6 units. Attention is drawn to these values primarily to point out how little is known about the potency of flavor effects on palatability (see Moskowitz, 1977) and to suggest that much more research needs to be done.

III. RELATIONSHIP OF INTRINSIC PALATABILITY TO FOOD CONSUMPTION

A. Objective

We showed above that intrinsic palatability can be measured under standardized conditions, that it is a stable property of food when tested under the same conditions in a given subject, and that it can be manipulated by addition of small amounts of nutritionally insignificant spices. In this section, we shall examine the effect of palatability changes on food intake.

FOOD	RATING	MEAN	STD	N
	1 2 3 4 5 6 7 8			
BANANA COLADA + 1/250 CUMIN		3.3	1.74	19
BANANA COLADA + 1/500 CUMIN		4.3	2.33	12
VANILLA CANNED LIQUID YOGURT		4.8	2.10	41
PLAIN CANNED SOLID YOGURT		5.0	1.95	13
DISTILLED WATER		5.0	1.23	294
6.1% SUCROSE		5.1	1.52	294
MILK		5.3	1.96	57
VANILLA FRESH LIQUID YOGURT + FRUIT AND TOFU		5.9	2.14	99
VEGETABLE CHUNKY SOUP		6.0	2.16	14
VANILLA CANNED MIXED YOGURT		6.0	1.31	15
VANILLA CANNED SOLID YOGURT		6.1	1.72	19
VANILLA FRESH SOLID YOGURT		6.1	2.00	19
VAN YOGURT AND FRUIT DIET W/OUT TOFU		6.1	1.90	28
BREAKFAST SQUARE		6.1	1.98	181
BANANA COLADA		6.7	1.48	29
VAN YOGURT AND FRIUT DIET W/OUT TOFU + 1/500 CINNAMON		6.7	1.94	27
CHOCOLATE SUSTACAL		6.8	1.59	10
ENGLISH MUFFIN		6.9	1.45	168
MACARONI AND BEEF		7.1	1.52	95
APPLE JUICE		7.3	1.47	266
TUNAFISH SALAD		7.6	0.91	35

Fig. 9. Mean intrinsic palatability ratings of various items tested, arranged in order of liking from top for least liked, to bottom for most liked. Line length indicates rating, which is also printed for convenience at the right, along with standard deviation (STD, across subjects) and number tested. Only the results of the first test on any subject are shown.

B. Methods

Eight men and eight women were selected on the basis of their responses to the taste test described above. Two of the items in the test were the banana colada frozen yogurt drink and the same drink adulterated with either 1/250 or 1/375 g cumin per gram of the drink. Subjects who gave a rating difference of at least 2 were subsequently also given intake tests. Those subjects who ate at least 250 g of an adaptation meal (yogurt and fruit diet without tofu) continued to the next phase. On two subsequent test meals, the selected subjects received either the unadulterated or adulterated banana colada diet in a counterbalanced order for each pair of subjects. The meals were given as lunches on nonconsecutive days (see Bobroff and Kissileff, 1985, for further details). Following each meal, subjects were given an extensive rating questionnaire to fill out on which the first question was "how much did you like or dislike the food you just ate?" Subjects whose ratings for the two diets did not remain at least 2 units apart or who reversed their ratings were not considered in the final analysis, because the problem was not whether palatability correlated with intake, but rather to quantify the relation that existed for those who maintained the palatability difference;

thus, the question was, how much difference did palatability make in their intakes?

C. Results

Intake of the cumin-adulterated banana colada meal (295 g) was 429 g ± 82.5 SED less than intake of the unadulterated meal (724 g). Ratings after the meal were 4.38 units ± 0.37 SED lower after the cumin-adulterated meal (2.93) than after the unadulterated meal (7.31). Both of these differences were highly significant [$F(1, 12) = 27.12$, $p < 0.001$; and $F(1, 12) = 153.12$, $p < 0.0001$, for intake and ratings, respectively; see Fig. 10]. Three additional computations of the relationship between intake difference and palatability difference gave similar results (see Bobroff and Kissileff, 1985, for details). Intake changed approximately 100 g for every unit difference in rating, whether expressed as the mean of the ratios of intake difference to palatability difference or as the slope of the regression line for all the data. If

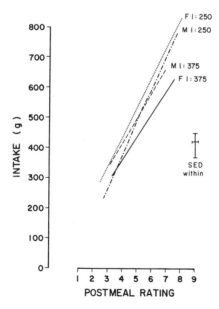

Fig. 10. Mean intake in relation to mean rating of like/dislike on questionnaire begun 5 min after the meal ended. Each line represents mean of 4 subjects in the group. "M" and "F" mean male and female. The numbers next to them are the concentration of cumin in the less liked of the two versions of the diet. Lengths of the horizontal and vertical bars indicate standard errors of the difference within subjects for each variable. (These, and not between-subject SDs, are the appropriate statistics for assessing significance of differences between diet conditions in this figure.)

expressed as a percentage of the mean intake under both conditions of palatability, intake changed by 19% per unit of palatability change. A similar computation carried out on the published data of Nisbett (1968) revealed similar results with ice cream adulterated with quinine to make it bitter. I believe that the present result represents the first attempt to quantify the relationship between intake and palatability for single-course meals. If it can be confirmed with other foods, it should be useful both for studying mechanisms controlling food intake as well as for designing and testing food products to raise or lower food intake.

In terms of variance, only 34.6% of the total was attributable to differences in palatability. However, within subjects, 65.8% of the variance was attributable to palatability. This result can be understood better by examining the individual responses (see Fig. 11). The figure reveals that the correlation between intake and palatability appears only when data from each individual at both palatability levels are pooled together. If the two sides of neutral, representing the two diets, are examined individually, there is no evidence of any relationship between palatability and intake within the same level of food adulteration, even though there is a range of 3 units between ratings within each level, a range that is statistically significant for ratings between items between subjects. However, the lines connecting the same individual's data show that the slopes are relatively consistent from subject to subject. These results indicate that palatability measurements can be useful in predicting, within a subject, how much a change in palatability will change intake, but palatability ratings apparently have little to do with setting the mean intake around which an individual's intake will vary. Thus, palatability ratings will be useful in predicting intake within subjects, but will be poor in predicting intake differences between subjects. If one wants to know quantitatively the intake of a particular food, brief-exposure taste tests will not be very helpful across subjects, and it would, therefore, be a good idea to do intake, as well as brief-exposure acceptability, tests.

The same results were also found when intake was regressed on ratings from the original, brief-exposure taste test on which the subjects were screened. Finally, when the taste-test ratings were plotted against the postmeal ratings, the correlations within groups given the same food was poor, but the relationship appears stronger when data points from each subject are connected by lines (see Fig. 12). Another point apparent from this figure is that the ratings before the meal (ordinate) were on the average approximately 1 unit higher than after the meal, a finding which is consistent with data of Rolls et al. (this volume) which shows that palatability drops when a food is consumed.

Hunger ratings were also obtained using a version of Silverstone's (1975) visual analog scale (mark on a 15-cm line being anchored at ends by descrip-

15. Palatability and Food Intake in Man

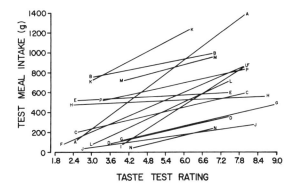

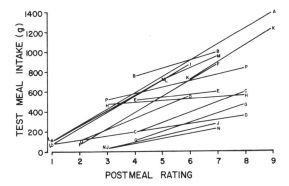

Fig. 11. Intakes of each subject plotted against ratings at either the screening test before any food was eaten as a meal (taste test ratings) or ratings made by questionnaire 5 min after meal was terminated (postmeal ratings). Each subject is shown by a different letter, and lines connecting each letter show the individual ratios (i.e., slopes) of the two variables. The left end of each line represents the ratings with the cumin-adulterated version of the test food. [From Bobroff and Kissileff, 1985. With permission from *Appetite* (in press). Copyright 1985 by Academic Press Inc. (London) Ltd.]

tors "hungriest I can imagine being" and "not hungry at all"). There were significant differences in hunger ratings between diets (see Fig. 13). When subjects stopped eating the cumin-adulterated diet, they were significantly more hungry than after the unadulterated diet. The corresponding results were also shown in the scaling of satiety. Subjects were more satiated by the more palatable, than by the less palatable diet. This result is not consistent with the idea that subjects stop eating because food drops in palatability below a level necessary to sustain eating. It is, however, consistent with the model that intrinsic palatability provides a constant input that sums with the excitation caused by food deprivation and with the inhibition caused by

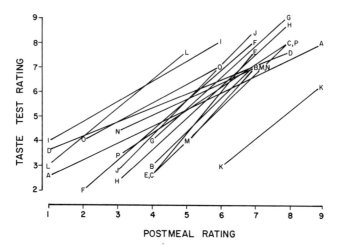

Fig. 12. Ratings of like/dislike on taste test (taste test rating) given when subjects were screened vs ratings of like/dislike for same food on questionnaire given 5 min after subject stopped eating it (postmeal rating) (see legend to Fig. 11).

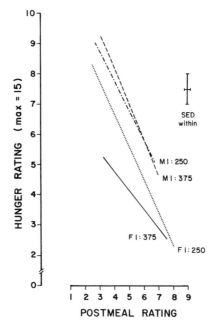

Fig. 13. Mean hunger ratings of the four groups described in the legend to Fig. 10 vs mean postmeal ratings.

postingestional effects to determine when an individual will stop eating (Kissileff and Thornton, 1982). The report of lower palatability after the meal, than before it, is postulated to be the result of a bias on the control of reporting of palatability rather than on the palatability itself. It is further suggested that the difference in ratings between eaten and uneaten foods seen in the experiments of Rolls *et al.* (1981) is the result of contrast effects making uneaten foods, which have not stimulated oropharyngeal receptors, temporarily more palatable than usual, following consumption of a different food. This hypothesis predicts that palatability of a food that is "sham eaten" will not drop but that palatability of a food that is not sham eaten will rise following sham eating of a different food.

The model mentioned above (i.e., Kissileff and Thornton, 1982) was derived from the fact that the rate of eating during a meal declines at a constant rate during the meal, giving rise to a decelerated increment in cumulative intake through the meal. It was specifically proposed that this rate of eating is the algebraic sum of a constant excitatory process reflected in the initial rate of eating and an inhibitory process that increases at a constant rate throughout the meal. Deprivation and palatability were the two major excitatory factors proposed. The present study provided a test of the model which predicted that the initial rate of eating would be significantly higher during a more, than during a less, palatable meal. Indeed it was found that the initial rate of eating was significantly higher after the unadulterated diet than after the adulterated diet, but the rates of deceleration (reflecting postingestional events) were not affected by palatability level (see Fig. 14).

IV. NEEDS FOR FUTURE RESEARCH

The studies described above raise many more problems than they have solved. Although it is almost trivially axiomatic that palatability influences food intake, there are many unanswered questions. For example, what are the qualities and quantities of stimulation, provided by a food, that influence palatability. Beyond a few simple, inverted U-shaped functions for preference aversion curves of solutions, there is neither quantitative description nor theoretical basis for persistent food preferences and aversions. Neophobias, other conditioned responses, and social–affective factors (e.g., Booth, 1978; Rozin and Kennel, 1983) are simply not adequate to explain food preferences that appear to be intrinsic to foods, (e.g., the liking of apple juice and dislike of certain formula diets). Sex differences in solution preferences appear to be emerging, and their possible extension to foods needs much more careful exploration than it has received in the past. More work needs to be done on the relation of brief exposure tests to intake. What kinds

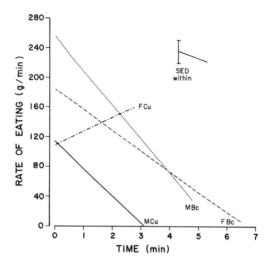

Fig. 14. Rate of eating adulterated (Cu) and unadulterated (Bc) test meals by men (M) and women (F), over course of meal. Initial rate of eating was determined by averaging the linear coefficient of the cumulative intake curve, and the slope was obtained by averaging twice the quadratic coefficient of the cumulative intake curve (i.e., the rate of deceleration) for each subject under each condition. The SED bar and slope show, respectively, the standard errors of the difference for the initial rate and half the rate of deceleration (i.e., quadratic coefficient) within subjects. It is clear from inspection, and statistical tests confirm, that initial rates were different for the two versions of the diet, but that rates of deceleration did not differ. Lengths of lines show mean meal durations for each condition.

of differences between subjects lead individuals, who give foods the same palatability ratings, to eat different amounts? Can such measures help in the successful selection of new food products? Finally, we know very little of the physiological mechanism by which a more palatable food leads to higher intake than a less palatable food. Are there specific neural pathways that code for palatability and lead to excitement of consumatory reflexes? Or, are such effects mediated by effects on the GI tract? For example, greater relaxation of the stomach or release of humoral agents could facilitate gastric emptying so that more palatable foods could leave the stomach faster, thus allowing more of them to be consumed (J. G. Kral, personal communication). Does palatability influence gut peptides, which conceivably could have a role in satiety? Do early metabolic responses, such as changes in respiratory quotient, which are in part diet specific (Nicolaidis, 1977), mediate effects of palatability on food intake? Is there a hormonal mediator whose release by a palatable food stimulates consumption, as suggested by Louis-Sylvestre and Le Magnen (1980) for insulin? Recognition that there are reliable and reproducible techniques for scaling palatability should lead to experiments in which palatability is manipulated under controlled conditions so that these mechanisms can be explored either independently or in combination.

V. SUMMARY AND CONCLUSIONS

We have incorporated previously developed test components into a rapid and efficient procedure for obtaining reliable data on palatability ratings of food items and solutions. We have been able to use this procedure to duplicate and even extend previously known psychophysical results with solutions. This success is attributed to careful control of the subject's physiological state and instructions to the subject.

A single test with an item seems to be just as reliable for assessing palatability as repeatedly testing the item. The time of testing per se did not appear to be a critical variable for the foods used. It is possible that testing in a relatively deprived (i.e., unfed) state leads to greater consistency in responding. We recognize the existence of, but do not know the basis for, the bimodal frequency distribution of liking of various foods.

Finally, palatability ratings predict intake differences reliably within subjects, but do so poorly across subjects. Therefore, it is suggested that acceptability testing of new foods should include intake as well as brief-exposure tests before marketing. Quantifying the relationship between intake and palatability is only in its infancy. Research needs to be done to characterize and describe relationships as well as to determine their mechanisms.

ACKNOWLEDGEMENT

This work was supported in part by Obesity Core Center Grant AM-26687.

REFERENCES

Bellisle, F., and Le Magnen, J. (1980). The analysis of human feeding patterns: The Edogram. *Appetite* **1**, 141–150.
Bellisle, F., and Le Magnen, J. (1981). The structure of meals in humans: Eating and drinking patterns in lean and obese subjects. *Physiol. Behav.* **27**, 649–658.
Birch, L. L., Billman, J., and Richards, S. S. (1984). Time of day influences food acceptability. *Appetite* **5**, 109–116.
Bobroff, E. M., and Kissileff, H. R. (1985). Effects of changes in palatability on food intake and the cumulative food intake curve in man. *Appetite* **6**. (In press.)
Booth, D. A. (1978). Acquired behavior controlling energy intake and output. *In* "The Psychi-

atric Clinics of North America" (A. J. Stunkard, ed.), Vol. 1, pp. 545–579. W. B. Saunders, Philadelphia.
Booth, D. A., Mather, P., and Fuller, J. (1982). Starch content of ordinary foods associatively conditions human appetite and satiation, indexed by intake and eating pleasantness of starch-paired flavors. *Appetite* 3, 163–184.
Bray, G. A. (1975). "Obesity in Perspective," Vol 2, part 2, pp. 7–11. National Institutes of Health, Washington, D.C.
Cabanac, M., and Duclaux, R. (1970). Specificity of internal signals in producing satiety for taste stimuli. *Nature (London)* 227, 966–967.
Drewnowski, A., and Greenwood, M. R. C. (1983). Cream and sugar: Human preferences for high-fat foods. *Physiol. Behav.* 30, 629–633.
Drewnowski, A., Grinker, J. A., and Hirsch, J. (1982). Obesity and flavor perception: Multidimensional scaling of soft drinks. *Appetite* 3, 361–368.
Drewnowski, A., Brunzell, J. B., Sande, K., Iverius, P. H., and Greenwood, M. R. C. (1983). Cream or sugar: Obese subjects' preferences for sweetened high fat foods. *Proc. Congr. Obesity, IVth,* New York. *Rec. Adv. Obesity Res.* 4, 30A.
Enns, M. P., Van Itallie, T. B., and Grinker, J. A. (1979). Contributions of age, sex and degree of fatness on preferences and magnitude estimations for sucrose in humans. *Physiol. Behav.* 2, 999–1003.
Hill, S. W., and McCutcheon, N. B., (1975). Eating responses of obese and nonobese humans during dinner meals. *Psychosom. Med.* 37, 395–401.
Houpt, T. R., Anika, S. M., and Houpt, K. A. (1979). Preabsorptive intestinal satiety controls of food intake in pigs. *Am. J. Physiol.* 236, R328–R337.
Jordan, H. A., Wieland, W. F., Zebley, S. P., Stellar, E., and Stunkard, A. J. (1966). Direct measurement of food intake in man: A Method for the objective study of eating behavior. *Psychosom. Med.* 28, 836–842.
Kissileff, H. R. (1984). Satiating efficiency and a strategy for conducting food loading experiments. *Neurosci. Biobehav. Rev.* 8, 129–135.
Kissileff, H. R., and Thorton, J. (1982). Facilitation and inhibition in the cumulative food intake curve in man. *In* "Changing Concepts of the Nervous System" (A. J. Morrison and P. Strick, eds.), pp. 585–607. Academic Press, New York.
Kissileff, H. R., Klingsberg, G., and Van Itallie, T. B. (1980). Universal eating monitor for continuous recording of solid or liquid consumption in man. *Am J. Physiol.* 238, R14–R22.
Kissileff, H. R., Pi-Sunyer, F. X., Thornton, J., and Smith, G. P. (1981). C-terminal octapeptide of cholecystokinin decreases food intake in man. *Am. J. Clin. Nutr.* 34, 154–160.
Le Magnen, J. (1971). Advances in studies on the physiological control and regulation of food intake. *In* "Progress in Physiological Psychology" (E. Stellar and J. M. Sprague, eds.), Vol. 4, pp. 203–261. Academic Press, New York.
Louis-Sylvestre, J., and Le Magnen, J. (1980). Palatability and preabsorptive insulin release. *Neurosci. Biobehav. Rev.* 4 (Suppl. 1), 43–46.
Moskowitz, H. R. (1977). Intensity and hedonic functions for chemosensory stimuli. *In* "The Chemical Senses and Nutrition" (M. R. Kare and O. Maller, eds.), pp. 71–99. Academic Press, New York.
Moskowitz, H. R., and Sidel, J. L. (1971). Magnitude and hedonic scales of food acceptability. *J. Food Sci.* 39, 677–680.
Nicolaidis, S. (1977). Sensory-neuroendocrine reflexes and their anticipatory and optimizing role on metabolism. *In* "The Chemical Senses and Nutrition" (M. R. Kare and O. Maller, eds.), pp. 123–143. Academic Press, New York.
Nisbett, R. E. (1968). Taste, deprivation, and weight determinants of eating behavior. *J. Personal. Soc. Psychol.* 10, 107–116.

Peryam, D. R., and Pilgrim, F. J. (1957). Hedonic scale method of measuring food preferences. *Food Technol. Suppl.* **11**, 9–14.

Pfaffmann, C., Dethier, V. G., and Hegsted, D. M. (1977). Concluding comments. *In* "The Chemical Senses and Nutrition" (M. R. Kare and O. Maller, eds.), pp. 463–475. Academic Press, New York.

Pi-Sunyer, F. X., Kissileff, H. R., Thornton, J. and Smith, G. P. (1982). C-Terminal octapeptide of cholecytokinin decreases food intake in obese men. *Physiol. Behav.* **29**, 627–630.

Price, J. M., and Grinker, J. (1973). Effects of degree of obesity, food deprivation, and palatability on eating behavior of humans. *J. Comp. Physiol. Psychol.* **85**, 265–271.

Rodin, J. (1975). Effects of obesity and set point on taste responsiveness and ingestion in humans. *J. Comp. Physiol. Psychol.* **89**, 1003–1009.

Rolls, B. J. (1986). *In* "Interaction of the Chemical Senses and Nutrition" (M. R. Kare and J. B. Brand, eds.), pp. 247–268. Academic Press, Orlando.

Rolls, B. J., Rowe, A., Rolls, E. T., Kingston, B., Megson, A., and Gunary, R. (1981). Variety in a meal enhances food intake in man. *Physiol. Behav.* **26**, 215–221.

Rozin, P., and Kennel, K. (1983). Acquired preferences for piquant foods by chimpanzees. *Appetite* **4**, 69–77.

Silverstone, T. (1975). Anorectic drugs. *In* "Obesity, Its Pathogenesis and Management" (T. Silverstone, ed.), pp. 193–227. Publishing Sciences Group, Acton, Mass.

Thompson, D. A., Moskowitz, H. R., and Campbell, R. G. (1976). Effects of body weight and food intake on pleasantness ratings for a sweet stimulus. *J. Appl. Physiol.* **41**, 77–83.

Witherly, S. A., Pangborn, R. M., and Stern, J. S. (1980). Gustatory responses and eating duration of obese and lean adults. *Appetite* **1**, 53–63.

Woo, R., Kissileff, H. R., and Pi-Sunyer, F. X. (1984). Elevated postprandial insulin levels do not induce satiety in normal-weight humans. *Am. J. Physiol.* **247**, R745–R749.

PART III

Discussion

McHugh: Dr. Kissileff, I have a question about your food stimuli. Dr. Rolls, in her work, cross-referenced sweet and savory; what about your carrier, tofu? Is that a sweet or savory substance?

Kissileff: Tofu, first of all, has no taste. It is, as a flavor, innert. The only flavors or tastes which were, therefore, in my mix were those that were derived from the fruits or yogurts. But Dr. Rolls has used yogurts also. Perhaps she could comment on how to classify their flavor.

Rolls: The way we use yogurt as a base, it is predominantly a sweet flavor. I think a very interesting question that has not been addressed very deeply is what happens when you provide, as a test stimulus, a food that is sweet, savory, bitter, etc.? How do these various qualities interact to yield a pleasant or an unpleasant rating?

Kissileff: One of the advantages of tofu is that it has nutrient properties. In a sense, it is a nutrient base into which one can introduce flavors.

Boyle: Dr. Rolls, there are a number of cases, I'm sure you are aware of, in which voluntary fluid intake is not adequate. Could one stimulate fluid intake by altering flavor?

Rolls: Yes, this is possible. We've reported on this with a low-calorie drink in humans and we've shown this to be possible in rats as well.

Kissileff: I have done some work with soup preloads, much as Dr. Rolls has done, and gotten somewhat different results. We modified the caloric density of soup either by adding water to it or cooking it to remove water. I found that consumption of a test meal following consumption of a high-caloric soup preload was lower than a low-caloric soup preload. The volume consumed in the preloads did not seem to make any difference. Isn't this different from what Dr. Roll's had found? Yes, it is, but I think it is important to realize that Dr. Rolls calorically diluted her soup not by adding water but by altering the maltodextrin or carrageenan levels. Thus, her soups were not as thin as the dilute soups I used.

Rolls: This opens many fascinating issues. When we did our macronutrient loads we tried to bring the weight up of our preloads. We tried to match these for calories but obviously were using very different bulk. What happened was that intake in the second course was affected by the bulk taken in during the first course. I think the whole issue of the effect of bulk, fiber, stomach fill, etc. on subsequent intake needs much more investigation.

Kissileff: I may have seen the results I did since, based on Dr. McHugh's work, a more concentrated calorie preload should stay in the stomach longer. I really don't see much of a conflict between our data and those of Dr. Rolls.

Rolls: The main point is that calories are not equal here, and we really need to know why not.

VanItallie: Originally, Dr. Porikos was scheduled to be here, but was not able to attend. I'd like to report on a study we carried out a few years ago. We used a platter or family-style paradigm to try to prevent hospitalized patients from restraining their food intake the way they usually do. People in the ward were served food on a large platter, family style and were encouraged to take as much food onto their plate as they chose. There was no constraint of portion size or not being able to have seconds of a favorite dish. We also put a lot of snacks in their refrigerators. We were careful, of course, to determine as best we could, exactly how much was eaten. We used healthy young individuals for this study. What we found was that these people gained a fair amount of weight. We certainly found, therefore, that offering a variety of highly palatable foods did lead to increased food intake. The other observation we made was that these people eat more than nonobese people contrary to what you often hear. If you rely on dietary histories, you may get the result that the obese do not eat more than the nonobese. But obese people apparently underreport what they eat. Having said all of this, I'd like to ask Dr. Rolls a question. How relevant is the work that you have done on sensory-specific satiety and related issues to the rather large prevalence of obesity in a country like the United States where a variety of foods is available? Statistics on obesity are rather staggering. There are 32 million adults overweight, according to the National Center for Health Statistics. Of these, 11 million are severely overweight. That's quite a large number. Do you think we're getting too much variety?

Rolls: Your work that you did with Dr. Porikos is about the only evidence we have that increasing food variety in humans will lead to increased intake and weight gain. Your study was for a limited period of time, a matter of days. It is quite possible that after these days, cognitive factors might come into play in humans telling them to slow down. So, we can't say that in humans variety per se inevitably leads to weight gain. But in rats, we obviously have many food variety experiments to look at. We vary many things. But what we did was try to keep palatability the same, tried to keep the basic proportions of macronutrients the same, but the food was really different in many sensory properties. We tried only to keep the basic proportions of fat, protein, and carbohydrates the same. But, these were real foods. We could get no effect just by altering the flavor of the chow.

Naim: But these foods you are referring to were different in composition as well as high in fat and sucrose. The basic problem with most of these feeding experiments with rats is that more than one parameter is being varied at a time. As a result, no firm conclusions can be made. As we reported here, we tried to vary only flavor while keeping macronutrient and micronutrient composition constant. When this is done using a low-fat semisynthetic diet in a multichoice arrangement, no hyperphagia is observed. Variety in sensory properties alone, therefore, was not sufficient to induce a dietary obesity in our studies. Let me ask a related question addressed to the human condition. In Japan, is it not true that most foods are low in fat and that the diet in general contains little variety, at least less variety than is in the modern Western diet? Is obesity a major problem in Japan?

VanItallie: There has been a striking increase in obesity in Japan paralleling an increase in meat intake. Until relatively recently, the Japanese have not had a high-fat intake. Now, however, their fat intake is on the rise. This has been blamed for the increase in obesity. As far as variety is concerned, it is a matter of judgment. When you go into a Sushi bar, you get a variety of fish, but it's still all fish.

Rolls: Monotony of a food may be important. I'm impressed by the tremendous variety of foods you have here in the United States, especially compared with England. Using liquid diets, we may be able to better control a test of the effect of flavor variety, but then you may have a problem with palatability.

PART IV

Interplay of Chemical Senses with Nutrient Metabolism

16

Taste and the Autonomic Nervous System*

RALPH NORGREN
Department of Behavioral Science and
The Milton S. Hershey Medical Center
Pennsylvania State University
Hershey, Pennsylvania

I. Introduction 323
II. The Afferent Limb 325
III. The Efferent Limb 329
IV. The Central Projections 334
References 340

I. INTRODUCTION

Gustatory afferent activity influences many aspects of feeding and drinking behavior. Psychophysics and electrophysiology attempt to understand how the gustatory apparatus codes and categories chemical stimuli, usually within the framework of human verbal descriptions of pure chemical stimuli. Perceived quality, however, represents only one dimension of the influence that taste exerts over energy and hydromineral balance (Norgren, 1984, 1985). Behavioral, endocrine, and autonomic responses to gustatory stimuli occur in chronically decerebrated rats, thus obviating the thalamocortical mechanisms that presumably underlie psychophysical assessments (Grill and Norgren, 1978a; Norgren and Grill, 1982). In addition, some of these responses to taste stimuli appear to be organized on sensory dimensions that differ from the four standard qualities derived from human judgments. For instance, in rats oral glucose appears to be the major unconditioned stimulus

*From *Chemical Senses*, Volume 10, No. 2, pp. 143–161. Reprinted by permission of IRL Press.

for the neurally mediated preabsorptive release of insulin, although glucose is less sweet to humans and less preferred by rodents than several other sugars (Grill et al., 1984). These observations serve only to emphasize that, as with other sensory systems, responses to gustatory afferent activity are organized at different levels of the nervous system, and perhaps on more than one sensory dimension.

Most nonbehavioral responses to gustatory stimuli involve endocrine or exocrine secretions that are mediated by the autonomic nervous system. Modern neuroanatomical methods for visualizing axonal pathways have revealed substantial parallels and even interdigitation of the central gustatory and autonomic systems. Before these central neural systems can be examined, however, the classical and modern views of the autonomic nervous system should be summarized to provide a common basis for comparison. The classical description of the autonomic nervous system was initiated by Langley (1903) at the turn of the century. Although elaborated in both anatomical and physiological detail, this description remained, in essence, unchanged through the first half of this century (Crosby et al., 1962). In this description, the autonomic nervous system consists of the motor neurons governing visceral, glandular, and cardiovascular responses by the opposing activity of the sympathetic and parasympathetic subdivisions. These subdivisions share several characteristics that differentiate the autonomic nervous system from the skeletal motor system, such as a two-neuron final common path to the effector organs and a relative independence of voluntary control, but differ from one another in other ways, such as the location of the preganglionic neurons within the central nervous system, their specificity of action, and their postganglionic neurotransmitters.

Although these characteristics remain important in the definition, investigations during the last two decades have documented enormous complexity in a system that had been at least conceptually simple. Much of this recent complexity is beyond the scope of the current discussion. Suffice it to say that the lack of voluntary control has been questioned, the specific–global response dichotomy attributed to the two subdivisions blurred, as well as their relative independence of one another both within the brain and the enteric nervous system (Brodal, 1981). The advances in understanding the autonomic nervous system that are most germaine to its interactions with the gustatory system include acceptance and careful description of an afferent as well as an efferent limb for both subdivisions, redefinition of the preganglionic parasympathetic motor pool in the medulla, and delineation of the anatomical connections between the limbic system areas known to influence autonomic activity and the final common motoneurons in the medulla and spinal cord. In what follows, these developments in the organization of the autonomic nervous system are discussed with reference to their anatomical and functional connections to the gustatory system. Much recent data, particularly with regard to the distribution of catecholamines and neuropeptides, will be given short shrift, not because they lack importance, but because their relationship to taste or taste-related functions remains dimly perceived (Armstrong et al., 1981, 1982; Helke et al., 1984; Higgins et al., 1984; Kawano and Chiba, 1984; Mantyh and Hunt, 1984).

II. THE AFFERENT LIMB

In its original incarnation, the autonomic nervous system was strictly motor, and some scientists adhere to the idea quite literally. "There is no satisfactory definition of visceral afferents [sic] that differentiates them from somatic afferents, thereby invalidating the use of the distinction." This quotation is taken from an anonymous review of a grant that proposed investigating the central organization of the gustatory and visceral afferent systems. Subsequently, I learned to phrase more carefully, as did Professor Amassian when he introduced a paper on the "central projections of the afferents distributed with the sympathetic outflow" (1951, p. 433). When referring to afferent axons from the body cavity, most scientists simply avoid the issue by using another, equally venerable classification, the doctrine of nerve components (Herrick, 1901, 1918). In this scheme, the autonomic nervous system is subsumed in the visceral efferent component, but a visceral afferent component is recognized as well.

The doctrine of nerve components, however, lacks the heuristic value of the autonomic nervous system, which is as much a physiological as an anatomical conception. The autonomic nervous system regulates the internal milieu; it is the nervous component of homeostasis. This regulation is more or less automatic. In nervous physiology, automatic means reflexive, and reflexes require afferent, as well as efferent, limbs. Most sensory information necessary for these regulatory reflexes also arises from the viscera. Virtually all sympathetic and parasympathetic nerves contain substantial numbers of afferent axons, often more than 50%. This afferent pool is far from neglected (see Mei, 1983, for a review), but often is related to the functions of specific organ systems, i.e., pulmonary stretch receptors, baroreceptors, atrial volume receptors, hepatic osmoreceptors. Visceral afferent axons seldom receive attention as a sensory system with common characteristics. Such an approach might establish whether or not a term such as the sympathetic afferent system has a physiological rationale.

As a group, afferent axons that reach the spinal cord via sympathetic and sacral parasympathetic nerves are less well described than their counterparts in the cranial parasympathetic nerves. To some extent, this may have resulted from their intertwining with the well-described somatosensory system. This intertwining, in fact, is physiological and provides the basis for visceral referred pain (Ammons et al., 1984; Rucker and Holloway, 1982). Unlike pain, referred or otherwise, much visceral afferent activity cannot be described, i.e., "does not reach consciousness." This characteristic supports the assumption that at some point within the CNS, this sympathetic afferent activity separates from the thalamocortical somatosensory system. In fact, recent evidence, both anatomical and physiological, indicates that visceral afferent axons entering the spinal cord relay independently from the somatosensory system, either within the cord itself, or in the caudal medulla (Nadelhaft and Booth, 1984; Neuhuber, 1982; Simon and Schramm, 1984; Wyss and Donovan, 1984).

The sensory neurons of the cranial parasympathetic nerves have received more attention than those in the sympathetic nerves, at least with respect to their central terminations. In all vertebrates that have been examined, the visceral sensory components of the vagus nerve terminate in the nucleus of the solitary tract or its homolog in the medulla (Ariens Kappers et al., 1936). Early on, the central destination of afferent axons in two other cranial parasympathetic nerves, the glossopharyngeal (IX) and the intermediate nerve of Wrisberg (VII), were the subject of some controversy, but the experimental evidence now agrees that the visceral afferent axons in these nerves also terminate in the solitary nucleus (see Norgren, 1984, for a review). The majority of the visceral afferent axons in VII and IX, and a small component in X, provide the gustatory innervation of the oral cavity and larynx. This fact, of course, establishes the major anatomical connection between taste and the autonomic nervous system. If my abbreviated argument for including afferent axons has any validity, then the gustatory system would be a sensory component of the autonomic nervous system.

Although the major terminus for gustatory and visceral afferent neurons in the cranial parasympathetic nerves has been long established, their central organization remains an active point of investigation. For the primary afferent neurons, two major conceptual problems can be identified. First, the range of adequate stimuli still need to be documented, as well as the nature and distribution of the receptors (Adachi and Niijima, 1982; Jeanningros, 1982; Mei, 1978; Vallet and Baertschi, 1980). Visceral afferent axons respond to chemical, mechanical, and thermal stimuli, often with great specificity. In many instances, however, the distribution or even gross location of the receptors is unknown. Even in the gustatory system, where the class of the adequate stimuli is known and the receptors more accessible, the response profiles of palatal taste buds have only begun to be investigated (Travers et al., 1986). Second, the functional organization of the primary afferent terminations within the solitary nucleus remains poorly understood. The small dimensions of both the nucleus and most of its neurons, as well as the difficulty of precisely stimulating many receptors, has hampered determination of somatotopy, or functional convergence within the nucleus.

The advent of techniques for visualizing transganglionic anterograde transport of horseradish peroxidase (HRP) prompted anatomical investigations of this second problem based upon injections of individual organs and nerve branches (Mesulem, 1978). Many studies are limited to only one or two nerve branches (Berger, 1980; Ciriello et al., 1981a,b; Hosoya and Sugiura, 1984; Leslie et al., 1982; Nomura and Mizuno, 1982; Wallach and Loewy, 1980; Whitehead and Frank, 1983). The more comprehensive series use different species to concentrate on different levels of the nucleus (Hamilton and Norgren, 1984; Kalia and Mesulam, 1980; Norgren and Smith, 1983). Nevertheless, taken together, these experiments provide a better picture of the somatotopic projection of oral and visceral afferent axons in the nucleus of the solitary tract.

Specific organs or receptor groups have restricted representations within the solitary nucleus that are arranged in a crude somatotopy, with the oral cavity most rostral and the gastric viscera, most caudal. Although restricted, the afferent distributions from visceral organs are neither independent of one another, nor confined to specific subnuclei (Kalia and Mesulam, 1980), but an organization is clear. Gustatory and somatosensory afferent axons from the oral cavity fill the anterior one-half of the nucleus (Fig. 1). Within this area, the modality representation may not be strictly somatotopic. Although the electrophysiological data are incomplete, it appears that gustatory axons from the oral cavity, i.e., the tongue and palate, concentrate most rostrally. Oral somatosensory axons from the lingual and glossopharyngeal nerves terminate caudal to the taste axons, but overlap presumed gustatory and somatosensory axons of the superior laryngeal nerve that synapse densely in the interstitial subnucleus. Based on our HRP data from the cervical vagus, the distributions of afferent axons from the oral and body cavitites are virtually nonoverlapping (Hamilton and Norgren, 1984). In cats, the afferent axons from the lungs and heart terminate preferentially in the subnuclei that surround the solitary tract with its enclosed interstitial subdivision, whereas those from the gastric viscera end more medially (Kalia and Mesulam, 1980). This last projection also appears in rats with the gastric branches denser rostrolaterally, and the hepatic and coeliac branches, caudomedially (Norgren and Smith, 1983).

When only the densest label is considered, each organ or nerve branch has a unique pattern that nevertheless provides an anatomical basis for some primary afferent convergence within the nucleus of the solitary tract. An important corollary to the variety of sensory information in the visceral afferent system is the presence of receptors in different locations that trans-

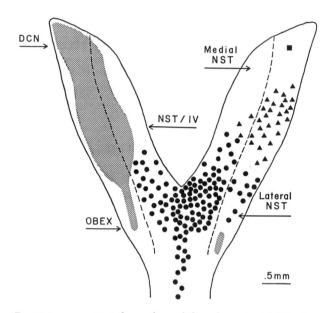

Fig. 1. Drawing representing the nucleus of the solitary tract bilaterally in horizontal section. Rostral is at the top. Shaded area represents the distribution of the nerves containing gustatory afferent axons, i.e., the chorda tympani and greater superficial petrosal branches of VII, the lingual–tonsilar branch of IX, and the superior laryngeal branch of X. Symbols represent the distribution patterns of the inferior alveolar (square) and lingual (triangles) branches of the trigeminal nerve and the cervical vagus (circles). DCN, Caudal limit of the dorsal cochlear nucleus; NST, nucleus of the solitary tract; NST/IV, point at which the NST meets the wall of the fourth ventricle. [Reprinted with permission from Hamilton and Norgren (1984), Fig. 10.]

duce the same physical stimulus. The most obvious example is the aortic and carotid baroreceptors, but many others exist. This raises the possibility of a functional as well as a topographical organization within this sensory system. Although the electrophysiological analysis of the central visceral afferent system now lags behind the anatomical data, a few experiments support this possibility. Hildebrandt (1974) recorded neurons in the solitary nucleus that responded to both carotid and aortic baroreceptors. Some neurons in the second-order visceral afferent relay in the pons respond to Na ions stimulating the tongue or the liver (Rogers et al., 1979). Recently, we found neurons in the anterior solitary nucleus that responded to gustatory stimuli applied either on the anterior tongue or the nasoincisor ducts (Travers et al., 1986). These data support the value of the anatomical evidence of converging primary afferent terminations. The nerves that innervate these gustatory receptor subpopulations, the chorda tympani and greater superficial petrosal branches of VII, have a common termination zone within the anterior solitary nucleus (Hamilton and Norgren, 1984). The neurons that responded to both receptor subpopulations were located within that zone. The same data, however, further complicate hypotheses concerning the functional organization of the solitary nucleus. Even though these neurons receive convergent input from the same class of receptors, i.e., taste buds, in many cases the most effective chemical stimulus differed depending upon which receptor field was stimulated. Sodium was most effective on the anterior tongue; sucrose, on the nasoincisor ducts.

The sensory organization of the nucleus of the solitary tract (NST) is becoming increasingly difficult to summarize. Superficially, it resembles the other major sensory systems in that the receptor sheet appears to be mirrored in the organization of the first central relay, i.e., somatotopy. This analogy is hampered not only by the absence of an easily specifiable receptor surface, but also by some degree of convergence from receptors at different locations on that surface. In any event, a succinct functional summary of the system would be scuttled by the neurochemical complexity that has been revealed during the last few years. For present purposes, however, one conclusion can be drawn that may prove useful in later functional analysis. The sensory representation of the oral and body cavities appears to be relatively independent within the nucleus of the solitary tract.

III. THE EFFERENT LIMB

As mentioned above, the autonomic nervous system was originally defined in efferent terms. Its dual, often functionally counterpoised subdivisions, sympathetic and parasympathetic, have clearly identified peripheral nerves and ganglia and, until recently, equally well-defined central preganglionic motoneuron pools. In the spinal cord, the vast majority of sympathetic motoneurons occur in the intermediolateral cell column that extends from the upper few lumbar segments through the thoracic level and, at least in some species, into the cervical cord. Recently, however, presumed sympathetic motoneurons have been observed in the intermediomedial column, the dorsal gray commissure, and even the ventral horn (Chung *et al.*, 1975, 1979; Hancock and Peveto, 1979; Luiten *et al.*, 1984; Petras and Faden, 1978). The preganglionic parasympathetic neurons in the spinal cord occur in an analogous location in the intermediolateral gray matter, but further caudally at the level of the L6–S1 vertebrae. The location and distribution of these neurons have been known for some time, and have been confirmed and refined with HPR histochemistry and electrophysiological techniques (deGroat *et al.*, 1982; Nadelhaft and Booth, 1984).

Due to their anatomical proximity and well-documented functional relationships, the gustatory system can be more readily connected to the autonomic outflow from the medulla than from the spinal cord. As it happens, the medullary preganglionic parasympathetic motoneurons have undergone as much or more redefinition than the autonomic neurons in the spinal cord. Classically, the two major sources for efferent axons in cranial parasympathetic nerves, the dorsal motor nucleus of the vagus (DMX) and the nucleus ambiguus (NA), were as functionally distinct as they were anatomically separate. The third source, the oculomotor parasympathetic neurons in and near the Edinger–Westphal nucleus, will not be discussed (see Burde and Loewy, 1980, for recent evidence). The DMX, sandwiched between the solitary and hypoglossal nuclei, contained true preganglionic parasympathetic neurons. Well ventral in the reticular formation, the NA controlled the muscular functions subserved by the glossopharyngeal and the vagus nerves, i.e., pharynx, larynx, and esophagus. In rodents at least, HRP applied to the vagus nerve revealed that the dorsal motor nucleus extends rostral to the limits observable in normal material. These neurons occur in the medial third of the NST almost to its rostral limit (Fig. 2A; Coil and Norgren, 1979; Dennison *et al.*, 1981). If the efferents that distribute with the lingual branch of the glossopharyngeal nerve are considered, even the rostral pole of the solitary nucleus contains a substantial population of preganglionic parasympathetic neurons (Contreras *et al.*, 1980).

As these recent anatomical studies extended the conventional boundaries of the dorsal motor nucleus, they also revealed a penumbra of vagal efferent neurons in the reticular formation surrounding the nucleus ambiguus. These ventral efferent neurons extend from just caudal to the facial nucleus, the so-called retrofacial area, around the nucleus ambiguus proper, and into the upper reaches of the ventral horn, the retroambiguus area (Kalia and Sullivan, 1982). Perhaps in keeping with this enlarged anatomical substrate, ample evidence now indicates that this ventral efferent column mediates not only skeletal motor activity, but also true parasympathetic responses. Gastric motility, acid secretion, cardiac inhibition, and neurally mediated insulin secretion each has been linked to neurons in or near the nucleus ambiguus by both anatomical and physiological experiments (Bereiter *et al.*, 1981; Geis and Wurster, 1980; Gunn *et al.*, 1968; Hopkins and Armour, 1982; Jordan *et al.*, 1982; Nosaka *et al.*, 1979; Pagani *et al.*, 1984; Sugimoto

---

Fig. 2. Preganglionic parasympathetic neurons in the medulla of the rat. (A) Neurons in the medial solitary nucleus (top right) and in and near the nucleus ambiguus (bottom left) retrogradely labeled after exposing the cervical vagus nerve to horseradish peroxidase. Labeled fascicles are the entering vagal afferent axons (top) and exiting efferents. (B) Labeled somata and dendrites of the inferior salivatory nucleus after exposing lanced otic ganglion to horseradish peroxidase. Dendrites extend from the rostral end of the solitary nucleus dorsally to the edge of the facial motor nucleus ventrally. [Reproduced with permission from Contreras *et al.* (1980), Fig. 9.]

16. Taste and the Autonomic Nervous System

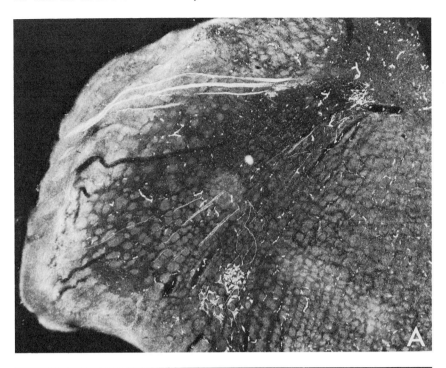

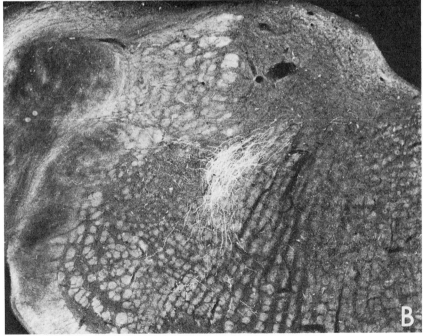

et al., 1979; Williford *et al.*, 1981). These responses extend rather than replace the classical skeletal motor functions attributed to the nucleus ambiguus. Nor does the region around the NA replace the dorsal motor nucleus as a major source of parasympathetic function. With a few exceptions, such as the neurons that innervate the thymus (Bulloch and Moore, 1981), there appears to be an alternate or parallel arrangement. The organs and functions subserved by the cranial nerve parasympathetic system, however, are so diverse that the actual division of labor has yet to be catalogued. Although the NA has acquired some functions previously reserved for the dorsal motor nucleus, the reverse is not the case. The dorsal motor nucleus does not innervate pharyngeal and laryngeal musculature.

The dorsal motor nucleus and nucleus ambiguus are not only functionally related, but also anatomically connected. Evidence from rats, cats, dogs, and monkeys indicates that the preganglionic parasympathetic neurons subserving the salivatory and lacrimal glands span the reticular formation between the rostral solitary nucleus and the periambiguus neurons (Chibuzo and Cummings, 1980; Chibuzo *et al.*, 1980; Perwaiz and Karim, 1982; Yu and Srinivasan, 1980; see Contreras *et al.*, 1980, for references prior to 1980). Neurons in the superior and inferior salivatory nuclei form a diffuse column that separates the medial, magnocellular reticular formation from the lateral, parvicellular zone. The salivatory neurons resemble the cells of the dorsal motor nucleus; they are small relative to skeletal motor neurons, spindle or pyramidal in shape, with few primary dendrites. Unlike DMS neurons, however, the salivatory preganglionic cells have extensive dendritic trees, oriented dorsoventrally, that often extend from the solitary nucleus down to the level of the nucleus ambiguus (Fig. 2B). In addition, unlike DMX, these neurons are dispersed in the reticular formation amid other, morphologically similar neurons that are presumably neither parasympathetic nor motor. Some of these other neurons may be interneurons involved in the coordination of oral behavior, such as licking, chewing, and swallowing (Travers and Norgren, 1983). In this sense, the term "superior and inferior salivatory nuclei" is a misnomer. In fact, a significant proportion of the preganglionic parasympathetic neurons in this area subserve nonsalivatory secretomotor functions, such as the mucous glands of the tongue, palate, and nasal passages. The neurons serving the lacrimal gland, on the other hand, are tightly clustered just lateral to the facial nucleus.

This brief summary of the efferent limb of the autonomic nervous system provides a basis for understanding how the gustatory system influences visceral and glandular responses, as well as behavior. If the parasympathetic neurons in the medulla were confined to the dorsal motor nucleus, much of the central gustatory system would appear functionally superfluous. The arc of preganglionic parasympathetic neurons extending from the solitary nu-

16. Taste and the Autonomic Nervous System   333

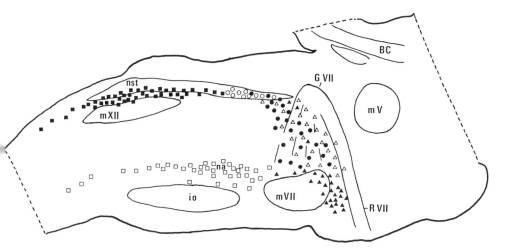

Fig. 3. Schematic representation on a parasagittal section of preganglionic parasympathetic neurons that distribute with the facial, glossopharyngeal, and vagus nerves in the rat. Rostral is to the right; dorsal is up. Efferents in the cervical vagus are concentrated in the dorsal motor nucleus of the vagus and the medial solitary nucleus (filled squares) and in and near the nucleus ambiguus (open squares). Those in the tympanic (open circles) and lingual–tonsilar (filled circles) branches of the glossopharyngeal nerve fill the medial edge of the solitary nucleus and the underlying reticular formation. Efferent neurons of the chorda tympani (open triangles) and greater superficial petrosal (filled triangles) branches of the facial nerve intermingle with those of the glossopharyngeal nerve in the reticular formation. The efferent neurons of the superior laryngeal (SLN) and pharyngeal branches of the vagus were excluded because they are primarily muscular. Some SLN cells located in the medial solitary nucleus are presumably preganglionic parasympathetic neurons serving the mucus glands of the larynx (Hamilton and Norgren, 1984). BC, Brachium conjunctivum; G VII, genu of the facial nerve; io, inferior olive; m V, trigeminal motor nucleus; m VII, facial motor nucleus; m XII, hypoglossal nucleus; na, nucleus ambiguus; nst, nucleus of the solitary tract; R VII, root of the facial nerve. [Redrawn with permission from Contreras *et al.* (1980), Fig. 10.]

cleus through the rostral medullary reticular formation and caudally past the nucleus ambiguus (Fig. 3) matches remarkably well both the local distribution of the central gustatory system and the reciprocal projections from forebrain structures that receive gustatory afferent input. The connections to the sympathetic pool in the spinal cord are more complicated, because additional interneurons must be invoked. Many of the germaine experiments involve cardiovascular functions, thus requiring parallel analogies for gustatory responses. Nevertheless, neurons that project to the intermediolateral columns are concentrated in the reticular formation near the nucleus ambiguus and, in the cat at least, within the solitary nucleus itself (Amendt *et al.*, 1979; Helke *et al.*, 1982; Loewy and Burton, 1978; Loewy and McKellar, 1981; Loewy *et al.*, 1979, 1981; Miura *et al.*, 1983; Yamada *et al.*, 1984; as examples).

IV. THE CENTRAL PROJECTIONS

Up to now this examination of the neural basis for gustatory interaction with the autonomic nervous system has been kept simple by treating only primary afferent and efferent neurons, largely independently, and ignoring the related neurochemistry. Venturing further into the CNS inevitably adds complexity (and uncertainty). In the present case, the anatomical data are extensive, but easily summarized. Functional or behavioral information is more limited, but provides some basis for inquiry. Electrophysiological data are sparse, and even the experiments in cognate areas, such as cardiovascular regulation, usually beg theoretical issues that might have broader applicability. Because much of this data has been reviewed recently from several perspectives (Norgren, 1983, 1984; Swanson and Mogenson, 1981), this analysis will concentrate as much on the implications of the research as on a summary of the results.

Unlike the primary afferent distributions, apparent species differences complicate a concise summary of the central ramifications of the gustatory and visceral afferent systems of the solitary nucleus (Norgren, 1981). These differences are documented in common laboratory species, such as rat, cat, and monkey, but more fragmentary evidence from other mammals and other vertebrates suggests further variations on the theme (Car *et al.*, 1975; Katz and Karten, 1979; Dubbledam *et al.*, 1976, 1979; Finger, 1978). In this synopsis rodents serve as the model, not because they are necessarily prototypical, but because they provide the largest data base.

The general plan of the interoceptive sensory systems that relay in the nucleus of the solitary tract is similar to that for the exteroceptive systems of the medulla. All have ascending components that establish the thalamocortical sensory elaborations, as well as local projections to the bulbar reticular formation and cranial motor nuclei. Beyond this superficial similarity, the two classes of sensory system diverge on several dimensions. First, the interoreceptive pathways to the thalamus are substantially ipsilateral and nonlemniscal, ascending in the central tegmental tract. Second, unlike the somatosensory system at least, the interoceptive systems of the solitary nucleus have a second-order relay in the pons, the parabrachial nuclei (Norgren and Leonard, 1971, 1973). Finally, unlike all other sensory systems except possibly olfaction, the largest distribution of the interoceptive sensory system is not on the cortex, but in the limbic system (Fig. 4; Norgren, 1976).

16. Taste and the Autonomic Nervous System   335

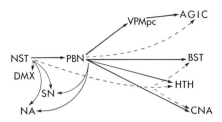

Fig. 4. Schematic representation in the sagittal plane of the central projections of the gustatory system in the rat. Rostral is right, dorsal is up. Heavy lines indicate major pathways to the forebrain; thin lines, local connections; dashed lines, related pathways without documented gustatory function. AGIC, Agranular insular (gustatory) cortex; BST, bed nucleus of the stria terminalis; CNA, central nucleus of the amygdala; DMX, dorsal motor nucleus of the vagus; HTH, hypothalamus including paraventricular nucleus; NA, nucleus ambiguus; NST, nucleus of the solitary tract; PBN, parabrachial nuclei; SN, salivatory nuclei; VPMpc, parvicellular division of the ventral posteromedial nucleus (thalamic gustatory relay). See Norgren (1985) for more details.

Local connections, of course, are a matter of definition. In the case of the gustatory and visceral afferent systems, it seems logical to include those axons arising from both the solitary and parabrachial nuclei that do not project beyond the pons or medulla. Based upon anatomical data alone, the major target of both the gustatory and visceral afferent neurons in the NST appears to be the ipsilateral parabrachial nuclei (PBN) (Norgren, 1978).

Electrophysiological data, however, indicate that only about 20% of NST gustatory neurons can be antidromically activated from the pons (Ogawa et al., 1984; Ogawa and Kaisaku, 1982). The remaining neurons contribute to a dense intranuclear axonal plexus or project into the parvicellular reticular formation between the NST and the facial motor nucleus, as well as to the region surrounding the nucleus ambiguus (Norgren, 1978; Loewy and Burton, 1978). Neurons in the parabrachial nuclei have similar local distributions, although the reciprocal projection to the NST is weak compared with that to the NA and the reticular formation (Norgren, 1976; Saper and Loewy, 1980). It also should be noted that the parabrachial nuclei receive substantial axonal projections from the medullary reticular formation, as well as the solitary nucleus (King, 1980). These reticular neurons, however, are spread across the parvicellular zone, rather than being restricted to areas rich in preganglionic parasympathetic efferents.

The local connections of both the solitary and parabrachial nuclei provide an obvious anatomical basis for gustatory influence on parasympathetic function. Although the mechanisms of this influence remain poorly documented electrophysiologically, behavioral evidence indicates that the forebrain is unnecessary for many autonomic responses. Chronically decerebrate rats, which lack any neural connections to and from the forebrain, do not thermoregulate effectively, but can survive for months, given appropriate care (Grill and Norgren, 1978b). Even the faulty thermoregulation may result from too much brain rather than not enough (Amini-Sereshki and Zarrindast, 1984). Appropriate care consists of maintaining nutrition and hydration by gavage and correcting hyperthermia if necessary. Otherwise the animals' regulatory systems manage on their own for many weeks, implying substantial autonomic competence. This is perhaps not surprising inasmuch as acutely decerebrate preparations are standard for studying autonomic reflexes. One of the few spontaneous behavioral sequences remaining in the chronically decerebrate rat, grooming, also presumes autonomic function, because effective grooming requires adequate salivary flow (Grill, 1980). In addition, specific autonomic reflexes, such as the sympathoadrenal response to glucoprivation and, more to the point, preabsorptive release of insulin elicited by intraoral gustatory stimuli, survive in the chronically decerebrate preparation (DiRocco and Grill, 1979; Grill and Berridge, 1981).

Similar observations of reasonably complex skeletal motor functions, such as walking, chewing, or swallowing, being organized in the spinal cord or caudal brainstem were cited as support for the Jacksonian hierarchical model of functional organization (Jackson, 1898; Grill, 1980). Mountcastle (1978) criticized the strict Jacksonian conception of rerepresentation, because the central organization of the skeletal motor system contains too many divergent, reciprocal, or reafferent connections to support such a model. In its stead, Mountcastle proposed a model for sensorimotor integration based upon distributed systems, in which similar functions were still dispersed across many neural levels, but the lines of authority extended in a complex web between these functional loci, rather than in a single, ladderlike hierarchy.

Mountcastle's distributed system model undoubtably is more accurate than Jacksonian hierarchy, but at present, it also is less heuristic. Although less is known about the autonomic sensorimotor system, its central organization suggests a functional arrangement that, if not hierarchical, may be more directly related to the first level of integration in the medulla and spinal cord.

In rodents, the central gustatory system has two distinct projections to the forebrain. A similar pattern probably exists for the other visceral afferent systems that first synapse in the NST, but the evidence to date is less compelling. As mentioned above, the gustatory neurons in the solitary nucleus do not themselves project to the forebrain. They ascend a few millimeters though the reticular formation and terminate in the parabrachial nuclei. In anesthetized animals, parabrachial neurons respond to gustatory stimuli with characteristics that differ only in degree from peripheral or solitary nucleus cells (Norgren and Pfaffmann, 1975; Perrotto and Scott, 1976; Travers and Smith, 1984; Van Buskirk and Smith, 1981). This appears to be the case with rabbits as well as rodents, but in awake, behaving rabbits parabrachial neurons that respond to gustatory and intraoral somatosensory stimuli alter those responses as a function of ingestive movements, associations, and the hedonic properties of the stimuli (DiLorenzo and Schwartzbaum, 1982; Schwartzbaum, 1983).

It is the parabrachial nuclei that give rise to two distinct projections to the forebrain. Parabrachial gustatory neurons ascend, largely ipsilaterally, to terminate in both the ipsilateral and contralateral thalamic gustatory relay, the parvicellular division of the ventral posteromedial nucleus (VPMpc) (Norgren and Leonard, 1971, 1973). The contralateral projection occurs primarily within the thalamus rostral to VPMpc, and it is not a precise mirror image of the ipsilateral terminal field (Norgren, 1976). From the thalamus, gustatory neurons project to a thin strip of agranular insular cortex subjacent to the primary somatosensory representation for the tongue (Norgren and Wolf, 1975; Kosar et al., 1986). Several laboratories also have documented a direct projection from neurons in and near the PBN to the vicinity of the gustatory cortex (Lasiter et al., 1982; Saper, 1982a; Shipley and Sanders, 1982). Unlike the thalamocortical route, however, electrophysiologial analysis of this projection is lacking, and thus its functional specificity remains uncertain. Aside from its ipsilateral character, the thalamocortical gustatory system seems to parallel the pattern of the exteroceptive senses. Of the species that have been investigated, only monkeys differ substantially. Old World monkeys, at least, have two cortical gustatory areas, but probably few if any gustatory neurons in the parabrachial nuclei (Beckstead et al., 1980; Block and Schwartzbaum, 1983; Pritchard et al., 1986; see Norgren, 1984, for a review).

The other rostral projections arising from the PBN distribute to the ventral forebrain. Based upon anatomical evidence, these parabrachial axons terminate densely in the lateral hypothalamus, the central nucleus of the amygdala, and the bed nucleus of the stria terminalis (Norgren, 1976). Equally dense projections pass through the lateral preoptic area and substantia innominata, but any inference of functional connections requires electrophysiological data. Less dense projections pass through or reach the substantia nigra, medial hypothalamic nuclei, paraventricular nucleus, and most of the remaining amygdaloid nuclei (Saper and Loewy, 1980). To some extent, the differences in the relative density of these distributions reflect differential projections from the medial, largely gustatory zones of the parabrachial nuclei compared with the more lateral, visceral afferent areas. In addition, it should be noted that similar, but less substantial projections to the ventral forebrain arise directly from the caudal or viscerosensory zone of the solitary nucleus (Ricardo and Koh, 1978; Ciriello and Calaresu, 1980a,b). The sensory nature of the parabrachial projections to the ventral forebrain has been documented at least for taste and baroreceptor afferent activity (Bloch and Schwartzbaum, 1983; Hamilton et al., 1981; Kannan et al., 1981; Norgren, 1974, 1976). These data prove that afferent activity critical to autonomic function reaches the ventral forebrain, but provide few hints about its organization or possible function within this extensive distribution.

The efferent projections back to the caudal brainstem from the gustatory cortex, lateral hypothalamus, paraventricular nucleus, amygdala, and bed nucleus of the stria terminalis may provide hints about functions. Each of these areas receives taste input and each projects directly back to the relay nuclei afferent to it. Gustatory cortex projects most strongly back to the thalamic taste relay, but also sends axons directly to both the parabrachial and solitary nuclei (Norgren and Grill, 1976; Saper, 1982b; Shipley, 1982). The amygdala, hypothalamus, paraventricular nucleus, and bed nucleus of the stria terminalis do not project to the gustatory thalamus, but do have reciprocal connections with the parabrachial and solitary nuclei (see Norgren, 1983, and Van Der Kooy et al., 1984, for other references). In addition to reciprocal connections to the sensory relay nuclei, these ventral forebrain areas also have direct axonal connections to the vicinity of the preganglionic parasympathetic neurons in the medulla and, in some cases, to the sympathetic motoneurons in the spinal cord as well (Berk and Finkelstein, 1982; Conrad and Pfaff, 1976; Schwanzel-Fukuda et al., 1984; Hosoya and Matsushita, 1981; Hosoya et al., 1984; Swanson, 1977; Swanson and Kuypers, 1980; Van Der Kooy et al., 1984). A few electrophysiological studies are beginning to answer the most obvious questions posed by the anatomy. Paraventricular neurons not only project directly to the NST, but influence presumed second-order visceral afferent neurons there (Rogers and Nelson, 1984). Lateral hypothalamic cells facilitate taste and orosensory neurons in the solitary nucleus, as well as identified preganglionic salivatory neurons in the medullary reticular formation (Matsuo et al., 1984; Matsuo and Kusano, 1984).

Based upon physiological responses to lesions and stimulation, the hypothalamus was long touted as the highest autonomic ganglion. The classical conception of hypothalamic anatomy, however, left it synaptically remote both from the visceral afferent information necessary for autonomic functions and from the final common path efferents in the medulla and spinal cord (see Ricardo, 1983, for a review). The research summarized here establishes that much of the sensory information relevant to autonomic function has direct access to the ventral forebrain, including the hypothalamus. These areas, in turn, project back directly to both gustatory and visceral afferent relays in the medulla and pons and to the final common pathways for autonomic function.

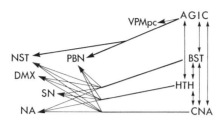

Fig. 5. Schematic representation in the sagittal plane of potential gustatory–autonomic reciprocal projections. Heavy lines indicate major long, descending forebrain projections to hindbrain; thinner lines, local forebrain interconnections. See Fig. 4 for abbreviations.

These long, ascending and descending projections provide an anatomical basis for hypothalamic influence on autonomic control, but also reduce the potential import of that influence. Classically, the hypothalamus was an important functional area in its own right, but also served as a funnel for integrating the neural influence of the limbic system. As far as its neuroendocrine and behavioral functions are concerned, the hypothalamus may still play this pivitol role. The autonomic pathways to and from the medulla, however, are organized in parallel. The central nucleus of the amygdala and the bed nucleus of the stria terminalis have reciprocal connections with one another and with the hypothalamus, but all three areas receive afferent projections directly from gustatory and visceral afferent relays and send reciprocal efferents back to the caudal brainstem. With respect to the autonomic nervous system, at least, these three ventral forebrain areas may function somewhat independently.

Taken with the thalamocortical gustatory system, which is also parallel to the ventral forebrain autonomic afferent and efferent projections, the central autonomic system appears to be, not hierarchical, but radial in arrangement (Fig. 5). Interconnections occur at the first central level in the medulla and spinal cord that clearly account for many reflex adjustments. Interconnections also occur among the forebrain areas involved, including the gustatory cortex, but the communication between the forebrain and hindbrain appears to be direct and parallel. If this admittedly simple conception has any validity, then hypotheses about the influence of specific forebrain regions on autonomic responses to gustatory stimuli may be easier to test than analogous hypotheses concerning behavior elicited by the same sapid chemicals.

ACKNOWLEDGMENTS

I thank Melanie Newman and Sharon Smith for bibliographic assistance. The author's research was supported by grants from NIH (NS 20397) and NSF (BNS (83-41375).

REFERENCES

Adachi, A., and Niijima, A. (1982). Thermosensitive afferent fibers in the hepatic branch of the vagus nerve in the guinea pig. *J. Auton. Nerv. Syst.* **5**, 101–109.
Amassian, V. E. (1951). Cortical representation of visceral afferents. *J. Neurophysiol.* **14**, 433–444.
Amendt, K., Czachurski, J., Dembowsky, K., and Seller, H. (1979). Bulbospinal projections to the intermediolateral cell column; a neuroanatomical study. *J. Auton. Nerv. System.* **1**, 103–117.
Amini-Sereshki, L., and Zarrindast, M. R. (1984). Brain stem tonic inhibition of thermoregulation in the rat. *Am. J. Physiol.* **247**, R154–R159.
Ammons, W. S., Blair, R. W., and Foreman, R. D. (1984). Greater splanchnic excitation of primate T1–T5 spinothalamic neurons. *J. Neurophysiol.* **51**, 592–603.
Ariens Kappers, C., Huber, G., and Crosby, E. (1936). "The Comparative Anatomy of the Nervous System of Vertebrates, Including Man." Macmillan, New York.
Armstrong, D. M., Pickel, V. M., Joh, T. H., Reis, D. J., and Miller, R. J. (1981). Immunocytochemical localization of catecholamine synthesizing enzymes and neuropeptides in area postrema and medial nucleus tractus solitarius of rat brain. *J. Comp. Neurol.* **196**, 505–517.
Armstrong, D. M., Ross, C. A., Pickel, V. M., Joh, T. H., and Reis, D. J. (1982). Distribution of dopamine-, noradrenaline-, and adrenaline-containing cell bodies in the rat medulla oblongata: Demonstrated by the immunocytochemical localization of catecholamine biosynthetic enzymes. *J. Comp. Neurol.* **212**, 173–187.
Beckstead, R., Morse, J., and Norgren, R. (1980). The nucleus of the solitary tract in the monkey: Projections to the thalamus and brain stem nuclei. *J. Comp. Neurol.* **190**, 259–282.
Bereiter, D. A., Berthoud, H.-R., Brunsmann, M., and Jeanrenaud, B. (1981). Nucleus ambiguus stimulation increases plasma insulin levels in the rat. *Am. J. Physiol.* **241**, E22–E27.
Berger, A. J. (1980). The distribution of the cat's carotid sinus nerve afferent and efferent cell bodies using the horseradish peroxidase technique. *Brain Res.* **190**, 309–320.
Berk, M. L., and Finkelstein, J. A. (1982). Efferent connections of the lateral hypothalamic area of the rat: An autoradiographic investigation. *Brain Res. Bull.* **8**, 511–526.
Block, C. H., and Schwartzbaum, J. S. (1983). Ascending efferent projections of the gustatory parabrachial nuclei in the rabbit. *Brain Res.* **259**, 1–9.
Brodal, A. (1981). "Neurological Anatomy in Relation to Clinical Medicine," 3rd ed. Oxford University Press, New York.
Bulloch, K., and Moore, R. Y. (1981). Innervation of the thymus gland by brain stem and spinal cord in mouse and rat. *Am. J. Anat.* **162**, 157–166.
Burde, R. M., and Loewy, A. D. (1980). Central origin of oculomotor parasympathetic neurons in the monkey. *Brain Res.* **198**, 434–439.
Car, A., Jean, A., and Roman, C. (1975). A pontine primary relay for ascending projections of the superior laryngeal nerve. *Exp. Brain Res.* **22**, 197–210.
Chibuzo, G. A., and Cummings, J. F. (1980). Motor and sensory centers for the innervation of mandibular and sublingual salivary glands: A horseradish peroxidase study in the dog. *Brain Res.* **189**, 301–313.
Chibuzo, G. A., Cummings, J. F., and Evans, H. E. (1980). Autonomic innervation of the tongue: A horseradish peroxidase study in the dog. *J. Auton. Nerv. Syst.* **2**, 117–129.
Chung, J. M., Chung, K., and Wurster, R. D. (1975). Sympathetic preganglionic neurons of the cat spinal cord: Horseradish peroxidase study. *Brain Res.* **91**, 126–131.

16. Taste and the Autonomic Nervous System   341

Chung, K., Chung, J. M., Lavelle, F. W., and Wurster, R. D. (1979). Sympathetic neurons in the cat spinal cord projecting to the stellate ganglion. *J. Comp. Neurol.* **185**, 23–30.
Ciriello, J., and Calaresu, F. R. (1980a). Autoradiographic study of ascending projections from cardiovascular sites in the nucleus tractus solitarii in the cat. *Brain Res.* **186**, 448–453.
Ciriello, J., and Calaresu, F. R. (1980b). Monosynaptic pathway from cardiovascular neurons in the nucleus tractus solitarii to the paraventricular nucleus in the cat. *Brain Res.* **193**, 529–533.
Ciriello, J., Hrycyshyn, A. W., and Calaresu, F. R. (1981a). Horseradish peroxidase study of brain stem projections of carotid sinus and aortic depressor nerves in the cat. *J. Auton. Nerv. Syst.* **4**, 43–61.
Ciriello, J., Hrycyshyn, A. W., and Calaresu, F. R. (1981b). Glossopharyngeal and vagal afferent projections to the brain stem of the cat: A horseradish peroxidase study. *J. Auton. Nerv. Syst.* **4**, 63–79.
Coil, J., and Norgren, R. (1979). Cells of origin of motor axons in the subdiaphragmatic vagus of the rat. *J. Auton. Nerv. Syst.* **1**, 203–210.
Conrad, L. C. A., and Pfaff, D. W. (1976). Efferents from medial basal forebrain and hypothalamus in the rat. *J. Comp. Neurol.* **169**, 221–262.
Contreras, R., Gomez, M., and Norgren, R. (1980). Central origins of cranial nerve parasympathetic neurons in the rat. *J. Comp. Neurol.* **190**, 373–394.
Crosby, E. C., Humphrey, T., and Lauer, E. W. (1962). "Correlative Anatomy of the Nervous System." Macmillan, New York.
deGroat, W. C., Booth, A. M., Milne, R. J., and Roppolo, J. R. (1982). Parasympathetic preganglionic neurons in the sacral spinal cord. *J. Auton. Nerv. Syst.* **5**, 23–43.
Dennison, S. J., Merritt, V. E., Aprison, M. H., and Felten, D. L. (1981). Redefinition of the location of the dorsal (motor) nucleus of the vagus in the rat. *Brain Res. Bull.* **6**, 77–81.
DiLorenzo, P. M., and Schwartzbaum, J. S. (1982). Coding of gustatory information on the pontine parabrachial nuclei of the rabbit: Magnitude of neural response. *Brain Res.* **251**, 229–244.
DiRocco, R. J., and Grill, H. J. (1979). The forebrain is not essential for sympathetic hyperglycemic response to glucoprivation. *Science* **204**, 1112–1114.
Dubbeldam, J. L., Karten, H. J., and Menken, S. B. J. (1976). Central projections of the chorda tympani nerve in the mallard, *Anas platyrhynchos* L. *J. Comp. Neurol.* **170**, 415–420.
Dubbeldam, J. L., Brus, E. R., Menken, S. B. J., and Zeilstra, S. (1979). The central projections of the glossopharyngeal and vagus ganglia in the mallard, *Anas platyrhynchos* L. *J. Comp. Neurol.* **183**, 149–168.
Finger, T. E. (1978). Gustatory pathways in the bullhead catfish. II. Facial lobe connections, *J. Comp. Neurol.* **180**, 691–706.
Geis, G. S., and Wurster, R. D. (1980). Horseradish peroxidase localization of cardiac vagal preganglionic somata. *Brain Res.* **182**, 19–30.
Grill, H. J. (1980). Production and regulation of ingestive consumatory behavior in the chronic decerebrate rat. *Brain Res. Bull.* **5**, 79–87.
Grill, H. J., and Berridge, K. C. (1981). Chronic decerebrate rats demonstrate preabsorptive insulin secretion and hyperinsulinemia. *Soc. Neurosci. Abst.* **7**, 29.
Grill, H. J., and Norgren, R. (1978a). The taste reactivity test. II. Mimetic responses to gustatory stimuli in chronic thalamic and chronic decerebrate rats. *Brain Res.* **143**, 281–297.
Grill, H. J., and Norgren, R. (1978b). Neurological tests and behavioral deficits in chronic thalamic and chronic decerebrate rats. *Brain Res.* **143**, 299–312.
Grill, H. J., Berridge, K. C., and Ganster, D. J. (1984). Oral glucose is the prime elicitor of preabsorptive insulin secretion. *Am. J. Physiol.* **246**, R88–R95.

Gunn, C. G., Sevelius, G., Puiggari, M. J., and Myers, F. K. (1968). Vagal cardiomotor mechanisms in the hindbrain of the dog and cat. *Am. J. Physiol.* **214**, 258–262.
Hamilton, R. B., and Norgren, R. (1984). Central projections of gustatory nerves in the rat. *J. Comp. Neurol.* **222**, 560–577.
Hamilton, R. B., Ellenberger, H., Liskowsky, D., and Schneiderman, N. (1981). Parabrachial area as mediator of bradycardia in rabbits. *J. Auton. Nerv. Syst.* **4**, 261–281.
Hancock, M. B., and Peveto, C. A. (1979). A preganglionic autonomic nucleus in the dorsal gray commissure of the lumbar spinal cord of the rat. *J. Comp. Neurol.* **183**, 65–72.
Helke, C. J., Neil, J. J., Massari, V. J., and Loewy, A. D. (1982). Substance P neurons project from the ventral medulla to the intermediolateral cell column and ventral horn in the rat. *Brain Res.* **243**, 147–152.
Helke, C. J., Shults, C. W., Chase, T. N., and ODonohue, T. L. (1984). Autoradiographic localization of substance P receptors in rat medulla: Effect of vagotomy and nodose ganglionectomy, *Neuroscience* **12**, 215–223.
Herrick, C. J. (1901). The cranial nerves and cutaneous sense organs of the North American siluroid fishes. *J. Comp. Neurol.* **11**, 177–249.
Herrick, C. J. (1918). "An Introduction to Neurology." W. B. Saunders, Philadelphia.
Higgins, G. A., Hoffman, G. E., Wray, S., and Schwaber, J. S. (1984). Distribution of neurotensin-immunoreactivity within baroreceptive portions of the nucleus of the tractus solitarius and the dorsal vagal nucleus of the rat. *J. Comp. Neurol.* **226**, 155–164.
Hildebrandt, J. R. (1974). Central connections of aortic depressor and carotid sinus nerves. *Exp. Neurol.* **45**, 590–605.
Hopkins, D. A., and Armour, J. A. (1982). Medullary cells of origin of physiologically identified cardiac nerves in the dog. *Brain Res. Bull.* **8**, 359–365.
Hosoya, Y., and Matsushita, M. (1981). Brainstem projections from the lateral hypothalamic area in the rat, as studied with autoradiography. *Neurosci. Lett.* **24**, 111–116.
Hosoya, Y., and Sugiura, Y. (1984). The primary afferent projection of the greater petrosal nerve to the solitary complex in the rat, revealed by transganglionic transport of horseradish peroxidase. *Neurosci. Lett.* **44**, 13–17.
Hosoya, Y., Matsushita, M., and Sugiura, Y. (1984). Hypothalamic descending afferents to cells of origin of the greater petrosal nerve in the rat, as revealed by a combination of retrograde HRP and anterograde autoradiographic techniques. *Brain Res.* **290**, 141–145.
Jackson, J. (1898). Relations of different divisions of the central nervous system to one another and to parts of the body. *Lancet* **1**, 79–87.
Jeanningros, R. (1982). Vagal unitary responses to intestinal amino acid infusions in the anesthetized cat: A putative signal for protein induced satiety, *Physiol. Behav.* **28**, 9–21.
Jordan, D., Khalid, M. E. M., Schneiderman, N., and Spyer, K. M. (1982). The location and properties of preganglionic vagal cardiomotor neurones in the rabbit. *Pflugers Arch.* **395**, 244–250.
Kalia, M., and Mesulam, M.-M. (1980). Brain stem projections of sensory and motor components of the vagus complex in the cat: II. Laryngeal, tracheobronchial, pulmonary, cardiac, and gastrointestinal branches. *J. Comp. Neurol.* **193**, 467–508.
Kalia, M., and Sullivan, J. M. (1982). Brainstem projections of sensory and motor components of the vagus nerve in the rat. *J. Comp. Neurol.* **211**, 248–264.
Kannan, H., Yagi, K., and Sawaki, Y. (1981). Pontine neurones: Electrophysiological evidence of mediating carotid baroreceptor inputs to supraoptic neurones in rats. *Exp. Brain Res.* **42**, 362–370.
Katz, D. M., and Karten, H. J. (1979). The discrete anatomical localization of vagal aortic afferents within a catecholamine-containing cell group in the nucleus solitarius. *Brain Res.* **171**, 187–195.

Kawano, H., and Chiba, T. (1984). Distribution of substance P immunoreactive nerve terminals within the nucleus tractus solitarius of the rat. *Neurosci. Lett.* **45,** 175–179.
King, G. W. (1980). Topology of ascending brainstem projections to nucleus parabrachialis in the cat. *J. Comp. Neurol.* **191,** 615–638.
Kosar, E. M., Grill, H. J., and Norgren, R. (1986). Gustatory cortex in the rat. I. Physiological properties and cytoarchitecture. *Brain Res.* (In press.)
Langley, J. N. (1903). The autonomic nervous system. *Brain* **26,** 1–26.
Lasiter, P. S., Glanzman, D. L., and Mensah, P. A. (1982). Direct connectivity between pontine taste areas and gustatory neocortex in rat. *Brain Res.* **234,** 111–121.
Leslie, R. A., Gwyn, D. G., and Hopkins, D. A. (1982). The central distribution of the cervical vagus nerve and gastric afferent and efferent projections in the rat. *Brain Res. Bull.* **8,** 37–43.
Loewy, A. D., and Burton, H. (1978). Nuclei of the solitary tract: Efferent projections to the lower brain stem and spinal cord of the cat. *J. Comp. Neurol.* **181,** 421–450.
Loewy, A. D., and McKellar, S. (1980). The neuroanatomical basis of central cardiovascular control. *Fed. Proc.* **39,** 2495–2503.
Loewy, A. D., and McKellar, S. (1981). Serotonergic projections from the ventral medulla to the intermediolateral cell column in the rat. *Brain Res.* **211,** 146–152.
Loewy, A. D., McKellar, S., and Saper, C. B. (1979). Direct projections from the A5 catecholamine cell group to the intermediolateral cell column. Brain Res. **174,** 309–314.
Loewy, A. D., Wallach, J. H., and McKellar, S. (1981). Efferent connections of the ventral medulla oblongata in the rat. *Brain Res. Rev.* **3,** 63–80.
Luiten, P. G. M., ter Horst, G. J., Koopmans, S. J., Rietberg, M., and Steffens, A. B. (1984). Preganglionic innervation of the pancreas islet cells in the rat. *J. Auton. Nerv. Syst.* **10,** 27–42.
Mantyh, P. W., and Hunt, S. P. (1984). Neuropeptides are present in projection neurones at all levels in visceral and taste pathways: From periphery to sensory cortex. *Brain Res.* **299,** 297–311.
Matsuo, R., and Kusano, K. (1984). Lateral hypothalamic modulation of the gustatory–salivary reflex in rats. *J. Neurosci.* **4,** 1208–1216.
Matsuo, R., Shimizu, N., and Kusano, K. (1984). Lateral hypothalamic modulation of oral sensory afferent activity in the nucleus tractus solitarius neurons of rats. *J. Neurosci.* **4,** 1201–1207.
Mei, N. (1978). Vagal glucoreceptors in the small intestine of the cat. *J. Physiol. (London)* **282,** 485–506.
Mei, N. (1983). Sensory structures in the viscera. *In* "Progress in Sensory Physiology," (H. Autrum, D. Ottoson, E. R. Perl, R. F. Schmidt, H. Shimazu, and W. D. Willis, eds.), Vol. 4, pp. 1–42. Springer-Verlag, New York.
Mesulam, M.-M. (1978). Tetramethyl benzidine for horseradish peroxidase neurohistochemistry: A non-carcinogenic, blue reaction-product with superior sensitivity for visualizing neural afferents and efferents. *J. Histochem. Cytochem.* **26,** 106–117.
Miura, M., Onai, T., and Takayama, K. (1983). Projections of upper structure to the spinal cardioacceleratory center in cats: An HRP study using a new microinjection method. *J. Auton. Nerv. Syst.* **7,** 119–139.
Mountcastle, V. B. (1978). An organizing principle for cerebral function: The unit module and the distributed system. *In* "The Mindful Brain" (G. M. Edelman and V. B. Mountcastle, eds.). M.I.T. Press, Cambridge, Mass.
Nadelhaft, I., and Booth, A. M. (1984). The location and morphology of preganglionic neurons and the distribution of visceral afferents from the rat pelvic nerve: A horseradish peroxidase study. *J. Comp. Neurol.* **226,** 238–245.

Neuhuber, W. (1982). The central projections of visceral primary efferent neurons of the inferior mesenteric plexus and hypogastric nerve and the location of the related sensory and preganglionic sympathetic cell bodies in the rat. *Anat. Embryol.* **164**, 413–425.

Nomura, S., and Mizuno, N. (1982). Central distribution of afferent and efferent components of the glossopharyngeal nerve: An HRP study in the cat. *Brain Res.* **236**, 1–13.

Norgren, R. (1974). Gustatory afferents to ventral forebrain. *Brain Res.* **81**, 285–295.

Norgren, R. (1976). Taste pathways to hypothalamus and amygdala. *J. Comp. Neurol.* **166**, 17–30.

Norgren, R. (1978). Projections from the nucleus of the solitary tract in the rat. *Neuroscience* **3**, 207–218.

Norgren, R. (1981). The central organization of the gustatory and visceral afferent systems in the nucleus of the solitary tract. In "Brain Mechanisms of Sensation" (Y. Katsuki, R. Norgren, and M. Sato, eds.), pp. 143–160. Wiley, New York.

Norgren, R. (1983). Afferent connections of cranial nerves involved in ingestion. *J. Auton. Nerv. Syst.* **9**, 67–77.

Norgren, R. (1984). Taste: Central neural mechanisms. In "Handbook of Physiology: The Nervous System Vol. III: Sensory Processes" (I. Darien-Smith, ed.), pp. 1087–1128. Am. Physiol. Soc., Washington, D.C.

Norgren, R. (1985). The sense of taste and the study of ingestion. In "Taste, Olfaction, and the Central Nervous System" pp. 233–249. (D. W. Pfaff, ed.). The Rockefeller University Press, New York.

Norgren, R., and Grill, H. J. (1976). Efferent distribution from the cortical gustatory area in rats. *Neurosci. Abstr.* **2**, 124.

Norgren, R., and Grill, H. J. (1982). Brain stem control of ingestive behavior. In "Physiological Mechanisms of Motivation" (D. Pfaff, ed.). pp. 99–131. Springer-Verlag, New York.

Norgren, R., and Leonard, C. M. (1971). Taste pathways in rat brainstem. *Science* **173**, 1136–1139.

Norgren, R., and Leonard, C. M. (1973). Ascending central gustatory pathways. *J. Comp. Neurol.* **150**, 217–238.

Norgren, R., and Pfaffmann, C. (1975). The pontine taste area in the rat. *Brain Res.* **91**, 99–117.

Norgren, R., and Smith, G. P. (1983). The central distribution of vagus subdiaphragmatic branches in the rat. *Soc. Neurosci. Abstr.* **9**, 611.

Norgren, R., and Wolf, G. (1975). Projections of thalamic gustatory and lingual areas in the rat. *Brain Res.* **92**, 123–129.

Nosaka, S., Yamamoto, T., and Yasunaga, K. (1979). Localization of vagal cardioinhibitory preganglionic neurons within rat brain stem. *J. Comp. Neurol.* **186**, 79–92.

Ogawa, H., and Kaisaku, J. (1982). Physiological characteristics of the solitario-parabrachial relay neurons with tongue afferent inputs in rats. *Exp. Brain Res.* **48**, 362–368.

Ogawa, H., Imoto, T., and Hayama, T. (1984). Responsiveness of solitario-parabrachial relay neurons to taste and mechanical stimulation applied to the oral cavity in rats. *Exp. Brain Res.* **54**, 349–358.

Pagani, F. D., Norman, W. P., Kasbekar, D. K., and Gillis, R. A. (1984). Effects of stimulation of nucleus ambiguus complex on gastroduodenal function. *Am. J. Physiol.* **246**, G253–G262.

Perrotto, R., and Scott, T. (1976). Gustatory nueral coding in the pons. *Brain. Res.* **110**, 283–300.

Perwaiz, S. A., and Karim, M. A. (1982). Localization of parasympathetic preganglionic neurons innervating submandibular gland in the monkey: An HRP study. *Brain Res.* **251**, 349–352.

Petras, J. M., and Faden, A. I. (1978). The origin of sympathetic preganglionic neurons in the dog. *Brain Res.* **144**, 353–357.

Pritchard, T., Hamilton, R. B., Morse, J., and Norgren, R. (1986). Projections of thalamic gustatory and lingual areas in the monkey, *Macaca fascicularis*. *J. Comp. Neurol.* (In press.)

Ricardo, J. A. (1983). Hypothalamic pathways involved in metabolic regulatory functions, as identified by track-tracing methods. *In* "Advances in Metabolic Disorders" (A. J. Szabo, ed.), Vol. 10, pp. 1–30. Academic Press, New York.

Ricardo, J., and Koh, E. (1978). Anatomical evidence of direct projections from the nucleus of the solitary tract to the hypothalamus amygdala and other forebrain structures in the rat. *Brain Res.* **153**, 1–26.

Rogers, R. C., and Nelson, D. O. (1984). Neurons of the vagal division of the solitary nucleus activated by the paraventricular nucleus of the hypothalamus. *J. Auton. Nerv. Syst.* **10**, 193–197.

Rogers, R. C., Novin, D., and Butcher, L. L. (1979). Electrophysiological and neuroanatomical studies of hepatic portal osmo- and sodium-receptive afferent projections within the brain. *J. Auton. Nerv. Syst.* **1**, 183–202.

Rucker, H. K., and Holloway, J. A. (1982). Viscerosomatic convergence onto spinothalamic tract neurons in the cat. *Brain Res.* **243**, 155–157.

Saper, C. B. (1982a). Reciprocal parabrachial–cortical connections in the rat. *Brain Res.* **242**, 33–40.

Saper, C. B. (1982b). Convergence of autonomic and limbic connections in the insular cortex of the rat. *J. Comp. Neurol.* **210**, 163–173.

Saper, C. B., and Loewy, A. D. (1980). Efferent connections of the parabrachial nucleus in the rat. *Brain Res.* **197**, 291–317.

Schwanzel-Fukuda, M., Morrell, J. I., and Pfaff, D. W. (1984). Localization of forebrain neurons which project directly to the medulla and spinal cord of the rat by retrograde tracing with wheat germ agglutinin. *J. Comp. Neurol.* **226**, 1–20.

Schwartzbaum, J. S. (1983). Electrophysiology of taste-mediated functions in parabrachial nuclei of behaving rabbit. *Brain Res. Bull.* **11**, 61–89.

Shipley, M. T. (1982). Insular cortex projection to the nucleus of the solitary tract and brainstem visceromotor regions in the mouse. *Brain Res. Bull.* **8**, 139–148.

Shipley, M. T., and Sanders, M. S. (1982). Special senses are really special: Evidence for a reciprocal, bilateral pathway between insular cortex and nucleus parabrachialis. *Brain Res. Bull.* **8**, 493–501.

Simon, O. R., and Schramm, L. P. (1984). The spinal course and medullary termination of myelinated renal afferents in the rat. *Brain Res.* **290**, 239–247.

Sugimoto, T., Itoh, K., Mizuno, N., Nomura, S., and Konishi, A. (1979). The site of origin of cardiac preganglionic fibers of the vagus nerve: An HRP study in the cat. *Neurosci. Lett.* **12**, 53–58.

Swanson, L. W. (1977). Immunohistochemical evidence for a neurophysin-containing autonomic pathway arising in the paraventricular nucleus of the hypothalamus. *Brain Res.* **128**, 346–353.

Swanson, L. W., and Kuypers, H. G. J. M. (1980). The paraventricular nucleus of the hypothalamus: Cytoarchitectonic subdivisions and organization of projections to the pituitary, dorsal vagal complex, and spinal cord as demonstrated by retrograde fluorescence double-labeling methods. *J. Comp. Neurol.* **194**, 555–570.

Swanson, L. W., and Mogenson, G. J. (1981). Neural mechanisms for the functional coupling of autonomic, endocrine, and somatomotor responses in adaptive behavior. *Brain Res. Rev.* **3**, 1–34.

Travers, J. B., and Norgren, R. (1983). Afferent projections to the oral motor nuclei in the rat. *J. Comp. Neurol.* **220**, 280–298.

Travers, S. P., and Smith, D. V. (1984). Responsiveness of neurons in the hamster parabrachial nuclei to taste mixtures. *J. Gen. Physiol.* **84**, 221–250.

Travers, S. P., Pfaffmann, C., and Norgren, R. (1986). Convergence of lingual and palatal gustatory neural activity in the nucleus of the solitary tract. *Brain Res.* (In press.)

Vallet, P., and Baertschi, A. J. (1980). Sodium chloride-sensitive receptors located in hepatic portal vein of the rat. *Neurosci. Lett.* **17**, 283–288.

Van Buskirk, R. L., and Smith, D. V. (1981). Taste sensitivity of hamster parabrachial pontine neurons. *J. Neurophysiol.* **45**, 144–171.

Van der Kooy, D., Koda, L. Y., McGinty, J. F., Gerfen, C. R., and Bloom, F. E. (1984). The organization of projections from the cortex, amygdala, and hypothalamus to the nucleus of the solitary tract in rat. *J. Comp. Neurol.* **224**, 1–24.

Wallach, J. H., and Loewy, A. D. (1980). Projections of the aortic nerve to the nucleus tractus solitarius in the rabbit. *Brain Res.* **188**, 247–251.

Whitehead, M. C., and Frank, M. E. (1983). Anatomy of the gustatory system in the hamster: Central projections of the chorda tympani and the lingual nerve. *J. Comp. Neurol.* **220**, 378–395.

Williford, D. J., Ormsbee, H. S., III, Norman, W., Harmon, J. W., Garvey, T. Q., III, DiMicco, J. A., and Gillis, R. A. (1981). Hindbrain GABA receptors influence parasympathetic outflow to the stomach. *Science* **214**, 193–194.

Wyss, J. M., and Donovan, M. K. (1984). A direct projection from the kidney to the brainstem. *Brain Res.* **298**, 130–134.

Yamada, K. A., McAllen, R. M., and Loewy, A. D. (1984). GABA antagonists applied to the ventral surface of the medulla oblongata block the baroreceptor reflex. *Brain Res.* **297**, 175–180.

Yu, W.-H. A., and Srinivasan, R. (1980). Origin of the preganglionic visceral efferent fibers to the glands in the rat tongue as demonstrated by the horseradish peroxidase method. *Neurosci. Lett.* **19**, 143–148.

17

Caudal Brainstem Integration of Taste and Internal State Factors in Behavioral and Autonomic Responses

HARVEY J. GRILL
Department of Psychology
University of Pennsylvania
Philadelphia, Pennsylvania

I.	Introduction	348
II.	The CBS Receives Input from Oral Exteroceptors That Evaluate the Sensory Characteristics of Food	348
III.	The CBS Is a Site of Metabolic Interoceptors	349
IV.	The CBS Contains Simple Reflex Connections between Oral Exteroceptor Input and Autonomic and Behavioral Effector Output	354
	A. Salivation	354
	B. Cephalic Insulin Response	355
	C. Sympathoadrenal Hyperglycemic Reflex	355
	D. Mastication	357
V.	The CBS Contains Connections between Exteroceptive Input and Behavioral Effector Output for the Production of Discriminative Responses to Taste	358
VI.	Interoceptive Input from Food Deprivation and Insulin-Induced Hypoglycemia Is Integrated with Oral Afferent Information within the CBS to Control the Ingestive Consummatory Behavior of Chronically Decerebrate Rats	365
VII.	Conclusion	369
	References	369

I. INTRODUCTION

A neural system that controls energy homeostasis should contain the following elements: contact *exteroceptors* that evaluate the sensory characteristics of food (the energy source), *interoceptors* that assess the organism's metabolic state, and *effectors* that initiate compensatory responses. Compensatory responses are behavioral in the presence of food. In the absence of food, compensatory responses are reflexive, making use of fuels that were previously stored. The behavioral consummatory response elicited by a particular food is *not* solely dependent on its sensory characteristics. Neural correlates of the animal's metabolic state and prior experience with the food appear to be integrated with oral afferent information in determining whether a food will be swallowed or spit out. *Integrators* are therefore a necessary element in the neural control of energy homeostasis (see Grill and Berridge, 1985, for more detail).

Research on the neural control of energy homeostasis has focused on the hypothalamus and other forebrain structures. As we now know, however, the hypothalamic hypothesis oversimplified the neural control of energy homeostasis. We now recognize that a more elaborate model involving intercommunication among multiple neural levels is required. Both peripheral and central nervous system mechanisms collectively participate to control energy homeostasis, and this control appears to be hierarchical (see Grill and Berridge, 1985, for more detail). The purpose of this chapter is to convince the reader that the brain caudal to the hypothalamus, which I refer to as the caudal brainstem (CBS), plays an important role in the neural control of energy homeostasis, and that other neural elements—both central and peripheral—interact with the CBS in a systematic fashion.

II. THE CBS RECEIVES INPUT FROM ORAL EXTEROCEPTORS THAT EVALUATE THE SENSORY CHARACTERISTICS OF FOOD

It is well established that oral gustatory receptors convey their evaluation of a food's taste to the nucleus of the solitary tract (NTS) via cranial nerves VII, IX, and X (see Norgren, this volume). Hamilton and Norgren (1984) have recently demonstrated the sites of specific afferent nerve terminals within the NTS. As shown in their figure, Fig. 1, VIIth nerve afferent terminals (chorda tympani and greater superficial petrosal) are located in the most rostral and lateral regions of the NTS, whereas IXth and Xth nerve terminals are found within the intermediate and lateral regions of the nucleus. NTS neurons receiving gustatory afferent information project rostrally

17. Caudal Brainstem and Energy Homeostasis   349

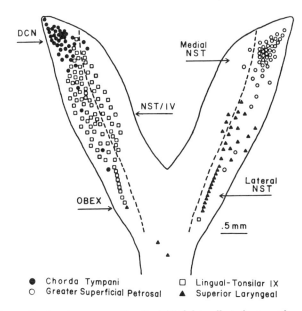

Fig. 1. Composite drawing representing the NTS bilaterally in horizontal section; rostral on top. Symbols represent the terminal distribution patterns of the chorda tympani, and greater superficial petrosal branches of VII, lingual–tonsilar branch of IX, and superior laryngeal branch of X. [Reprinted with permission from Hamilton and Norgren (1984).]

upon neurons in the parabrachial nuclei of the pons (Norgren and Leonard, 1971). Oral tactile and thermal afferent neurons in the Vth nerve project within the CBS to the principal sensory nucleus of V as well as to the rostral NTS within the CBS.

III. THE CBS IS A SITE OF METABOLIC INTEROCEPTORS

Glucose is the primary substrate for central nervous system (CNS) metabolism. When this substrate is insufficient for the metabolic needs of CNS neurons, compensatory behavioral, autonomic, and autonomic–endocrine responses are elicited to elevate circulating blood glucose levels (e.g., Smith and Epstein, 1969; Himsworth, 1970). While CNS efferent neurons provide the final path for the production of the majority of these compensatory responses, it remained to be shown that CNS metabolic interoceptors participated in triggering compensatory responses. The existence of CNS interoceptors was initially supported by the ability of lateral intracerebroventricular (ICV) injections of phlorizin, a glucose transport inhibitor (Betz et al., 1975), to stimulate compensatory feeding behavior; plasma glucose levels were not measured in this study (Glick and Mayer, 1968). Subsequently, inhibition of brain glycolysis by lateral ICV injections of 2-deoxy-D-glucose (2-DG) (Wick et al., 1957) was shown to evoke both behavioral and autonomic compensatory responses (Miselis and Epstein, 1975; Berthoud and Mogenson, 1977; Coimbra et al., 1979).

Which receptive regions of the CNS are exposed to a drug injected intracerebroventricularly? Due to the normal caudad movement of cerebrospinal fluid, lateral ICV injections make the glucodynamic drug available to the entire ventricular system (Bradbury, 1979). On the other hand, peripheral detection is eliminated in these experiments by the ventricular application of small drug dosages or the ineffectiveness of peripheral administration. Despite the fact that ICV injections of glucodynamic drugs have access to the caudal brainstem surrounding the fourth ventricle, the zeitgeist of the hypothalamic hypothesis directed the interpretation of the site of drug action to the existence of hypothalamic interoceptors. Such cells had been hypothesized based on electrophysiological demonstrations of glucoresponsive neurons in the hypothalamus (Chhina et al., 1971; Desiraju et al., 1968; Oomura, 1976). The existence of hypothalamic neurons responsive to glucose, however, neither demonstrated that these neurons were the exclusive site of interoception, nor proved that they alone were responsible for eliciting compensatory behavioral and autonomic responses in the ICV studies.

It has recently been shown that metabolically sensitive neurons in the caudal brainstem provide an afferent limb for compensatory behavioral and autonomic responses in the intact rat. R. Ritter and colleagues demonstrated that injections of 5-thioglucose (5-TG), an antimetabolite of glucose, restricted to the 4th ventricle by means of a cerebral aqueduct plug, produced both feeding and sympathoadrenal hyperglycemic compensatory responses (Ritter et al., 1981). In contrast, lateral ventricular injections of 5-TG, restricted to the forebrain ventricles by the same cerebral aqueduct plug, stimulated neither feeding nor sympathoadrenal hyperglycemia (Ritter et al., 1981). These data suggest that the effects of lateral ICV-administered phlorizin on feeding (Glick and Mayer, 1968) might also be mediated by caudal brainstem mechanisms. To investigate this possibility and to extend the investigation of caudal brainstem interoception, initiated by the important study of Ritter and co-workers, we have applied both phlorizin and 5-TG to the 4th ventricle of intact rats (Flynn and Grill, 1984).

Rats, previously implanted with cannulas aimed at the 4th ventricle, were given, in a random order, 4th-ICV injections of saline, 150 µg 5-TG, 210 µg 5-TG, 6.5 µg phlorizin (3.0 mM), and 13.0 µg phlorizin (6.0 mM). The effects of these treatments on food intake were measured at 3, 6, and 24 hr after injection. The range of phlorizin doses tested was restricted by the drug's limited solubility. All injections were of a 5.0 µl volume.

To verify that any effect of phlorizin on food intake was mediated by its central actions and not by a systemic action of the drug, food intake was also measured in response to systemic phlorizin injection in the same dosage. Rats were injected intraperitoneally (ip) with 13.0 µg phlorizin or saline (0.15 M), and food intake was measured 3 and 6 hr later.

After the food intake experiments were completed, other experiments evaluated changes in plasma glucose and insulin concentrations produced by 4th-ICV injections in the *absence* of food. Food was removed prior to tail blood collection and returned after the collection of the last sample. Rats were injected on alternate days with 4th-ICV 5-TG (150 µg, 210 µg), phlorizin (13 µg), or saline. Tail blood samples were collected prior to the 4th-ICV injection, and 30, 60, 120, and 180 min, and 6 hr after the injection.

Histological analyses revealed that the cannulas were positioned on the midline, just anterior to the primary cerebellar fissure, as shown in Fig. 2. Ink (5 µl) injected prior to sacrifice was restricted to the 4th ventricle in all cases; there were no traces of ink rostral to the recess of the inferior colliculus. As shown in Fig. 3, cumulative food intake 3 and 6 hr following 4th-ICV 13.0 µg phlorizin, and 150 µg 5-TG and 210 µg 5-TG was significantly greater than that consumed following 4th-ICV saline. Phlorizin- and 5-TG-treated rats ate similar amounts of food. In contrast, systemically injected phlorizin (13.0 µg) did not affect food intake.

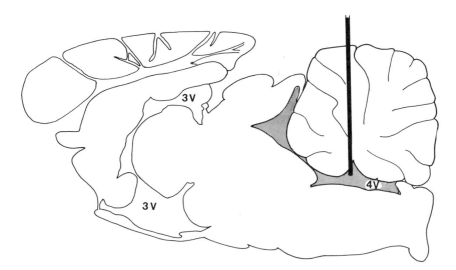

Fig. 2. Schematic midsagittal section of the rat brain, indicating the position of the 4th-ICV cannulas. Drugs injected into the 4th ventricle were confined to the 4th ventricle (darkened region) in all rats.

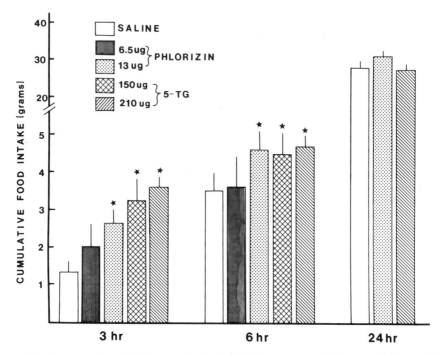

Fig. 3. Cumulative 3-, 6-, and 24-hr food intake (mean ± standard error of the mean) following 4th-ICV saline, 6.5 μg phlorizin, 13.0 μg phlorizin, 150 μg 5-TG, and 210 μg 5-TG. (Asterisk indicates significantly different from saline.)

Figure 4 reveals that 4th ventricular injections of 5-TG (210 μg) elicited sympathoadrenal hyperglycemia. In contrast to the sympathetic-arousing effect of 5-TG, 4th-ICV phlorizin (13 μg) injection did *not* elicit sympathoadrenal hyperglycemia. Using our paradigm, the inability of phlorizin to stimulate hyperglycemia cannot be attributed to cannula placement or patency since injections of 5-TG through the same cannula elicited hyperglycemia. In addition, insulin secretion was not altered by the ICV injection of phlorizin. So the absence of hyperglycemia following phlorizin injections cannot be attributed to a concomitant stimulation of insulin secretion. These data indicate that phlorizin stimulates a caudal brainstem system controlling behavioral but not autonomic compensatory responses. Similarly, S. Ritter and Strang (1982) reported that 4th ventricular injection of subtoxic alloxan doses stimulates feeding but not sympathoadrenal hyperglycemia in rats. Thus, the action(s) of phlorizin and subtoxic alloxan appear(s) specific for those mechanisms controlling compensatory ingestive behavior.

The neural systems controlling compensatory behavioral and autonomic

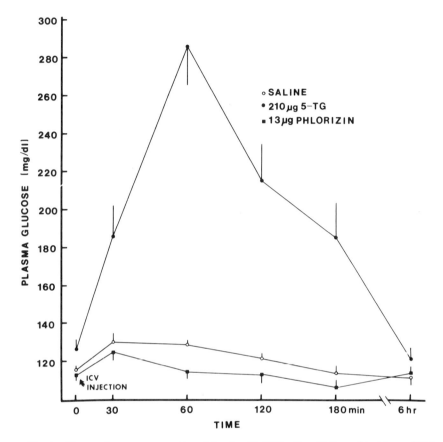

Fig. 4. Plasma glucose concentrations following 4th-ICV saline, 13.0 μg phlorizin, and 210 μg 5-TG.

responses differ on other measures. First, in response to the metabolic challenge provided by 2-DG, the neural mechanism mediating sympathoadrenal hyperglycemia develops prior to that of compensatory feeding in the rat (Houpt and Epstein, 1973). Second, the neural system mediating feeding in response to systemic 2-DG is more susceptible to the effects of toxic 4th-ICV doses of alloxan than is that system mediating sympathoadrenal hyperglycemia (S. Ritter et al., 1982). Finally, compensatory behavioral and autonomic responses are dissociable by intravenous infusions of different metabolic fuels (see Stricker et al., 1977, for more details). It appears that different neural systems or metabolic events mediate compensatory behavioral and autonomic responses.

Our findings using 4th-ICV injections of 5-TG replicate the report of R.

Ritter and associates that 4th-ICV 5-TG injections, restricted to the caudal brainstem by a cerebral aqueduct plug, stimulate feeding and hyperglycemia (R. Ritter et al., 1981). In addition, we found that 4th-ICV injections of phlorizin that were confined to the caudal brainstem also stimulated feeding. Therefore, feeding stimulated by lateral ICV phlorizin cannot be exclusively attributed to phlorizin's action on forebrain neurons (Glick and Mayer, 1968). Our data, taken together with the indentification of caudal brainstem metabolic receptors by R. Ritter et al. (1981), argue that lateral ICV application of phlorizin is acting on caudal brainstem mechanisms to stimulate feeding. These 4th ventricular injection studies demonstrate that the detection of metabolically relevant signals takes place within the caudal brainstem of intact rats.

IV. THE CBS CONTAINS SIMPLE REFLEX CONNECTIONS BETWEEN ORAL EXTEROCEPTOR INPUT AND AUTONOMIC AND BEHAVIORAL EFFECTOR OUTPUT

Oral sensation elicits a variety of behavioral and autonomic reflexes like mastication, salivating, swallowing, and cephalic insulin secretion. As stated above, the CBS receives primary oral sensory information from the cranial nerves. The final, common-path motor neurons for these reflexive responses are located in the CBS. Our present concern is whether the reflexive connection between oral input and the final, common-path output is contained within the CBS or requires the participation of the forebrain.

A. Salivation

The preganglionic parasympathetic motor neurons innervating the salivary glands are found within the rostral medullary reticular formation in a narrow column beginning ventrally at the exit of the facial root and proceeding dorsally to the rostral nucleus of the solitary tract (Contreras et al., 1980). A variety of studies have shown that taste quality and intensity directly determine the secretory output of the salivary glands. For example, acidic tastes stimulate a much greater parotid salivary response than sugar, salt, or bitter tastes in a variety of species. Kawamura and Yamamoto (1978) have determined that the relationship between taste quality and intensity and salivary output does *not* depend upon the forebrain. The saliva secretory responses of acutely decerebrate rabbits and anesthetized control rabbits to taste stimuli were similar.

Salivation is subject to classical conditioning. Using a classical conditioning paradigm, Pavlov was able to elicit salivary secretion with stimuli that were ordinarily ineffective.

B. Cephalic Insulin Response

In a variety of species, oral contact with food elicits an immediate elevation of plasma insulin levels in the absence of a rise in glycemia (e.g., Strubbe and Steffens, 1975). This immediate cephalic or preabsorptive insulin response persists for approximately 8–10 min after the beginning of a sham-fed meal (Berthoud and Jeanrenaud, 1982) and is neurally mediated. The preganglionic parasympathetic neurons in the dorsal motor nucleus of the vagus project via the vagus nerve to a ganglion in the vicinity of the pancreas (Laughton and Powley, 1979). Surgical or pharmacological interference with these neurons blocks the cephalic insulin response (Strubbe and Steffens, 1975; Berthoud et al., 1980).

Like salivation, the range of stimuli that elicits the cephalic insulin response also appears to be affected by conditioning (e.g., Berridge et al., 1981).

It has been assumed that the ability of a taste to elicit cephalic insulin secretion correlates with its preference—that is, its capacity to elicit and sustain ingestive behavior. For example, rats and humans prefer "sweet" stimuli to nonsweet stimuli. Since the nutritive value of sweet stimuli ranges from high (glucose) to none (saccharin) and since a preference hierarchy for sugars can be constructed, we examined the capacity of 10 novel, "sweet" stimuli to elicit a cephalic insulin response in the intact rat (Grill et al., 1984).

Seven sugars (sucrose, maltose, glucose, fructose, lactose, galactose, mannose), two sugar alcohols (mannitol, sorbitol) of equal molarity, and a nonnutritive sweetener (saccharin) were administered to the mouths of rats via indwelling oral cannulas. Taste order was randomized. Rats were fitted with gastric drainage fistulas so that taste stimuli would not be digested and absorbed. Rats had no previous experience with each taste that might have led to a conditioned cephalic response. Although all taste solutions were ingested, only oral glucose evoked a rapid, statistically significant elevation of insulin levels. This insulin secretion was independent of a rise in glycemia, as seen in Figs. 5–7. The preeminence of oral glucose as an elictor of cephalic insulin secretion is especially striking, considering that glucose is neither the most intense (as measured electrophysiologically in peripheral gustatory nerve) nor the most preferred (as measured in two bottle-preference tests) taste stimulus examined. These results suggest the existence of a gustatory and/or gastric chemoreceptor that is most responsive to glucose.

C. Sympathoadrenal Hyperglycemic Reflex

Cannon (1932) cleverly demonstrated that when glucose utilization falls, the adrenal medulla is neurally stimulated to secrete epinephrine. This

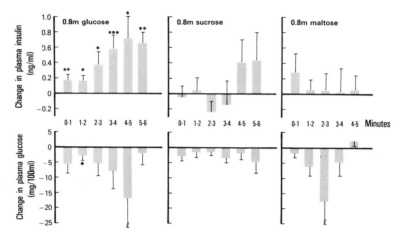

Fig. 5. Relative mean (±SE) preabsorptive insulin and plasma glucose responses to oral 0.8 M glucose, sucrose, and maltose. Mean plasma insulin baselines (ng/ml): 1.64 (glucose), 2.27 (sucrose), and 1.89 (maltose). Mean plasma glucose baselines (mg/100 ml): 146.94 (glucose), 153.08 (sucrose), and 150.24 (maltose). *Changes from baseline significant at $p < 0.05$; **significant at $p < 0.02$; ***Significant at $p < 0.01$.

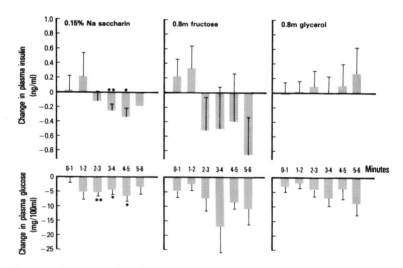

Fig. 6. Relative mean (±SE) preabsorptive insulin and plasma glucose responses to oral sodium saccharin (0.15%), fructose (0.8 M), and glycerol (0.8 M). Mean plasma insulin base lines (ng/ml): 1.78 (saccharin), 2.09 (fructose), and 1.95 (glycerol). Mean plasma glucose base lines (mg/100 ml): 142.55 (saccharin), 141.20 (fructose), and 148.75 (glycerol). (Significance as in Fig. 5.)

17. Caudal Brainstem and Energy Homeostasis 357

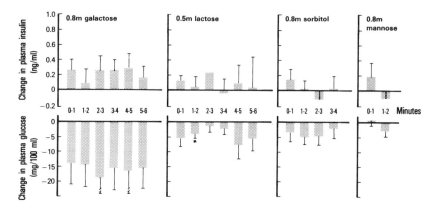

Fig. 7. Relative mean (±SE) preabsorptive insulin and plasma glucose responses to 0.8 M galactose, sorbitol, and mannose, and to 0.5 M lactose. Mean plasma insulin baselines (ng/ml): 1.11 (galactose), 0.66 (lactose), 1.57 (sorbitol), and 2.29 (mannose). Mean plasma glucose baselines (mg/100 ml): 149.96 (galactose), 144.23 (lactose), 134.80 (sorbitol), and 151.45 (mannose). (Significance as in Fig. 5.)

blood-borne epinephrine acts on the liver to promote glycogenolysis, and a new pool of glucose is thus liberated. Where are the interoceptors located that triggered this autonomic–endocrine compensatory reflex? Again we must be reminded of the hegemony of the hypothalamic hypothesis and the associated tendency to posit all interoceptors as hypothalamic. Himsworth (1970) and others had concluded that the interoceptors triggering sympathoadrenal hyperglycemia were located in the hypothalamus. DiRocco and I (1979) demonstrated that chronically decerebrate rats would display sympathoadrenal hyperglycemia when challenged with 2-DG, as seen in Fig. 8. Since these rats did not have a neurally connected hypothalamus, these data meant that interoceptors triggering sympathoadrenal hyperglycemia were located either within the CBS itself and/or in the periphery connected to the CBS. As stated above, there is now support for CBS interoceptors triggering sympathoadrenal hyperglycemia in the intact rat. It should be noted, however, that the decerebrate data do not rule out the possibility that a separate pool of interoceptors exists within the forebrain.

D. Mastication

Sherrington (1917) put forth a chain-reflex view of mastication that was dependent on sensory feedback from the oral cavity. The chain reflex was initiated by the placement of food in the mouth, which reflexively initiated jaw closure. The contraction of the jaw-closing muscles as well as the pres-

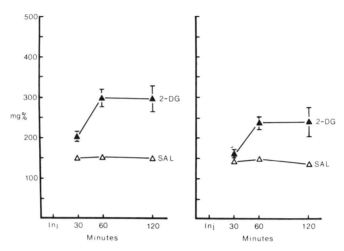

Fig. 8. Plasma glucose concentrations of decerebrate (right) and control (left) rats 30, 60, and 120 min after intraperitoneal injection (Inj) of 2-DG and physiological saline.

sure created by jaw closure on food provided sensory feedback, initiating a reflex jaw opening. The food in the mouth then initiated a jaw closure and so on. Subsequent to these and other observations, it has become increasingly clear that mastication is made up of a primary, basic pattern of cyclic jaw movements, predominantly opening and closing. This pattern of cyclic activity measured in the nerves to the jaw muscles itself persists after muscle paralysis (Lund and Dellow, 1971). This observation is not compatible with a reflex theory of mastication because muscle paralysis would prevent the proprioceptive feedback required to initiate jaw opening and so prevent the cyclic activity. A central pattern generator can function independent of sensory feedback. Current evidence suggests that a central pattern generator for mastication is located in the pontine reticular formation of the CBS. Dellow and Lund (1971) have shown that the pattern of cyclic jaw muscle activity produced by oral stimulation is similar for acutely decerebrate and intact rabbits.

V. THE CBS CONTAINS CONNECTIONS BETWEEN EXTEROCEPTIVE INPUT AND BEHAVIORAL EFFECTOR OUTPUT FOR THE PRODUCTION OF DISCRIMINATIVE RESPONSES TO TASTE

The final or consummatory acts of feeding behavior, i.e., ingestion or rejection, involve highly stereotyped responses guided principally by taste stimuli. My laboratory views these ingestion and rejection oral behavior patterns as the basic functional units of feeding behavior. Connections between the basic sensory input (gustatory and somatosensory) and the motor output of discriminative responses to taste might exist within the caudal brainstem itself. Sherrington (Miller and Sherrington, 1916) described reflex components of ingestion and rejection in acutely decerebrate cats. It was not clear from these data or that of others (Macht, 1951) whether the same discriminative taste stimuli that elicit ingestion and rejection responses in the intact cat were effective in the decerebrate cat. Therefore, we developed a chronically, supracollicularly decerebrate rat preparation to examine the sufficiency of the isolated caudal brainstem for producing discriminative responses to taste. Histological and general behavioral data from the chronically decerebrate rat are shown in Figs. 9 and 10.

Chronically decerebrate rats never initiate spontaneous meals. To circumvent this lack of spontaneous feeding, we (Grill and Norgren, 1978a) developed a methodology for presenting tastes directly into the rat's mouth via oral cannulas anchored to the skull (see Fig. 11). In an apparatus like the one shown in Fig. 12, responses elicited by remotely presenting taste stimuli into the mouths of freely moving rats were videotaped and subsequently analyzed by advancing single frames of videotape. A lexicon of response topographies, called *taste reactivity,* was described first in intact rats (see Grill and Norgren, 1978a, for more detail). In response to sucrose, an *ingestion-response sequence* of rhythmic mouth movements, rhythmic tongue protrusions, and larger lateral tongue protrusions accompanied swallowing, as seen in Fig. 13. Movement was restricted to oral region in response to sucrose; rats did not move about or in any way increase their activity in response to sucrose. Taste reactivity to quinine differed strikingly. As seen in Fig. 14, intact rats gape, chin rub, head shake, face wash, forelimb shake, and paw rub in response to orally delivered quinine. Quinine produced a *rejection-response sequence* that included locomotion, rearing, and rejection of the orally presented taste.

In response to a variety of 50-μl orally delivered taste stimuli, chronically decerebrate rats demonstrated the same ingestion and rejection behavioral sequences as intact rats did (see Fig. 15) (Grill and Norgren, 1978b). Furthermore, the threshold of individual taste reactivity response components of decerebrates was identical to that of intact rats. For example, the gape response threshold was 0.03 mM quinine for both chronically decerebrate and pair-fed intact rats. These data can be generalized by comparison with behavioral observations on the taste-elicited, oral–facial responses of anencephalic and intact human newborns (Steiner, 1977). The isolated caudal brainstem is therefore sufficient for the production of discriminative responses in response to taste stimulation.

A given taste does not always elicit the same discriminative response, be it

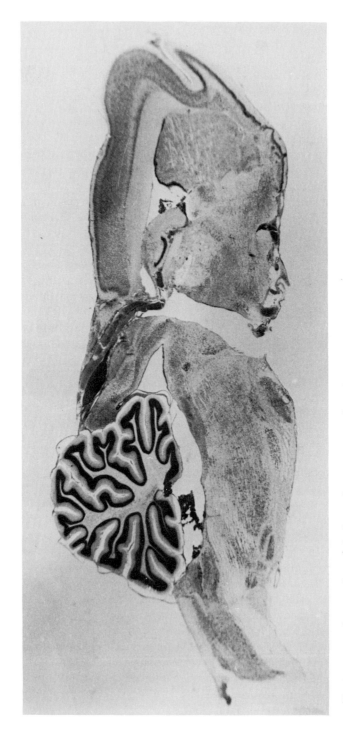

Fig. 9. A sagittal section through the brain of a representative decerebrate rat preparation (Cresylecht violet stain). This animal survived 37 days after the second stage of the decerebration. The plane of section was highly similar in all rats. The tissue posterior to the transection appeared normal in the light microscope. In many preparations the space normally occupied by the thalamus and hippocampus was replaced by a fluid-filled cyst. [From Grill and Norgren (1978b)].

17. Caudal Brainstem and Energy Homeostasis 361

Fig. 10. Chronically decerebrate rats exhibit no spontaneous activity other than grooming, but often overreact with well-coordinated movements to seemingly inappropriate stimuli. Tail pinch facilitates a brisk, well-coordinated sequence of cage climbing (a–e). Decerebrate rats maintain their fur: face washing (f), grooming of the flanks (g and h), and anal (i) and genital grooming involve complicated postures, which are executed in a coordinated fashion by these rats.

ingestion or rejection. In other words, knowledge of the input (the taste stimulus) is insufficient to predict the output (the response that follows). Rather, the neural mechanisms effecting ingestion and rejection responses and autonomic responses are affected by the integration of the peripheral gustatory signals with neural control signals deriving from the organism's present physiological state and past experience to produce responses appropriate to the prevailing physiological and environmental conditions. For example, signals arising from the digestive tract, such as the osmotic signals

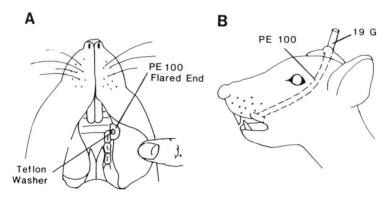

Fig. 11. Diagram of the intraoral catheter. The intraoral end is placed just rostral to the first maxillary molar. The tubing is led out subcutaneously to the skull and secured to a short piece of 19-gauge (19 G) stainless-steel tubing with dental acrylic. (A) Ventral view. (B) Lateral view.

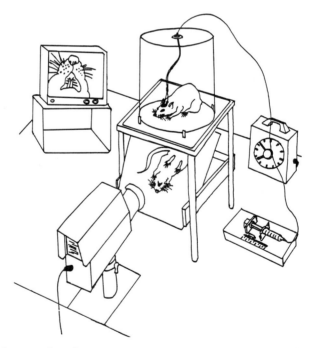

Fig. 12. Apparatus for videotaping taste-reactivity responses to taste stimuli injected into the mouth via chronic intraoral catheters. Video-taping is done via a mirror located beneath the Plexiglas floor in the upper right-hand corner of the figure.

Ingestion Sequence

Aversion Sequence

Fig. 13. Taste-elicited fixed action patterns. Top: Ingestive responses are elicited by oral infusions of glucose, sucrose, isotonic sodium chloride, and include rhythmic mouth movements, tongue protrusions, lateral tongue protrusions, and paw licking. Bottom: Aversive responses are elicited by infusions of quinine, caffeine, and sucrose octaacetate solutions and include gapes, chin rubs, head shakes, paw wipes, forelimb flailing, and locmotion (not shown). [From Berridge, K., Grill, H., and Norgren, R. (1981). Relation of consummatory responses and learned taste aversions. *J. Comp. Physiol. Psych.* **95**, 363–382. Copyright (1981) by the American Psychological Association. Reprinted by permission of the authors.]

described by Davis and Levine (1977), may act directly or indirectly upon the caudal brainstem circuitry effecting ingestive consummatory behavior to alter the response elicited by a particular taste (see also Rogers and Novin, 1983). Examples of neural control signals are the neural correlates of food and sodium depletion and repletion as well as those of taste–illness association. These neural control signals impart greater complexity to behaviors guided by taste. In Jackson's hierarchical neurology, a neural control system is located rostral to the circuitry effecting the most basic behavioral responses. A distributed-systems neurology allows for control mechanisms to be located within the same neural level as the circuitry that produces the

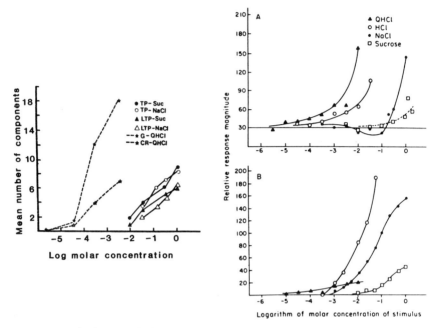

Fig. 14. The figure on the left shows the mean number of individual ingestive and aversive response components seen in response to intraoral infusions of various tastants as a function of concentration. (Symbols used here correspond to response components, not tastants.) The two figures on the right, similar to those of Pfaffmann (1977), show the relationship between taste stimulus concentration and electrophysiological responses of two peripheral gustatory nerves. Note that the glossopharyngeal nerve (A) is more sensitive to quinine than the chorda tympani (B) and that the slope of the electrophysiological function and the slopes of aversive-response functions are very similar for concentrations between −5 and −3 log molar. Similarly, there is a correspondence between the chorda tympani's electrophysiological response to sucrose and the ingestive-behavior response functions for concentrations −2 to 0 log molar.

basic behavioral responses (Mountcastle, 1978). Our approach is to manipulate a given physiological or associative condition known to alter the discriminative responses to taste or intake volume of the intact rat, and then apply those manipulations to neurologically damaged preparations in order to determine the sufficient neural substrate for the integration of that particular neural control signal. A series of experiments were performed to examine whether the caudal brainstem, in the absence of neural participation of the forebrain, is sufficient to integrate neural control signals (e.g., neural correlates of caloric deficit) with taste afferent input to alter ingestive consummatory responses.

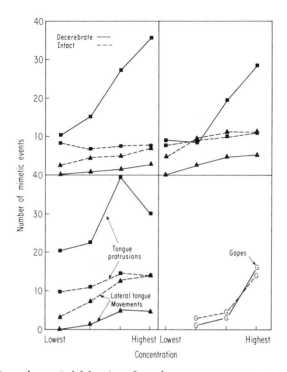

Fig. 15. Decerebrates (solid lines) performed more tongue protrusions before the appearance of the first lateral tongue movement than intact rats (dashed lines). The number of initial mouth movements before the first tongue protrusion was very similar for both preparations and is not shown. For the highest concentrations of sucrose (upper left-hand quadrant), NaCl (upper right-hand quadrant), and HCl (lower left-hand quadrant) stimuli, decerebrates executed approximately the same number of lateral tongue movements as intact rats performed in response to the lowest concentration of these stimuli. Nevertheless, the slopes of the lateral tongue movement–concentration function were roughly parallel for both preparations. The number of gape responses to quinine (lower right-hand quadrant) was very similar for decerebrate and intact rats.

VI. INTEROCEPTIVE INPUT FROM FOOD DEPRIVATION AND INSULIN-INDUCED HYPOGLYCEMIA IS INTEGRATED WITH ORAL AFFERENT INFORMATION WITHIN THE CBS TO CONTROL THE INGESTIVE CONSUMMATORY BEHAVIOR OF CHRONICALLY DECEREBRATE RATS

The effect of simple food deprivation on food consumption was tested first. In order to measure the food intake of the aphagic, chronically decerebrate rat, we took advantage of the taste-reactivity methodology just described. Instead of presenting a variety of taste stimuli to the oral cavity in brief, 50-µl aliquots, a single taste stimulus was now continuously infused. Intake of orally infused sucrose (0.03 M; 0.6 ml/min) was examined under two conditions: sated (1 hr following a tube-fed meal) and deprived (24 hr after the same meal). Sucrose infusion stopped as soon as a drop of fluid was observed to drip from the rat's mouth (rejection) and started again 30 sec later. Termination of intake was defined as two consecutive instances of sucrose rejection. As shown in Fig. 16, food-deprived chronic decerebrates, like control rats, ingested 2–3 times the volume of sucrose they had consumed in the sated condition (Grill and Norgren, 1978c; Grill, 1980). To be certain that the effects of food deprivation were specific to sucrose (a potent food cue) and did not reflect a general facilitation of fluid intake, the effects of deprivation on another taste stimulus, water, was examined. Neither food-deprived control nor decerebrate rats increased their water intake above their sated levels.

Many physiological consequences of food deprivation and repletion may be integrated with food taste signals to alter the ingestive behavior of chronically decerebrate rats. They include alterations in blood-borne levels of metabolic fuels and digestive hormones and the state of the digestive tract. To limit this list of candidate signals, an additional set of experiments was performed. The effect of insulin (5 or 10 U/kg) or saline control injections on the oral sucrose intake of sated decerebrate and control rats was examined.

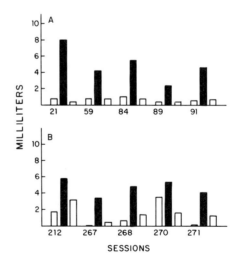

Fig. 16. Intraoral sucrose (0.03 M) intakes by chronic decerebrate (A) and pair-fed control rats (B) under 2 metabolic conditions. Rats tested were either food deprived (solid bars) or food sated (open bars). Code numbers of individuals are depicted below the bars.

Insulin treatment produced comparable levels of hypoglycemia in decerebrate and pair-fed control rats. As can be seen in Fig. 17, Flynn and I (1983) found that chronically decerebrate rats made hypoglycemic by the insulin injection increased their sucrose but not their water intake, as did their pair-fed controls.

As just stated, the correlates of food deprivation and repletion that control the ingestive consummatory behavior of decerebrate rats include the absorption of metabolic fuels, osmotic alterations, the mechanical state of the gastrointestinal tract, and gastrointestinal hormone secretion. Another approach to defining what correlates of a tube-fed meal might reduce the sucrose intake of chronically decerebrate rats has involved experiments on lateral hypothalamic-lesioned (LHX) rats. Fifty microliter presentations of sucrose elicit ingestive taste reactivity in intact or decerebrate rats, whether these rats are tested food-deprived or tube-fed. In both preparations, repeated presentations of sucrose (continuous infusions) result in a more rapid cessation of intake in the tube-fed condition than in the deprived condition. The taste-reactivity responses of aphagic LHX rats in the food-deprived state were found to be identical to those of decerebrate and intact rats. In contrast, Fluharty and I (1980) have shown that the tube-fed LHX rat rejects all

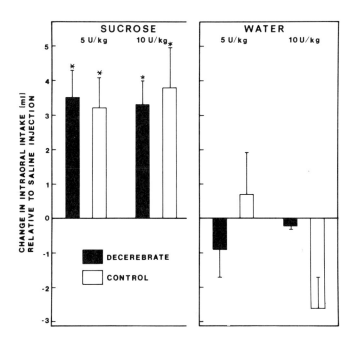

Fig. 17. Change in intake (milliliters) relative to saline injection condition. $*p < 0.05$: paired t test comparisons, significantly different from saline of that group. Sucrose = $0.03\ M$. Intake values are in milliliters.

orally delivered tastes, even concentrated sucrose. We took advantage of the LHX rat's dramatic sensitivity to tube feeding to highlight the potential correlates of repletion for decerebrate rats. The tube feeding-induced suppression of sucrose intake and ingestive taste-reactivity responses in the LHX rat were *not* duplicated by either inserting the feeding tube without delivering diet, delivering a nonabsorbable substance (mannitol) of an equivalent osmolarity and volume, or delivering the water content of the tube-fed meal. As seen in Fig. 18, equicaloric, equivolemic, and equiosmotic glucose was the only intubated substance that duplicated the dramatic effects of tube-fed meals on the taste reactivity and intraoral intake of sated LHX rats. Parallel experiments with decerebrate rats will pursue this suggestion of the role of caloric feedback in altering intake and taste reactivity.

In a recent experiment we have shown that decerebrates, like controls, are sensitive to elevated levels of cholecystokinin (CCK), a digestive hormone that is released in the course of normal digestion. When treated with CCK, 24-hr food-deprived chronically decerebrate rats, like their pair-fed controls, reduced both their intake of intraoral sucrose and decreased the magnitude of ingestive taste-reactivity that accompanies sucrose ingestion (Grill *et al.*, 1983).

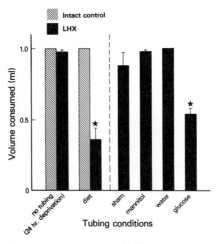

Fig. 18. The effect of six randomly presented intragastric intubation conditions on intraoral sucrose intake. The conditions are 12 ml of liquid diet, 0.2 M mannitol, 1.4 M glucose, or 8 ml of distilled water (the portion of 12 ml diet that is water); passing the tube but not delivering any liquid (sham), on neither passing the tube nor any liquid (24-hr deprived). Ninety minutes after each intubation condition, 1 ml of 1.0 M sucrose (1 ml/min) was intraorally infused into the mouths of lateral hypothalamic-lesioned and yoked control rats. The volume of the sucrose ingested is shown as the amplitude of each histogram. Analysis of variance reveals that sucrose intake following diet and glucose tubing conditions is significantly lower (*$p > 0.01$) than other conditions for the lesioned rat.

VII. CONCLUSION

Hunger has been operationally defined as the increase in food intake that correlates with the magnitude of food deprivation (Silverstone, 1976). As noted above, the control of feeding exerted by the correlates of deprivation ad repletion was thought to require the participation solely of the forebrain and the hypothalamus in particular. The data just described, however, make a case for the caudal brainstem being sufficient for interoception, for reflex connections between these interoceptors and compensatory responses, for reflex connections between gustatory afferent input and the effectors producing salivation, mastication, cephalic insulin secretion, and discriminative behavioral responses, and finally, for integrating some aspects of deprivation/repletion-derived control of intake and taste reactivity.

As noted above, chronically decerebrate rats never spontaneously feed. These rats die without oral or gastric infusion of nutrients. Obviously then, the case made for certain neural controls on feeding behavior contained within the caudal brainstem must be viewed within the context of a larger, integrated nervous system. A hierarchical perspective assumes some degree of redundancy and would therefore presume that some mechanisms and receptors are found within more than one level of the neuraxis. From this perspective, forebrain mechanisms can act directly on caudal brainstem neural circuits to inhibit or facilitate particular responses that would regulate energy input and expenditure. From a functional perspective, the forebrain may be required to initiate action (appetitive behavior) and mediate between competing behaviors prior to the decision to feed.

ACKNOWLEDGMENTS

My thanks to my collaborators Ralph Norgren, Kent Berridge, Bill Flynn, Gary Schwartz, and Richard DiRocco. I am grateful to Gary Schwartz for his critical reading of the manuscript. This work was supported by AM 21397.

REFERENCES

Berridge, K., Grill, H. J., and Norgren, R. (1981). Relation of consummatory responses and preabsorptive insulin release to palatability and learned taste aversions. *J. Comp. Physiol. Psychol.* **95**, 363–382.

Berthoud, H. R., and Jeanrenaud, B. (1982). Sham feeding-induced cephalic phase insulin release in the rat. *Am. J. Physiol.* **238** (*Endocrinol. Metab.* 5), E336–E340.

Berthoud, H. R., and Mogenson, G. J. (1977). Ingestive behavior after intracerebral and intracerebroventricular infusions of glucose and 2-deoxy-D-glucose. *Am. J. Physiol.* **233**, R127–R133.

Berthoud, H. R., Trimble, E. R., Siegel, E. G., Bereiter, D. A., and Jeanrenaud, B. (1980). Cephalic phase insulin secretion in normal and pancreatic islet-transplanted rats. *Am. J. Physiol.* **238**, E336–E340.

Betz, A. L., Drewes, R., and Gilboe, D. D. (1975). Inhibition of glucose transport into brain by phlorizin, phloretin, and glucose analogues. *Biochim. Biophys. Acta* **406**, 505–515.

Bradbury, M. (1979). "The Concept of a Blood–Brain Barrier." Wiley, New York.

Cannon, W. B. (1932). "The Wisdom of the Body." W. W. Norton, New York.

Chhina, G. S., Anand, B. K., and Rao, P. S. (1971). Effect of glucose on hypothalamic feeding centers in deafferented animals. *Am. J. Physiol.* **221**, 662–667.

Coimbra, C. C., Gross, J. I., and Migliorini, R. H. (1979). Intraventricular 2-deoxyglucose, insulin, and free fatty acid mobilization. *Am. J. Physiol.* **207**, E317–E329.

Contreras, R. J., Gomez, M. M., and Norgren, R. (1980). Central origins of cranial nerve parasympathetic neurons in the rat. *J. Comp. Neurol.* **190**, 373–394.

Davis, J. D., and Levine, M. W. (1977). A model for the control of ingestion. *Psychol. Rev.* **84**, 379–412.

Desiraju, T., Benerjee, M. G., and Anand, B. K. (1968). Activity of single neurons in the hypothalamic feeding centers: Effect of 2-deoxy-D-glucose. *Physiol. Behav.* **3**, 757–760.

DiRocco, R. J., and Grill, H. J. (1979). The forebrain is not essential for sympathoadrenal hyperglycemic response to glucoprivation. *Science* **204**, 1112–1114.

Fluharty, S. J., and Grill, H. J. (1980). Taste reactivity of lateral hypothalamic lesioned rats: Effects of deprivation and tube feeding. *Neurosci. Abstr.* **6**, 28.

Flynn, F. W., and Grill, H. J. (1983). Insulin elicits ingestion in decerebrate rats. *Science* **221**, 188–190.

Flynn, F. W., and Grill, H. J. (1984). Fourth ventricular phlorizin dissociates feeding from hyperglycemia in rats. *Brain Res.* (in press).

Glick, Z., and Mayer, J. (1968). Hyperphagia caused by cerebral ventricular infusion of phlorizin. *Nature (London)* **219**, 1374.

Grill, H. J. (1980). Production and regulation of ingestive consummatory behavior in the chronic decerebrate rat. *Brain Res. Bull.* **5**, 79–87.

Grill, H. J., and Berridge, K. C. (1985). Taste reactivity as a measure of the neural control of palatability. *Prog. Psychobiol. Physiol. Psychol.*, in press.

Grill, H. J., and Norgren, R. (1978a). The taste reactivity test. I. Mimetic responses to gustatory stimuli in neurologically normal rats. *Brain Res.* **143**, 263–279.

Grill, H. J., and Norgren, R. (1978b). The taste reactivity test. II. Mimetic responses to gustatory stimuli in chronic thalamic and chronic decerebrate rats. *Brain Res.* **143**, 281–297.

Grill, H. J., and Norgren, R. (1978c). Neurological tests and behavioral deficits in chronic thalamic and chronic decerebrate rats. *Brain Res.* **143**, 299–312.

Grill, H. J., Ganster, D., and Smith, G. P. (1983). CCK-8 decreases sucrose intake in chronic decerebrate rats. *Neurosci. Abstr.* **9**, 903.

Grill, H. J., Berridge, K. C., and Ganster, D. (1984). Oral glucose is the prime elicitor of preabsorptive insulin secretion. *Am. J. Psychol.* **246**, 88–95.

Hamilton, R. B., and Norgren, R. (1984). Central projections of gustatory nerves in the rat. *J. Comp. Neurol.* **222**, 560–577.

Himsworth, R. L. (1970). Hypothalamic control of adrenalin secretion in response to insufficient glucose. *J. Physiol. (London)* **206**, 411–417.

Houpt, K. A., and Epstein, A. N. (1973). Ontogeny of controls of food intake in the rat: GI fill and glucoprivation. *Am. J. Physiol.* **225**, 58–66.

Kawamura, Y., and Yamamoto, T. (1978). Studies on neural mechanisms of the gustatory–salivary reflex in rabbits. *Jpn. J. Physiol.* **285**, 35–47.
Laughton, W. L., and Powley, T. L. (1979). Four central nervous system sites project to the pancreas. *Neurosci. Abstr.* **4**, 415.
Lund, J. P., and Dellow, P. G. (1971). The influence of interative stimuli on rhythmical masticatory movements in rabbits. *Arch. Oral Biol.* **16**, 215–223.
Macht, M. B. (1951). Subcortical localization of certain "taste" responses in the cat. *Fed. Proc.* **10**, 88.
Miller, F. R., and Sherrington, C. S. (1916). Some observations on the buccopharyngeal stage of reflex deglutition in the cat. *Q. J. Exp. Physiol.* **9**, 147–186.
Miselis, R. R., and Epstein, A. N. (1975). Feeding induced by intracerebroventricular 2-deoxy-D-glucose in the rat. *Am. J. Physiol.* **229**, 1438–1447.
Mountcastle, V. B. (1978). An organizing principle for cerebral function: The unit module and the destribution system. *In* "The Mindful Brain" (V. B. Mountcastle and G. M. Edelman, eds.). MIT Press, Cambridge, Mass.
Norgren, R., and Leonard, C. M. (1971). Taste pathways in rat brainstem. *Science* **173**, 1136–1139.
Oomura, Y. (1976). Significance of glucose, insulin, and free fatty acid on the hypothalamic feeding and satiety neurons. *In* "Hunger: Basic Mechanisms and Clinical Implications" (D. Novin, W. Wyrwicka, and G. A. Bray, eds.). Raven Press, New York.
Pfaffmann, C. (1977). Biological and behavioral substrates of the sweet tooth. *In* "Taste and Development: The Genesis of Sweet Preference" (J. M. Weiffenbach, ed.). U.S. Department of Health, Education, and Welfare, Bethesda, Maryland.
Ritter, R. C., Slusser, P. G., and Stone, S. (1981). Glucoreceptors controlling feeding and blood glucose: Location in the hindbrain. *Science* **213**, 451–453.
Ritter, S., and Strang, M. (1982). Fourth ventricular alloxan injection causes feeding but not hyperglycemia in rats. *Brain Res.* **249**, 198–201.
Ritter, S., Murane, J. M., and Landenheim, E. E. (1982). Glucoprivic feeding is impaired by lateral or fourth ventricular alloxan injection. *Am. J. Physiol.* **243**, R312–R317.
Rogers, R. C., and Novin, D. (1983). The neurological aspects of hepatic osmoregulation. *In* "The Kidney in Liver Disease" (M. Epstein, ed.), 2nd ed., pp. 337–350. Elsevier Biomedical, New York.
Sherrington, C. S. (1917). Reflexes elicitable in the cat from pinna vibrissae and jaws. *J. Physiol. (London)* **51**, 404–431.
Silverstone, J. T. (1976). The CNS and feeding: Group report. "Dahlem Workshop on Appetite and Food Intake." Abakon Verlagsgesellschaft, Berlin.
Smith, G. P., and Epstein, A. N. (1969). Increased feeding in response to decreased glucose utilization in the rat and monkey. *Am. J. Physiol.* **217**, 1083–1087.
Steiner, J. E. (1977). Facial expressions of the neonate infant indicating the hedonics of food-related chemical stimuli. *In* "Taste and Development: The Genesis of Sweet Preference" (J. M. Weiffenbach, ed.). U.S. Dept. of Health, Education and Welfare, Bethesda, Maryland.
Stricker, E. M., Rowland, N., Saller, C., and Friedman, M. I. (1977). Homeostasis during hypoglycemia: Central control of adrenal secretion and peripheral control of feeding. *Science* **196**, 79–81.
Strubbe, J. H., and Steffens, A. B. (1975). Rapid insulin release after ingestion of a meal in the unanesthetized rat. *Am. J. Physiol.* **229**, 1019–1022.
Wick, A. N., Dury, D. R., Nakada, H. I., and Wolfe, J. B. (1957). Localization of the primary metabolic block produced by 2-deoxyglucose. *J. Biol. Chem.* **224**, 963–969.

18

Possible Participation of Oro-, Gastro-, and Enterohepatic Reflexes in Preabsorptive Satiation

MAURICIO RUSSEK AND RADU RACOTTA
Department of Physiology
National School of Biological Sciences
National Polytechnic Institute
Mexico City, Mexico

I.	Introduction	374
II.	Regulation of Glycemia	375
III.	Hepatic Receptors and Control of Food Intake	377
IV.	The Hepatic Hypothesis of Feeding	379
V.	Preabsorptive Satiation	381
VI.	Oropharyngeal Receptors	383
VII.	Gastric Distention Receptors	384
VIII.	Gastrointestinal Chemoreceptors	385
IX.	Gastrointestinal Hormones	386
X.	Possible Relation between Satiation and Lipostasis	387
XI.	Conclusions	387
	References	388

It is known that animals stop eating before a substantial amount of the food ingested has been absorbed. This preabsorptive satiation must be effected by nervous and/or hormonal signals originating in the oropharyngeal cavity and gastroduodenal wall. These signals could act directly on the central nervous system (CNS), inhibiting feeding behavior, or on a peripheral organ, which in turn would inform the CNS. Our hypothesis is that nervous signals from oral, gastric, and duodenal chemoreceptors play an important role in preabsorptive satiation via a reflex secretion of noradrenaline and adrenaline from intrahepatic sympathetic terminals and chromaffin cells. The local action of these catecholamines on hepatic receptors elicits a strong inhibition of food intake with the characteristics of postabsorptive satiety. Thus their action depends on the extent of previous depletion (i.e., hepatic glycogen level) and on the palatability of the food offered, and does not affect other behaviors (e.g., drinking of water, or self-stimulation in areas unrelated to feeding).

Gastric distention (and other stimuli such as tastes, smells, sounds, and images) may, by conditioning, acquire the capacity to produce reflex intrahepatic catecholamine secretion and, therefore, to elicit conditional preabsorptive satiation.

Thus, by assuming that the amount of intrahepatic catecholamine secretion depends on the quantity and quality of the food ingested and that the satiating effect of the food depends on the amount of glycogen remaining in the liver from the previous meal, we have a mechanism for adjusting food intake to the level of depletion, before any food has been absorbed.

I. INTRODUCTION

All living cells are thermodynamically open systems which require a continuous supply of energy to preserve their structure and function. In most living organisms, the input of energy from the environment is intermittent; therefore, in order to assure a constant supply of energy, they must store it in some way. Vertebrates store metabolic energy in pouches of their gastrointestinal tract, where ingested substances are processed to simpler molecules, and in internal metabolic reserves, where the absorbed molecules are stored as large polymers (polysaccharides and proteins) or water-insoluble substances (triglycerides). This precludes dangerous changes in cytoplasmic osmolarity when the amount of reserves increases or decreases. Thus, all cells of vertebrate animals receive a constant supply of metabolic fuels derived from either one or both of the above-mentioned stores.

All cells are provided with the enzymatic machinery to obtain their energy requirements from glucose, but most animal tissues can also utilize other metabolites, e.g., amino acids, free fatty acids, and ketone bodies.

Mammalian red cells can metabolize only glucose because they lack mitochondria. Nervous tissue normally consumes mostly glucose and a small mount of amino acids; after a definite fasting period, however, it adapts so as to fulfill approximately half of its energy requirements with ketone bodies derived from fatty acids in the liver. On the other hand, muscle and adipose tissue can derive most of their energy directly from fatty acids.

In view of the importance of glucose, especially for the metabolism of nervous tissue, it is not surprising that its blood level (glycemia) is regulated very efficiently against perturbations that tend to decrease it (e.g., increased glucose consumption elicited by exercise) and less efficiently against perturbations that tend to increase it (e.g., intestinal absorption).

II. REGULATION OF GLYCEMIA

The reader is referred to the following sources: Racotta and Ciures (1970), Newsholme and Start (1973), Lundquist and Tygstrup (1974), Hanson and Mehlman (1976) and Veneziale (1981).

In mammals eating an omnivorous diet (rat, dog, man), in nonruminant herbivores (rabbit), and even in carnivores (cat), basal postabsorptive glycemia is rather similar (60–80 mg/100 ml) and kept fairly constant except during absorption. Its *regulation* is achieved by the *control* of metabolic reserves, in the following manner (Fig. 1).

After a meal, glycemia shows a variable tendency to increase, depending on the amount of carbohydrate in the diet. The regulatory response that counteracts this increase in glycemia is an increase in the storage of glucose in the form of hepatic and muscular glycogen, and of its transformation into fatty acids, stored as triglycerides in adipose tissue. This regulatory response is promoted by an increase in insulin secretion, which also stimulates the storage of amino acids as protein. Insulin also increases the permeability of muscle fibers and adipocytes to glucose, and stimulates glucose catabolism in these tissues.

Thus, during the absorption of a meal, this regulatory response prevents an unduly large increase in glycemia, and allows all metabolic reserves to be replenished and most tissues to obtain their energy only from glucose (with the notable exception of the liver, which always consumes mainly fatty acids).

Once absorption has ceased or has substantially diminished, glycemia decreases, sometimes passing through a rebound period of hypoglycemia, before it stabilizes to its basal level. This basal glycemia (set point) is maintained during fasting periods of considerable length, showing only a minimal, gradual decrease. This regulation is achieved, in the initial phase of the fasting period, by the control of hepatic glycogenolysis and, in muscle and adipose tissue, by a gradual change of energy source from glucose to fatty acids. Thus, despite a constant decline in liver glycogen along with the concomitant decrease in glucose output by the liver, glycemia decreases during fasting at a much slower rate than expected because a reduction in consumption accompanies the reduced production.

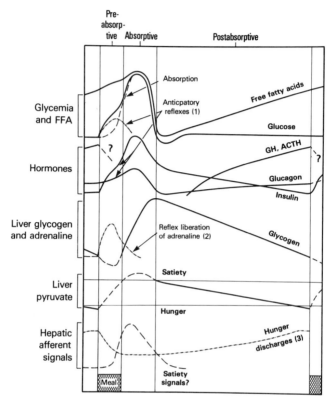

Fig. 1. Events occurring during a meal and in the intermeal period: (———), based on well-known evidence; (–·–·–·), based on recent experimental evidence in (1) Nicolaidis (1969, 1978), (2) Martinez et al. (1974), and (3) Niijima (1969); (— — —), hypothetical. [Redrawn from Russek (1981). With permission from *Appetite* **2**, 137–162. Copyright 1981, by Academic Press Inc. (London) Ltd.]

All of these processes are effected by a continuous decline in insulin level and an increase in glucagon, ACTH, and glucocorticoid levels throughout the postabsorptive period.

The increase in circulating fatty acids resulting from the lipolysis elicited by glucagon, ACTH, and sympathetic activity together with the decrease in insulin compels muscle and other insulin-dependent tissues to gradually reduce their glucose consumption in favor of fatty acid utilization. The increase in glucagon and glucocorticoids stimulates hepatic gluconeogenesis from lactate, glycerol, and amino acids, which also helps to spare glycogen and retard its depletion.

In spite of all these regulatory responses, however, hepatic glycogen declines steadily during the postabsorptive period and can be fully restored only by the ingestion and subsequent absorption of food. This usually occurs before hepatic glycogen reaches its minimum concentration. If food is not available, a much larger increase in gluconeogenesis, mostly from amino acids, occurs, in order to fulfill the glucose requirements of the body. From the metabolic point of view, this is an expensive process, because ammonia liberated from the amino acids used in gluconeogenesis must be detoxified (i.e., transformed to urea), and eventually, protein reserves must be restored from the amino acids ingested in the next meal. Both of these processes, gluconeogenesis and protein anabolism, consume large amounts of metabolic energy, which may explain why the search for and eventual ingestion of food (hunger) are initiated before the depletion of liver glycogen (only hours after a meal), in spite of the fact that the animal has lipid and protein reserves enough to survive for days or weeks (the larger the species, the longer the survival time without food).

Thus, we believe that hunger is a mechanism to replenish mainly carbohydrate reserves and to avoid the large energy cost of gluconeogenesis (i.e., a short-term control of food intake). The amount of food ingested in each meal may be modulated by lipid reserves, which also depend on feeding for their replenishment (i.e., lipostatic, long-term control of food intake).

III. HEPATIC RECEPTORS AND CONTROL OF FOOD INTAKE

Rephrasing the last paragraph, we can thus postulate that the control of food intake is the effector mechanism of a system that reglates the availability of hepatic carbohydrate. The primary physiological control for feeding should therefore be dependent on the detection of carbohydrate stores. This detection is provided by hepatic metabolic receptors. A parallel system regulates adipose reserves mainly by controlling energy output (basal metabolic rate and activity) and possibly also by modulating the size of the meals initiated by the first system.

The existence of hepatic receptors influencing feeding was postulated 20 years ago (Russek, 1963) on the basis of data obtained in dogs (Russek and Piña, 1962). More convincing evidence for the hepatic control of feeding in the dog (Russek, 1970; Russek et al., 1968, 1980) and data suggesting a similar mechanism in cats (Russek and Morgane, 1963) and rats (Racotta et al., 1984; Racotta and Russek, 1977; Russek, 1971; Russek and Stevenson, 1972; Russek et al., 1967) were obtained later. Some evidence suggesting the participation of these receptors in the regulation of glycemia was also obtained in cats (Rodriguez-Zendejas et al., 1968). Confirmatory evidence for the participation of hepatic receptors in the control of food intake has been obtained by other research groups in rabbits (Novin, 1976; Novin and VanderWeele, 1977; Novin et al., 1973, 1974; VanderWeele et al., 1982), rats (Booth and Jarman, 1976; Campbell and Davis, 1974a,b; Friedman and Stricker, 1976; Rowland and Nicolaidis, 1974; Stricker et al., 1979), sheep (Anit and Forbes, 1980; Islas-Chaires and Russek, 1984), and hamsters (N. Rowland, personal communication).

The principal finding reported in all of these studies is that the intraperitoneal or intraportal injections of glucose, lactate, or adrenaline produced a strong inhibition of food intake (from 70% to more than 90%) in animals made hungry by fasting periods of varying duration, whereas the intravenous, intramuscular, or subcutaneous administration of the same doses of these substances had no significant effect on feeding.

This differential effect is due to the fact that these substances are rapidly taken up and stored or catabolized by all tissues, so the different routes of administration produce very different hepatic concentrations.

Other substances, such as isoproterenol and glucagon, may also act on feeding via the liver, but because these substances may have a longer half-life in blood, their effects may be similar irrespective of the route of administration (Stunkard et al., 1955; Balagura and Hoebel, 1967). We suggest that the apparent hypothalamic anorectic effect of isoproterenol might be due to leakage into the blood and action on the liver (Leibowitz, 1970).

Some substances (e.g., 2-deoxyglucose), besides having a long half-life in blood, might also have both an hepatic and a central effect that can occlude one another; in this instance, the only difference observed with diverse methods of administration is in the latency of the effect (Rowland and Nicolaidis, 1974).

The failure of some authors to obtain inhibition of feeding with intraportal glucose and adrenaline in dogs (Bellinger et al., 1976) and with intraportal glucose in rats (Strubbe and Steffens, 1977) can be explained by their use of insufficient doses for the conditions under which tests were performed (Russek et al., 1980). More difficult to account for is the failure to obtain feeding suppression in pigs (Stephens and Baldwin, 1974) with intraportal injections of glucose in doses similar to the ones that we found effective in dogs. Among the possible explanations are those suggested by the authors themselves, i.e., the use of a glucose solution half the concentration of ours, and the fact that the pigs were still growing and that they belonged to a breed selected for its maximum food consumption and rapid growth rate. Another possibility is that despite the fasting period, their liver glycogen was still high (see later).

Other evidence suggesting an hepatic control of food intake is the fact that fructose, which is easily metabolized by liver but does not cross the blood–

brain barrier, suppresses insulin-induced feeding, whereas ketone bodies, which enter the brain and are utilized by it, inhibit the catecholamine response to hypoglycemia but have no significant effect on feeding (Stricker et al., 1977).

In other studies, it has been shown that the effects on feeding of intraportal 2-deoxyglucose, glucose, glucagon, and adrenaline are substantially reduced after vagotomy and/or coeliacotomy (Martin et al., 1978; Novin and VanderWeele, 1977; Schneider et al., 1976; Tordoff et al., 1982). For glucose and adrenaline, whose anorectic effect can only be portohepatic, as stated previously, this finding is unequivocal evidence that at least part of this effect is elicited through nervous afferents. Quite astonishing and very difficult to explain is the recently reported finding that the elimination of the direct hepatic branch of the vagus in rats reduces the small anorectic effect of glucagon more than does the section of the rest of vagus nerve (Geary and Smith, 1983); exactly the opposite happens with the much more pronounced anorexia elicited by adrenaline (Tordoff et al., 1982).

This same operation reduces the inhibition of insulin-induced feeding observed 3 hr after the end of a fructose perfusion (Friedman, 1980; Friedman and Granneman, 1983). The hepatic vagal branch, which does not exist in all species (e.g., the rabbit), represents only a small fraction of the vagal innervation entering the liver, most of which enters via the stomach and coeliac ganglia together with the sympathetic innervation (Sutherland, 1965; Warwick and Williams, 1973; for other references, see Tordoff et al., 1982).

A plausible interpretation of these results is that the hepatic vagal branch contains the main vagal effector innervation of liver (whereas the gastric–coeliac route contains most of the afferents), and that these effector fibers may stimulate the synthesis of glucagon receptors in the hepatocyte membranes and/or the induction of some enzyme specifically involved in fructose utilization (e.g., fructokinase) and in the anorectic effect of glucagon (e.g., pyruvate kinase). This effect could thus be due to an induction of these enzymes during the previous hours, so that acute blockage with atropine (Geary and Smith, 1983) would not produce the same result as chronic section of the nerve. Further research is needed to elucidate this.

IV. THE HEPATIC HYPOTHESIS OF FEEDING

After many years and several modifications of our hypothesis, what we now believe regarding the hepatic control of food intake is as follows (Russek, 1981).

Hepatic receptor discharges (Niijima, 1969, 1982) are inversely proportional to the hepatocyte concentration of some key metabolite. Pyruvate is a good candidate (Racotta *et al.*, 1984) because it represents the metabolic integration of the hepatocyte uptake of glucose, lactate, glycerol, and glycogenic amino acids (especially alanine) *and* the rate of glycogenolysis (Fig. 2). In accord with this, hepatic pyruvate is low in starved and diabetic hyperphagic rats (Table I). Therefore the frequency of these discharges will be low during intestinal absorption (absorptive satiety) and in periods of elevated liver glycogen levels (postabsorptive satiety). As liver glycogen decreases, pyruvate concentration is reduced and the hepatic receptor discharges increase.

These steadily increasing hepatic discharges might be responsible for the continuous decrease in insulin levels and for the increases in glucagon,

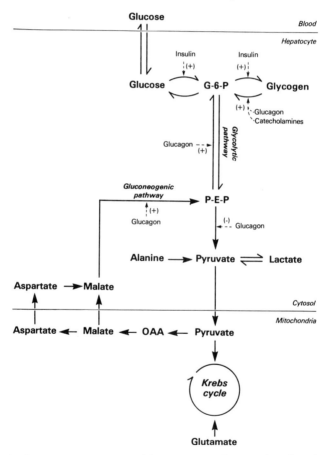

Fig. 2. A schematic representation of the main sites of action of insulin, glucagon, and catecholamines on some metabolic pathways of the hepatocyte. The actions are effected through membrane or intracellular receptors and second messengers.

TABLE I   Metabolites in Rat Liver and Blood[a]

Substance	Liver (mg/100 g)			Blood (mg/100 ml)		
	Fed	Starved	Diabetic	Fed	Starved	Diabetic
Pyruvate	1.14–2.22	0.35–0.70	0.18–0.77	0.74–1.8	0.46	2.1
Lactate	13.5–23.7	4.1–8.6	13.6–27.1	8.6–26.6	5.4	—
Glucose	123–128	65	637	97–103	74	500
Glycogen	3924–5470	2540	227–1190			

[a] Data extracted from Bergmeyer (1974).

ACTH, growth hormone (GH), and in sympathetically controlled lipolysis, occurring during the postabsorptive period (Rodriquez-Zendejas et al., 1968; Sakaguchi and Hayashi, 1981; Sakaguchi and Yamaguchi, 1979). Therefore, the metabolic effects elicited by these hormonal changes, directed toward sparing glycogen, are closely coordinated with the production of glucose by the liver.

When hepatic glycogen, and thus the concentration of pyruvate, decreases to a certain level in spite of increases in lactate, amino acids, and glycerol derived from muscle and adipose tissue, the correlated hepatic receptor discharges begin to facilitate feeding behavior. We call this state "hepatic hunger" or "real hunger."

Animals having food available *ad libitum*—a rare instance in nature— begin to ingest food long before this state of "real hunger" has been reached because of the conditioning of hunger to external stimuli, or to internal stimuli other than hepatic (i.e., gastric or intestinal emptiness). The animal thus attempts to avoid the reaction of "real hunger."

Conditional hunger (appetite) usually occurs even before absorption is completed, at a moment when liver glycogen is almost at a saturation level (when the animal is metabolically satiated). The portal administration of glucose or other metabolites does not affect this type of hunger (Janowitz, 1967; Novin et al., 1974; VanderWeele et al., 1976), which can be inhibited only by external conditional satiety stimuli (Booth and Davis, 1973; Booth et al., 1976a,b; Russek and Pina, 1962) by preabsorptive stimuli (Liebling et al., 1974; Novin, 1976; Davis et al., 1976; Kraly and Smith, 1978; Deutsch et al., 1980; Deutsch and Gonzalez, 1981), or by the glycogenolytic effect of adrenaline or glucagon (Russek et al., 1967; Russek and Racotta, 1980; Geary et al., 1981).

V. PREABSORPTIVE SATIATION

A meal, whether initiated by real or conditional hunger, is ended when only a small portion of the ingested food has been absorbed. The inhibitory process responsible for the inhibition of ingestion has been called *preabsorptive satiation* or simply *satiation*, as opposed to *satiety*, which would be the absence of real hunger.

We hypothesize that satiation results from metabolic changes that simulate satiety, which are elicited by hormones reflexly secreted as a consequence of preabsorptive afferentation. More specifically, we have postulated that the stimulation of oropharyngeal and gastrointestinal receptors produces a reflex secretion of pancreatic glucagon and/or intrahepatic catecholamines, i.e., noradrenaline from hepatic sympathetic terminals and adrenaline from chromaffin cells in the portal spaces (Russek and Racotta, 1980). The hepatic glycogenolysis elicited by these hormones restores hepatic pyruvate, which we believe to be the cause of *satiety*.

In accord with this hypothesis is the similarity between the behavioral changes accompanying "adrenaline-induced anorexia" and those observed during normal satiation. For instance, the inhibition of food intake elicited by intraperitoneal adrenaline (1) depends on the palatability of the food offered to the animal (Russek, 1965); (2) decreases self-stimulation of the lateral hypothalamus only when this self-stimulation elicits stimulus-bound feeding (Russek, 1979); (3) decreases spontaneous activity (Russek and Bruni, 1972), leading to paradoxical sleep (Danguir and Nicolaidis, personal communication); and (4) does not inhibit water intake (Russek, unpublished).

We have obtained histological evidence for the existence of hepatic chromaffin cells and indirect evidence that during a meal there is secretion of adrenaline from these cells and of intrahepatic noradrenaline from sympathetic terminals (Martinez *et al.*, 1974; Russek and Racotta, 1980). There is also evidence that during the course of a meal, glucagon is secreted (Nicolaidis, 1969, 1978; Berthoud and Jeanrenaud, 1982) and glycogenolysis occurs (Geary *et al.*, 1981; Langhans *et al.*, 1982a), and that the stimulation of the ventromedial nucleus of the hypothalamus elicits, through sympathetic efferents, hyperglycemia resulting from hepatic glycogenolysis (Shimazu *et al.*, 1966; Booth *et al.*, 1969). Moreover, it has been shown in the rat fed *ad libitum* that antibodies against glucagon stimulate feeding (Langhans *et al.*, 1982b), and that guanethidine, an adrenergic depletor that acts only in the periphery, and dichloroisoproterenol, an inhibitor of hepatic glycogenolysis, increase meal size (Oceguera *et al.*, 1983).

These findings suggest that both glucagon and catecholamines are involved in the generation of satiation in an animal fed *ad libitum*, which, as we discussed previously, is metabolically satiated (i.e., its liver glycogen is in a saturated condition). However, after a certain fasting period, when liver glycogen has decreased to a certain critical level, the satiating effect of glucagon disappears or is reversed, whereas that of adrenaline is only somewhat reduced (VanderWeele *et al.*, 1979; Russek and Racotta, 1980). This correlates with the fact that, in the fasting rat, adrenaline increases liver lactate (and supposedly pyruvate), whereas glucagon has no effect (Racotta et al., 1984). It has been shown that glucagon inhibits pyruvate kinase much more than adrenaline (Veneziale, 1981; Claus and Pitkis, 1982), suggesting that as real hunger develops, the relative contribution of glucagon to satiation decreases and finally disappears.

We shall now briefly describe the participation of oropharyngeal, gastric, and intestinal receptors, in the generation of satiation, and discuss whether their action can be explained by the mechanism proposed earlier or by other neural or hormonal actions.

VI. OROPHARYNGEAL RECEPTORS

A normal rat, when offered food after several hours deprivation, begins eating at a high rate, which then gradually declines. If the animal is ingesting liquid food, after taking some 15 ml in about 15 min, it stops eating for a while, a sign that satiation has inhibited its feeding behavior. If the rat has an open gastric fistula, the decrease from the initial high eating rate is much slower, reaching about half the maximum about 30 min after feeding began; the rat then continues to sham feed at this rate for several hours (Smith et al., 1974). Thus, continuous stimulation of oropharyngeal receptors can mobilize satiation to only about half the necessary strength to stop a hungry rat from feeding.

On the other hand, a hungry rat with a pyloric clamp that precludes stomach emptying will stop feeding after ingesting a meal of normal size (Kraly and Smith, 1978). This observation suggests that the simultaneous stimulation of oropharyngeal and gastric receptors is sufficient to elicit the full, normal strength of satiation. This does not mean that intestinal receptors play no role in satiation, but rather that there are redundant mechanisms capable of eliciting equivalent inhibition of feeding that occlude each other, as we shall discuss later.

The mechanism by which oropharyngeal receptors participate in satiation may be related to a reflex secretion of hormones. It is well known that sham feeding or the stimulation of oropharyngeal receptors with glucose or saccharine elicits a reflex secretion of insulin that is accompanied by a mild hyperglycemia, but no glucagon secretion (Nicolaidis, 1969, 1978; Berthoud et al., 1980; Berthoud and Jeanrenaud, 1982). There is some evidence that glucose is a more potent stimulus of this "cephalic phase" of insulin than other sugars or sweet substances (Grill et al., 1984). The fact that glycemia increases in spite of insulin secretion in the absence of glucagon suggests that some other mechanism counteracts the hypoglycemic effect of insulin. We suggest that this mechanism could be a reflex intrahepatic liberation of catecholamines, which would explain not only the mild hyperglycemia, but also the partial satiation.

Another phenomenon involving the oropharyngeal receptors that may play a role in satiation is the change in the subjective appreciation of food flavors and odors that has been called *alliesthesia* (Cabanac et al., 1973; Cabanac and Fantino, 1977). Alliesthesia consists of a decrease in or reversal of the pleasantness of different concentrations of sucrose and of different food odors after intubation of glucose into the stomach or, more effectively, into the duodenum. Thus, it is actually an action of the afferent signals from duodenal receptors upon the processing by the central nervous system of the signals from oropharyngeal receptors. Apparently, duodenal glucoreceptors elicit greater negative alliesthesia after stimulation by sweet substances than after stimulation by proteins or peptides, suggesting some specificity for this mechanism. This specificity is demonstrated in the person who is allowed to eat a food with a particular flavor to satiation. Negative alliesthesia affects that flavor to a much greater extent than other flavors (Rolls et al., 1981). Therefore, we think that alliesthesia is a consequence and not a cause of satiation and that its purpose is not to terminate a meal, but to make an omnivorous animal switch to different flavors before satiation is complete, increasing the probability of selecting a more balanced diet.

The contribution of oropharyngeal receptors to satiation is not indispensable. Rats or men fed by self-injection of food into the stomach develop satiation even more quickly than normals and tend to maintain a lower body weight (Epstein, 1967a,b; Fantino, 1980). Thus, the net effect of oropharyngeal receptors is to compel subjects to vary food intake according to palatability, which usually results in an increased feeding (Jacobs and Sharma, 1969). The mild satiation that these receptors seem to elicit is not sufficient to counteract their stimulating effect on feeding.

VII. GASTRIC DISTENTION RECEPTORS

If oropharyngeal receptors are not necessary to elicit satiation and the blockage of stomach emptying does not preclude it (Kraly and Smith, 1978), then gastric receptors should be the main source of information on which this process depends. This is apparently true under normal feeding conditions (Davis et al., 1976; Deutsch et al., 1980). The reason may be that the discharges from gastric distention receptors correlate with the total amount (volume) of food accumulated in the stomach, which gives the most accurate information on total calories ingested, provided that the caloric density of the food remains constant; or alternatively, that duration of gastric distention varies as a function of caloric density, as seems to be the case (McHugh and Moran, 1979).

Thus, the threshold distention required to produce satiation should adapt to the caloric content of the food. In fact, this is what is actually observed (Adolph, 1947; Booth and Davis, 1973; Jonowitz and Grossman, 1949; Levitsky and Collier, 1968). Thus gastric distention appears to be a conditional stimulus for satiation, whose effect must be continually reinforced by some other unconditional stimulation.

The mechanism we propose for the reconditioning of gastric distention, in order for it to adapt to a change in caloric density, is as follows. The first meal on a diluted diet is of the same average size as the preceding meals. However, the stimulation of gastric and/or intestinal chemoreceptors informs the animal of the decreased concentration of metabolites. More importantly, when the animal, driven by conditional stimuli eats again, liver glycogen will be lower than normal (lower level of satiety), so that a stronger distention (larger meal size) will be needed to elicit enough satiation. In subsequent meals, the animal will sense the necessary distention that (when the ingested food is absorbed) will produce the same saticty as before.

It is quite likely that the early secretion of glucagon observed during a normal meal [which is absent during sham feeding (Berthoud and Jeanrenaud, 1982)] and the early rapid liberation of hepatic catecholamines (Martinez et al., 1974; Russek and Racotta, 1980) are, at least in part, conditional reflexes elicited by gastric distention receptors and might be, as already stated above, an important component of the mechanism of satiation. If this is so, what is the unconditional stimulus that produces the reflex secretion of these hormones? More generally, what are the primary unconditional stimuli eliciting satiation? By a process of elimination, they must be the stimuli acting on the gastrointestinal chemoreceptors.

VIII. GASTROINTESTINAL CHEMORECEPTORS

There is electrophysiological evidence of gluco and amino acid receptors in the intestinal and gastric walls (Mei, 1978; Sharma and Nasset, 1962; Sharma et al., 1975), which are most likely responsible for the satiation elicited by duodenal administration of glucose (Novin et al., 1974; VanderWeele et al., 1976), and for the alliesthesia that accompanies it (Cabanac and Fantino, 1977).

In the rabbit, duodenal glucose elicits satiation only when the animals are fed *ad libitum*, that is, when the hepatic glycogen level is high. After a certain fasting period (i.e., after specific decrease in liver glycogen) has occurred the satiating effect of duodenal glucose is lost, and, at the same time, the anorexigenic effect of glucagon disappears (VanderWeele et al., 1976, 1979).

These observations are concordant with the hypothesis that duodenal glucose produces satiation by the reflex secretion of glucagon, whose effect depends on the level of liver glycogen. Apparently, in these experiments the liberation of intrahepatic catecholamines was not sufficient to elicit satiation, perhaps because prior stimulation of orophayngeal and gastric chemoreceptors is needed for catecholamine release to attain its normal intensity. However, in fasting pigs, in which intraportal glucose had no effect on food intake (Stephens and Baldwin, 1974), duodenal glucose elicited some satiation (Stephens, 1980). A possible explanation for these results is that the period of fasting used in these experiments was too short to substantially reduce the level of liver glycogen, and therefore the pigs responded in the same manner as *ad libitum* rabbits. Another possibility is that in omnivorous animals such as the pig, duodenal glucoreceptors produce a much stronger liberation of intrahepatic catecholamines than they do in herbivorous animals such as the rabbit.

Another mechanism by which gastrointestinal chemoreceptors could induce satiation is by the direct secretion of gut hormones. Their possible participation of satiation is discussed in the following section.

IX. GASTROINTESTINAL HORMONES

It has been shown that gastrin, bombesin, secretin, somatostatin, gastric inhibitory peptide, and cholecystokinin (CCK) elicit inhibition of feeding without any apparent toxic effects. The most potent of these is CCK, which elicits an apparently normal satiation at comparatively smaller doses than the other GI hormones (Smith et al., 1974; Smith and Gibbs, 1981). However, there is evidence that the CCK normally produced by a meal is insufficient to elicit satiation and that satiation elicited by a normal meal or by the intestinal administration of food does not need CCK to attain its usual intensity (Koopmans and Heird, 1977; Glick, 1979). These same experiments eliminate the possibility that other duodenal hormones play a significant role in the mechanism of satiation.

It would be interesting to study whether intestinal glucagon liberated from caudal ileum and colon by glucose and fat (Sundby and Moody, 1980) plays a role in prolonging satiation during the last phases of absorption and/or in the anorexia observed in animals with intestinal bypass (Sclafani and Koopmans, 1980). Also, still to be elucidated is whether bombesin contained in the stomach wall is liberated during feeding, and whether this polypeptide plays some role in the satiation induced by gastric receptors.

X. POSSIBLE RELATION BETWEEN SATIATION AND LIPOSTASIS

It has been shown that an important factor in the regulation of fat reserves is a compensatory modification of metabolism, effected by major changes in triiodothyronine (T_3) production, with only small changes in thyroxine (T_4) levels (Hayashi, 1983; Vagenakis, 1977). The discrepancy in levels suggests that the changes in T_3 are due to changes in the conversion of T_4 to T_3 and not to alterations in thyroid function. Another interesting observation is that, when fasting has caused a substantial decrease in T_3, the ingestion of carbohydrates or the injection of catecholamines results in rapid increases in T_3 concentration. Both of these effects are blocked by propranolol, a blocker of β-adrenergic receptors, and potentiated by caffeine, an inhibitor of cAMP catabolism (Bukowiecki et al., 1983; Rothwell et al., 1983), suggesting that sugars increase T_3 levels through the action of catecholamines at β-adrenoceptors having cAMP as a second messenger.

The main site of the T_4 to T_3 conversion is the liver. The conversion is modulated by the cytoplasmic ratio of NADH/NAD, which depends to a large extent on the pyruvate/lactate ratio.

Given all of these findings, we can postulate that the changes in T_3 levels (which control energy output in order to regulate weight) depend on the level of hepatic pyruvate; thus the short-term control of feeding is linked to long-term regulation of fat reserves. Moreover, we could speculate that the stimulation of gut glucoreceptors, through a reflex secretion of intrahepatic catecholamines, participates both in the development of satiation and in the transformation of T_4 to T_3 that will increase metabolism during the next intermeal period, as an anticipatory response against an excess weight gain.

XI. CONCLUSIONS

We have analyzed some of the nervous and hormonal responses to the stimulation of oropharyngeal and gastrointestinal receptors, and the possible participation of these responses in the generation of (preabsorptive) satiation.

From the observations presented, it could be suggested that glucoreceptors along the gut from mouth to caudal ileum, possibly potentiating each other, seem to play a dominant role in satiation. We have postulated that an important component of the effector mechanism mobilized by glucoreceptor stimulation may be the reflex secretion of pancreatic glucagon, intrahepatic catecholamines, and perhaps the direct liberation of intestinal glucagon. The satiation elicited by these glycogenolytic hormones depends on the hepatic glycogen level; thus this satiation gradually decreases during fasting. The effects of glucagon and noradrenaline decrease more during fasting than those of adrenaline perhaps because of a reduction in α-adrenergic and glucagon receptors in the hepatocyte membranes. Thus, the three hormones may have a synergistic role in the production of satiation in an animal fed *ad libitum*, whose eating we believe is driven by conditional hunger on a background of metabolic satiety (i.e., high liver glycogen); however, in a fasting animal (i.e., having low liver glycogen), adrenaline may have greater relative importance in the production of satiation.

Gastric distention receptors elicit a powerful satiating effect, which seems to be a conditional reflex that adapts to changes in the caloric density of the ingested food.

Whether this effect is due to a conditional reflex secretion of glycogenolytic hormones, or to conditional satiety elicited by the direct action of nervous discharges from these receptors upon the central nervous system, will have to be elucidated by further research.

There is some evidence that the catecholamine secretion elicited by gut glucoreceptors involved in satiation also plays a role in lipostasis by potentiating the conversion of T_4 to T_3 by the liver.

Intestinal amino acid and fat receptors might also be involved in the mechanism of satiation, but apparently not through the secretion of catecholamines (protein and fat have no effect on the T_4 to T_3 conversion). However, they could act via glucagon and/or some combination of gastrointestinal hormones.

ACKNOWLEDGMENT

The authors are fellows of the DEDICT-COFAA-IPN. This work was partially supported by Grant PCSABNA-022625 of the Consejo Nacional de Ciencia y Tecnología (CONACyT) of Mexico.

REFERENCES

Adoph, E. F. (1947). Urges to eat and drink in rats. *Am. J. Physiol.* **151**, 110–125.
Anit, M. H., and Forbes, J. M. (1980). Feeding in sheep during intraportal infusions of short-chain fatty acids and the effect of liver denervation. *J. Physiol (London)* **298**, 407–414.
Balagura, S., and Hoebel, B. G. (1967). Self-stimulation of the lateral hypothalamus modified by insulin and glucagon. *Physiol. Behav.* **2**, 337–340.
Bellinger, L. L., Trietley, G. J., and Bernardis, L. L. (1976). Failure of portal blood glucose and adrenaline infusions or liver denervation to affect food intake in dogs. *Physiol. Behav.* **16**, 299–304.
Bergmeyer, H. V. (1974). "Methods of Enzymatic Analysis," 2nd ed., Vol. 4, pp. 2267–2300. Academic Press, New York.

Berthoud, H. R., and Jeanrenaud, B. (1982). Sham feeding-induced cephalic phase insulin release in rat. *Am. J. Physiol.* **242**, E280–E285.
Berthoud, H. R., Trimble, E. R., Siegel, E. G., Bereiter, D. A., and Jeanrenaud, B. (1980). Cephalic phase insulin secretion in normal and pancreatic islet-transplanted rats. *Am. J. Physiol.* **238**, E336–E340.
Booth, D. A., and Davis, J. D. (1973). Gastrointestinal factors in the acquisition of oral sensory control of satiation. *Physiol. Behav.* **11**, 23–29.
Booth, D. A., and Jarman, S. P. (1976). Inhibition of food intake in the rat following complete absorption of glucose delivered into the stomach, intestine, or liver. *J. Physiol. (London)* **259**, 501–522.
Booth, D. A., Coons, E. E., and Miller, N. E. (1969). Blood glucose response to electrical stimulation of hypothalamic feeding areas. *Physiol. Behav.* **4**, 991–1001.
Booth, D. A., Lee, M., and McAleavey, C. (1976a). Acquired sensory control of satiation in man. *Br. J. Psychol.* **67**, 137–147.
Booth, D. A., Toates, F. M., and Platt, S. U. (1976b). Control system for hunger and its implications in animals and man. In "Hunger: Basic Mechanisms and Clinical Implications" (D. Novin, W. Wyrwicka, and G. A. Bray, eds.), pp. 127–142. Raven Press, New York.
Bukowiecki, L., Lupien, J., Follea, N., and Jahjah, L. (1983). Effects of sucrose, caffeine, and cola beverages on obesity, cold resistance and adipose tissue cellularity. *Am. J. Physiol.* **244**, R500–R507.
Cabanac, M., and Fantino, M. (1977). Origin of olfacto-gustatory alliesthesia: Intestinal sensitivity to carbohydrate concentration. *Physiol. Behav.* **18**, 1039–1045.
Cabanac, M., Pruvost, M., and Fantino, M. (1973). Alliesthésie négative pour des stimulus sucrés après diverses ingestions de glucose. *Physiol. Behav.* **11**, 345–348.
Campbell, C. S., and Davis, J. D. (1974a). Licking rate of rats induced by intraduodenal and intraportal glucose infusion. *Physiol. Behav.* **12**, 357–365.
Campbell, C. S., and Davis, J. D. (1974b). Peripheral control of food intake: Interaction between test diet and postingestive chemoreception. *Physiol. Behav.* **12**, 377–384.
Claus, T. H., and Pilkis, S. J. (1982). Hormonal control of gluconeogenesis. In "Biochemical Actions of Hormones" (G. Litwack, ed.), Vol. VIII, pp. 209–271. Academic Press, New York.
Davis, J. D., Collins, B. J., and Levine, M. W. (1976). Peripheral control of meal size: Interaction of gustatory stimulation and postingestional feedback. In "Hunger: Basic Mechanisms and Clinical Implications" (D. Novin, W. Wyrwicka, and G. A. Bray, eds.), pp. 395–408. Raven Press, New York.
Deutsch, J. A., and Gonzalez, M. F. (1981). Gastric fat content and satiety. *Physiol. Behav.* **26**, 673–676.
Deutsch, J. A., Gonzalez, M. F., and Young, W. G. (1980). Two factors control meal size. *Brain Res. Bull.* **5** (Suppl. 4), 55–67.
Epstein, A. N. (1967a). Oropharingeal factors in feeding and drinking. In "Handbook of Physiology" Sect. 6: "Alimentary Canal "Vol. I: "Food and Water Intake" (C. F. Code and W. Heidel, eds.), pp. 197–218. American Physiological Society, Washington, D.C.
Epstein, A. N. (1967b). Feeding without oropharingeal sensations. In "Chemical Senses and Nutrition" (M. R. Kare and O. Maller, eds.), pp. 263–280. The Johns Hopkins University Press, Baltimore.
Fantino, M. (1980). Masse corporelle et comportement alimentaire. Doctoral Thesis, Université Claude Bernard, Lyon, France, pp. 103–118.
Friedman, M. I. (1980). Hepatic-cerebral interactions in insulin induced eating and gastric acid secretion. *Brain Res. Bull.* **5** (Suppl. 4), 63–68.

Friedman, M. I., and Granneman, J. (1983). Food intake and peripheral factors after recovery from insulin-induced hypoglycemia. *Am. J. Physiol.* **244**, R374–R382.
Friedman, M. I., and Stricker, E. M. (1976). The physiological psychology of hunger: A physiological perspective. *Psychol. Rev.* **83**, 407–431.
Geary, N., and Smith, G. P. (1983). Selective hepatic vagotomy blocks pancreatic glucagon's satiety effect. *Physiol. Behav.* **31**, 391–394.
Geary, N., Langhans, W., and Scharrer, E. (1981). Metabolic concomitants for glucagon-induced suppression of feeding in the rat. *Am. J. Physiol.* **241**, R330–R335.
Glick, Z. (1979). Intestinal satiety with or without upper intestinal factors. *Am. J. Physiol.* **236**, R142–R146.
Grill, H. J., Berridge, K. C., and Ganster, D. F. (1984). Oral glucose is the prime elicitor of preabsorptive insulin secretion. *Am. J. Physiol.* **246**, R88–R95.
Hanson, R. W., and Mehlman, M. A. (eds.) (1976). "Gluconeogenesis: Its Regulation in Mammalian Species." Wiley, New York.
Hyashi, M. (1983). Restraint induced thermogenesis blunted by fasting in rats. *Am. J. Physiol.* **244**, E323–E328.
Islas-Chaires, M., and Russek, M. (1984). Efecto de la adrenalina sobre la ingestion de alimento en rumiantes. *Cong. Nal. Cien. Fisiol. (Mex.)* **XXVII**, Morelia, Mich., Mexico.
Jacobs, H. L., and Sharma, K. N. (1969). Taste versus calories: Sensory and metabolic signals in the control of food intake. *Ann. N.Y. Acad. Sci.* **157**, 1084–1112.
Janowitz, H. D. (1967). Role of the gastrointestinal tract in regulation of food intake. *In* "Handbook of Physiology," Sect. 6: "Alimentary Canal," Vol. I: "Food and Water Intake" (C. F. Code and W. Heidel, eds.), pp. 219–224. American Physiological Society, Washington, D.C.
Janowitz, H. D., and Grossman, M. I. (1949). Effect of variations in nutritive density on intake of food of dogs and rats. *Am. J. Physiol.* **158**, 184–193.
Koopmans, H. S., and Heird, W. C. (1977). Hunger satiety dissociated from pancreatic enzyme flow. *Int. Conf. Physiol. Food and Fluid Intake*, **VI**, Jouy en Josas, France.
Kraly, F. S., and Smith, G. P. (1978). Combined pregastric and gastric stimulation by food is sufficient for normal meal size. *Physiol. Behav.* **21**, 405–408.
Langhans, W., Geary, N., and Scharrer, E. (1982a). Liver glycogen content decreases during meals in rats. *Am. J. Physiol.* **243**, R450–R453.
Langhans, W., Zieger, V., Scharrer, E., and Geary, N. (1982b). Stimulation of feeding in rats by intraperitoneal injection of antibodies to glucagon. *Science* **218**, 894–895.
Leibowitz, S. F. (1970). Reciprocal hunger-regulating circuits involving alpha- and beta-adrenergic receptors located, respectively, in the ventromedial and lateral hypothalamus. *Proc. Natl. Acad. Sci. U.S.A.* **67**, 1063–1070.
Levitsky, D. A., and Collier, G. (1968). Effects of diet and deprivation on meal eating behavior in rats. *Physiol. Behav.* **3**, 137–140.
Liebling, D. S., Elisner, J. D., Gibbs, J., and Smith, G. P. (1974). Intraduodenal infusion of liquid food decreases feeding in rats with open gastric fistula. *Fed. Proc.* **33**, 259 (abstract).
Lundquist, F., and Tygstrup, N. (eds.) (1974). "Regulation of Hepatic Metabolism." Academic Press, New York.
McHugh, P. R., and Moran, T. H. (1979). Calories and gastric emptying: A regulatory capacity with implications for feeding. *Am. J. Physiol.* **236**, R254–R260.
Martin, J. R., Novin, D., and VanderWeele, D. A. (1978). Loss of glucagon suppression of feeding following vagotomy in rats. *Am. J. Physiol.* **234**, E314–E318.
Martinez, I., Racotta, R., and Russek, M. (1974). Hepatic chromaffin cells. *Life Sci.* **15**, 267–271.

Mei, N. (1978). Vagal glucoreceptors in the small intestine of the cat. *J. Physiol. (London)* **282**, 485–506.
Newsholme, E. A., and Start, C. (1973). "Regulation in Metabolism." Wiley, London.
Nicolaidis, S. (1969). Early systemic responses to orogastric stimulation in the regulation of food and water balance: Functional and electro-physiological data. *Ann. N.Y. Acad. Sci.* **157**, 1176–1200.
Nicolaidis, S. (1978). Role des reflexes anticipateurs oro-végétatifs dans la régulation hydrominérale et énergétique. *J. Physiol. (Paris)* **71**, 1–19.
Niijima, A. (1969). Afferent impulse discharges from glucoreceptors in the liver of the guinea pig. *Ann. N.Y. Acad. Sci.* **157**, 690–700.
Niijima, A. (1982). Glucose-sensitive afferent nerve fibres in the hepatic branch of the vagus nerve in the guinea pig. *J. Physiol. (London)* **322**, 315–323.
Novin, D. (1976). Visceral mechanisms in the control of food intake. *In* "Hunger: Basic Mechanisms and Clinical Implications" (D. Novin, W. Wyrwicka, and G. A. Bray, eds.), pp. 357–367. Raven Press, New York.
Novin, D., and VanderWeele, D. A. (1977). Visceral involvement in feeding: There is more to regulation than the hypothalamus. *Prog. Psychobiol. Physiol. Psychol.* **7**, 193–241.
Novin, D., VanderWeele, D. A., and Rezek, M. (1973). Infusion of 2-deoxy-D-glucose into the hepatic-portal system causes eating: Evidence for peripheral glucoreceptors. *Science* **181**, 858–860.
Novin, D., Sanderson, J. D., and VanderWeele, D. A. (1974). The effect of isotonic glucose on eating as a function of feeding condition and infusion site. *Physiol. Behav.* **13**, 3–7.
Oceguera, M. G., De la Cruz, F., Chambert, G., and Russek, M. (1983). Effects of adrenergic blockers and depletors on food intake in rats. *Appetite* **4**, 187–193.
Racotta, R., and Ciures, A. (1970). A model of the fuel-tension regulatory system. *An. Inst. Biol. Univ. Nal. Antón (MX) Ser. Biol. Exp.* **41**, 55–88.
Racotta, R., and Russek, M. (1977). Food intake of rats after intraperitoneal and subcutaneous administration of glucose, glycerol, and sodium lactate. *Physiol. Behav.* **18**, 267–273.
Racotta, R., Islas-Chaires, M., Vega, C., Soto-Mora, M., and Russek, M. (1984). Glycogenolytic substances, hepatic and systemic lactate, and food intake in rats. *Am. J. Physiol.* **246**, R247–R250.
Rodriguez-Zendejas, A. M., Vega, C., Soto-Mora, L. M., and Russek, M. (1968). Some effects of intraperitoneal glucose and of intraportal glucose and adrenaline. *Physiol. Behav.* **3**, 259–264.
Rolls, B. J., Rolls, E. T., Rowe, E. A., and Sweeney, K. (1981). Sensory specific satiety in man. *Physiol. Behav.* **27**, 137–142.
Rothwell, N. J., Saville, M. E., and Stock, M. J. (1983). Metabolic responses to fasting and refeeding in lean and genetically obese rats. *Am. J. Physiol.* **244**, R615–R620.
Rowland, N., and Nicolaidis, S. (1974). Response glycemique et de prise d'aliments a l'injection intra-auriculaire et portale de 2-desoxy-D-glucose chez le rat. *C. R. Acad. Sci. (Paris)* **279**, 1093–1096.
Rowland, N., and Stricker, E. M. (1979). Differential effects of glucose and fructose infusions on insulin-induced feeding in rats. *Physiol. Behav.* **22**, 387–389.
Russek, M. (1963). Participation of hepatic glucoreceptors in the control of intake of food. *Nature (London)* **197**, 79–80.
Russek, M. (1965). The influence of food palatability on adrenaline-induced anorexia and its conditioning. *Int. Cong. Physiol. Sci.* **122**, 1123.
Russek, M. (1970). Demonstration of the influence of an hepatic glucosensitive mechanism of food-intake. *Physiol. Behav.* **5**, 1207–1209.

Russek, M. (1971). Hepatic receptors and the neurophysiological mechanisms controlling feeding behavior. *Neurosci. Res.* **4**, 213–282.

Russek, M. (1979). The effect of various hunger manipulations on self-stimulation and the feeding elicited by it. *Physiol. Behav.* **22**, 661–667.

Russek, M. (1981). Current status of the hepatostatic theory of food intake control. *Appetite* **2**, 137–162.

Russek, M., and Bruni, E. (1972). On the mechanism of the anorexigenic effect of amphetamine and adrenaline. *An. Inst. Biol. Univ. Nal. Autón. (Mex.)* **40**, 1–6.

Russek, M., and Morgane, P. J. (1963). Anorexic effect of intraperitoneal glucose in the hypothalamic hyperphagic cat. *Nature (London)* **199**, 1004–1005.

Russek, M., and Piña, S. (1962). Conditioning of adrenaline induced anorexia. *Nature (London)* **193**, 1296–1297.

Russek, M., and Racotta, R. (1980). A possible role of adrenaline and glucagon in the control of food intake. In "Frontiers in Hormone Research" (T. B. Van Wimersma, ed.), Vol. 6, pp. 120–137. Karger, Basel.

Russek, M., and Stevenson, J. A. F. (1972). Correlation between the effects of several substances on food intake and on the hepatic concentration of reducing sugars. *Physiol. Behav.* **8**, 245–249.

Russek, M., Mogenson, G. J., and Stevenson, J. A. F. (1967). Calorigenic, hyperglycemic, and anorexigenic effects of adrenaline and noradrenaline. *Physiol. Behav.* **2**, 429–433.

Russek, M., Rodriguez-Zendejas, A. M., and Piña, S. (1968). Hypothetical liver receptors and the anorexia caused by adrenaline and glucose. *Physiol. Behav.* **3**, 249–257.

Russek, M., Lora-Vilchis, M. C., and Islas-Chaires, M. (1980). Food intake inhibition elicited by intraportal glucose and adrenaline in dogs on a 22 h fasting/2 h feeding schedule. *Physiol. Behav.* **24**, 157–161.

Sakaguchi, T., and Hayashi, Y. (1981). Reflex secretion of insulin evoked by hepatic portal infusion of D-glucose anomers. *Biomed. Res.* **2**, 222–224.

Sakaguchi, T., and Yamaguchi, K. (1979). Changes in efferent activities of the gastric vagus nerve by administration of glucose in the portal vein. *Experientia* **35**, 875–876.

Schneider, K., Rezek, M., and Novin, D. (1976). Effects of visceral sympathectomy on 2-DG induced eating. *Physiol. Behav.* **16**, 55–58.

Sclafani, A., and Koopmans, H. S. (1980). Food intake and body weight following jejunoileal bypass in obese and lean rats. *Brain Res. Bull.* **5**, (Suppl. 4), 69–73.

Sharma, K. N., and Nasset, E. S. (1962). Electrical activity in mesenteric nerves after perfusion of gut lumen. *Am. J. Physiol.* **202**, 725–730.

Sharma, K. N., Dua-Sharma, S., and Jacobs, H. L. (1975). Electrophysiological monitoring of multilevel signals related to food intake. In "Neural Integration of Physiological Mechanisms and Behavior" (G. J. Mogenson and F. R. Calaresu, eds.), pp. 194–212. Toronto Univ. Press, Toronto.

Shimazu, T., Fukuda, A., and Ban, T. (1966). Reciprocal influences of the ventromedial and lateral hypothalamic nuclei on blood glucose levels and liver glycogen content. *Nature (London)* **210**, 1178–1179.

Smith, G. P., and Gibbs, J. (1981). Brain-gut peptides and the control of food intake. In "Neurosecretion and Brain Peptides" (J. B. Martin, S. Reichilin, and K. L. Bick, eds.), pp. 389–395. Raven Press, New York.

Smith, G. P., Gibbs, J., and Young, R. C. (1974). Cholecystokinin and intestinal satiety in the rat. *Fed. Proc.* **33**, 1146–1149.

Stephens, D. B. (1980). The effects of alimentary infusions of glucose, amino acids, or neutral fat on meal size in hungry pigs. *J. Physiol. (London)* **299**, 455–465.

Stephens, D. B., and Baldwin, B. A. (1974). The lack of effect of intrajugular or intraportal injections of glucose or amino acids on food intake in pigs. *Physiol. Behav.* **12**, 923–929.
Stricker, E. M., Rowland, N., Saller, C. F., and Friedman, M. (1977). Homeostasis during hypoglycemia: Central control of adrenal secretion and peripheral control of feeding in rats. Science **196**, 79–81.
Strubbe, J. H., and Steffens, A. B. (1977). Blood glucose levels in portal and peripheral circulation and their relation to food intake in the rat. *Physiol. Behav.* **19**, 303–307.
Stunkard, A. J., Van Itallie, T. B., and Reis, B. B. (1955). The mechanism of satiety: Effect of glucagon on gastric hunger contractions in man. *Proc. Soc. Exp. Biol. Med.* **89**, 258–261.
Sundby, F., and Moody, A. J. (1980). Gut glucagon like immunoreactants: Isolation, structure, and possible role. *In* "Gastrointestinal Hormones" (G. B. J. Glass, ed.), pp. 307–313. Raven Press, New York.
Sutherland, S. D. (1965). The intrinsic innervation of the liver. *Rev. Intern. d'Hépatol.* **15**, 569–578.
Tordoff, M. G., Novin, D., and Russek, M. (1982). Effects of hepatic denervation on the anorexic response to epinephrine, amphetamine, and lithium chloride. *J. Comp. Physiol. Psychol.* **96**, 361–375.
Vagenakis, A. B. (1977). Thyroid hormone metabolism in prolonged starvation in man. *In* "Anorexia Nervosa" (R. A. Vigersky, ed.), pp. 243–253. Raven Press, New York.
VanderWeele, D. A., Skoog, D. R., and Novin, D. (1976). Glycogen levels and peripheral mechanisms of glucose-induced suppression of feeding. *Am. J. Physiol.* **231**, 1655–1659.
VanderWeele, D. A., Geiselman, D. J., and Novin, D. (1979). Pancreatic glucagon, food deprivation and feeding in intact and vagotomized rabbits. *Physiol. Behav.* **23**, 155–158.
VanderWeele, D. A., Haraczkiewicz, E., and Van Itallie, T. B. (1982). Elevated insulin and satiety in obese and normal-weight rats. *Appetite* **3**, 99–109.
Veneziale, C. M. (ed.) (1981). "The Regulation of Carbohydrate Formation and Utilization in Mammals." Univ. Park Press, Baltimore.
Warwick, R., and Willilams, P. L. (eds.) (1973). "Gray's Anatomy of the Human Body," 35th ed., p. 1306. Longman, London.

19

Effects of Protein and Carbohydrate Ingestion on Brain Tryptophan Levels and Serotonin Synthesis: Putative Relationship to Appetite for Specific Nutrients

JOHN D. FERNSTROM
Department of Psychiatry
Western Psychiatric Institute and Clinic and
The Center for Neuroscience
University of Pittsburgh School of Medicine
Pittsburgh, Pennsylvania

I.	Introduction	395
II.	Diet, Brain Tryptophan Uptake, and Serotonin Synthesis	396
III.	Diet, Brain Tryptophan and Serotonin, and Appetite	404
IV.	Summary and Conclusions	412
	References	413

I. INTRODUCTION

The rate of synthesis of serotonin in brain neurons is very sensitive to tryptophan supply. Serotonin synthesis changes almost *immediately* following an injection of tryptophan (Hess and Doepfner, 1961; Ashcroft et al., 1965; Fernstrom and Wurtman, 1971a). And quite small doses of tryptophan, amounting to a fraction of the tryptophan an animal would normally consume each day in food, elicit a significant stimulation of serotonin synthesis (Fernstrom and Wurtman, 1971a). No other neurotransmitter shows this extraordinary sensitivity of synthesis rate to changes in precursor supply [Catecholamine and acetylcholine synthesis rates do respond to changes in the local levels of their respective precursors, but the responses are not as rapid and do not occur at small precursor doses; they are clearly pharmacologic (M. H. Fernstrom, 1981; J. D. Fernstrom, 1983).] The biochemical basis for the influence of precursor supply on serotonin synthesis is known: the enzyme catalyzing the rate-limiting step in serotonin biosynthesis, tryptophan hydroxylase, is normally unsaturated with substrate. Hence, variations in local tryptophan level can influence enzyme saturation, and thus the rate of tryptophan hydroxylation and the overall rate of serotonin production (see J. D. Fernstrom, 1983).

Given the sensitivity of serotonin synthesis rate to variations in brain tryptophan supply, two questions can immediately be posed. First, does anything normally take place in the body that influences tryptophan access to brain, and thereby also serotonin production? And second, if serotonin production is routinely influenced by changes in brain tryptophan level elicited by normal vagaries in body function, does the brain do anything differently as a consequence of such changes in serotonin synthesis?

Over the past 15 years my associates and I have studied the relationship in brain of serotonin synthesis to tryptophan supply. The results of this work have been a set of findings connecting the ingestion of proteins and carbohydrates by rats to the uptake of tryptophan into brain and the synthesis of serotonin. These findings seem clearly to indicate that normal meals can influence the synthesis of an important brain neurotransmitter. Thus, an answer to the first question is in hand, at least under certain conditions (see discussion below). An answer to the second question has been more difficult to obtain. Relevant to the content of this symposium are recent efforts to connect the relationship between diet and brain serotonin synthesis to a regulatory mechanism for governing appetite for carbohydrates and protein. This is a reasonable working hypothesis, one that can be (and has been) tested. What has been the outcome of this line of investigation to date? Some of the results are suggestive, though far from complete.

The discussion that follows first considers the relationship between food intake and serotonin synthesis in brain. The logic in approaching this work is reviewed, in part to illuminate the thought process under which it developed, but also to provide a context in which to view subsequent work by others regarding the effect of nutrient selection on brain serotonin synthesis and on appetite. Accordingly, the second part of this chapter reviews available data linking the dietary–neurochemical relationship to appetite for specific nutrients.

II. DIET, BRAIN TRYPTOPHAN UPTAKE, AND SEROTONIN SYNTHESIS

The first studies connecting the event of eating something to a change in brain tryptophan level and serotonin production followed from a simple experiment in endocrinology. The following question led to the experiment: If an injection of L-tryptophan rapidly elevated brain tryptophan level and stimulated serotonin production, was there a way to elicit a change in brain tryptophan without injecting tryptophan itself? If so, would the change be sufficient to alter serotonin production? Insulin was known to influence blood levels of amino acids. Hence, if insulin could be shown to alter the blood level of tryptophan, perhaps it would as a result influence tryptophan uptake into brain and thus serotonin synthesis. An insulin experiment was therefore conducted. The animals were fasted overnight to reduce serum insulin levels to low values. The next morning, the rats received an injection of insulin, and were killed at intervals thereafter. In these early studies (Fernstrom and Wurtman, 1971b), insulin injection was found to elevate serum tryptophan, brain tryptophan, and brain serotonin levels within 2 hr of injection. It was surprising that insulin raised blood tryptophan; this hormone usually lowers blood levels of individual amino acids. However, the effect was clear, and the interpretation was that insulin elevated brain tryptophan level indirectly via its effect on blood tryptophan level.

As an endocrine experiment, this study lacked one control. The presumption was that insulin elevated brain tryptophan and serotonin via its effects on blood tryptophan level. However, just as likely might be some nonspecific effect on brain serotonin mediated by the sizable reduction that occurred in blood glucose levels (the insulin dose was 2 IU/kg, which caused a sizable decline in blood glucose). As a control for this potential problem, we decided to attempt to elicit the effects by giving the animal sugar instead of insulin. In a fasted animal, the sugar would elicit insulin secretion, but blood glucose levels would not decline. Hence, any effect on brain tryptophan and serotonin, if it still occurred, could not be attributed to hypoglycemia. Moreover, if an effect were obtained, it would show the release of normal quantities of insulin to be capable of eliciting the changes observed when the hormone was injected. Instead of intubating the rats with glucose, we decided to feed them a carbohydrate food following an overnight fast. The fast was needed to ensure low insulin levels (as would be required for any glucose tolerance test). The idea of giving them a food was appealing, since the animals would dose themselves with their own desired amount of carbohydrate. In practice, they gave themselves a much greater amount of sugar than the experimenter would have chosen. The outcome of this experiment was gratifying. Following an overnight fast, the animals were presented with carbohydrate. They consumed sufficient quantities to produce a robust elevation in serum insulin levels, without causing any reduction in blood glucose levels. Associated with these changes was (1) a rise in blood tryptophan levels, (2) a rise in brain tryptophan levels, and (3) a rise in the brain levels of serotonin and its principal metabolite, 5-hydroxyindoleacetic acid (5-HIAA). These increments all occurred within 2 hr after food presentation, and were sustained for at least an hour thereafter (Fig. 1).

At this point, the strictly endocrine approach was modified when it was realized that the changes were caused by the simple act of eating a normal food substance. Studies turned to an exploration of effects of single meals on brain tryptophan and serotonin levels. But the model was retained, since the explanation of the events was tied to it. That is, the reason that brain tryptophan and serotonin rose following carbohydrate ingestion was that insulin was secreted and caused a sufficient rise in blood tryptophan level to induce the changes seen in brain. To elicit the rise in serum insulin, a fast was needed to ensure low circulating insulin levels at the start of the experiment. This is not a small point. If insulin levels were high at the outset of the experiment, it is doubtful that the ingestion of carbohydrates would cause

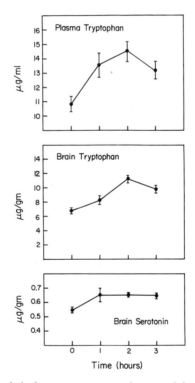

Fig. 1. Effect of carbohydrate ingestion on plasma and brain tryptophan, and brain serotonin levels. Groups of overnight-fasted rats received a protein-free diet composed largely of carbohydrates, and were killed at the indicated times thereafter. Serum and brain tryptophan levels were significantly elevated 1, 2, and 3 hr, and brain serotonin 2 and 3 hr after food presentation ($p < 0.05$). (Adapted from Fernstrom and Wurtman, 1971b.)

any further, sizable increment in blood insulin levels. As a result, it is likewise doubtful that any effects on blood tryptophan, brain tryptophan, or brain serotonin would be seen. This issue is relevant to the appetite studies described below. An unstated assumption in the model, which holds that protein and/or carbohydrate intakes in the rat are regulated from meal to meal, is that the ingestion of carbohydrates will raise brain tryptophan level and stimulate serotonin production, even though the animal is not fasted. This is not likely to be the case, since after a few "meals," the rat will have high circulating insulin levels. Following the ingestion of yet another small meal, regardless of composition, insulin levels would probably not change sufficiently to influence further brain tryptophan levels (or serotonin synthesis).

The emphasis was thus shifted to food-induced changes in brain tryptophan, rather than insulin-induced changes in this parameter. The notion immediately arose that the only food substance other than carbohydrate that was likely to influence blood tryptophan was *protein*. Naively, we assumed that if carbohydrate ingestion raised blood tryptophan, and this alone accounted for the increment in brain tryptophan, the ingestion of protein along with carbohydrates should cause even greater increments in blood tryptophan (due to the addition of new tryptophan molecules to the blood from the food), and thus in brain tryptophan levels. As indicated above, the experiment was done using rats fasted overnight. The result obtained was unexpected. The ingestion of a protein-containing meal did indeed cause larger increments in blood tryptophan level than were caused by the ingestion of carbohydrate alone. However, despite the large rise in blood tryptophan, the ingestion of the protein meal not only did *not* cause a rise in brain tryptophan comparable to that seen following carbohydrate ingestion, it caused *no* increase in brain tryptophan *at all* (Fernstrom et al., 1973). Clearly, the hypothesis that brain tryptophan simply reflected blood tryptophan levels was no longer tenable.

But what could explain this paradoxical effect? The solution was fortunately straightforward. A long line of research has previously established that amino acids are transported into a variety of tissues, including brain, by carrier-mediated mechanisms (e.g., see Pardridge, 1977). These transport carriers were found not to be specific for each amino acid. Instead, only three or four carriers were identified, each one specific to a *group* of amino acids. In the case of tryptophan, the carrier is that which transports all large neutral amino acids (LNAAs): tryptophan, tyrosine, phenylalanine, leucine, isoleucine, and valine. The most important features of the transport carriers are that they are saturable and *competitive*. Hence, if blood tryptophan levels rise, but those of the other competitors do not, more tryptophan should enter brain. If blood tryptophan remains constant, but the levels of its competitors rise, transport of tryptophan into brain should fall. If blood tryptophan rises, and the levels of its competitors *also* rise, no change in brain tryptophan uptake should be obtained. It was this latter notion that was embraced as a hypothesis to explain why the ingestion of a protein-containing meal did not raise brain tryptophan levels. Perhaps the blood levels of competitors rose in parallel with that of blood tryptophan, preventing any increments in brain tryptophan because with the ingestion of protein, the meal would be adding new molecules of tryptophan's competitors to blood. In fact, dietary proteins contain much more of tryptophan's competitors than they do of tryptophan itself. Hence, the influx of competitors would be considerable. These would first offset the insulin-induced fall in their blood levels, and then increment the blood pools over fasting values.

The testing of this notion was straightforward. If brain tryptophan did not rise after protein was consumed because large amounts of the competing amino acids entered the blood from the gut, then brain tryptophan *should* rise if a "protein" were ingested that lacked these amino acids. Again, the same paradigm was used because it was well characterized. That is, the studies were carried out using animals fasted overnight. The experiment was to feed rats a diet that looked just like a normal, protein (casein)-containing food in its amino acid content. But the diet was made up using free amino acids rather than the protein. With this design, amino acids could be deleted from the food, and the effect of such diets studied. In particular, we deleted the large neutral amino acid competitors from one diet. Groups of fasted animals were thus fed a complete diet, or one that lacked the competitors, or one that lacked *not* the large neutral amino acids, but instead the *basic* amino acids. This latter diet would serve as a control for amino acid deletion. When the diets were fed, the results were clear (Fig. 2). Blood tryptophan level rose rapidly in rats consuming the complete diet (i.e., that which

Fig. 2. Effect of the ingestion of various amino acid-containing diets on plasma and brain tryptophan, and brain 5-hydroxyindoles. Groups of eight rats were killed 1 or 2 hr after diet presentation. Vertical bars represent standard errors of the mean; open circles, fasting controls; closed squares, complete amino acid mixture (i.e., equivalent to casein in amino acid composition) diet; closed circles, mixture diet minus tyrosine, phenylalanine, leucine, isoleucine, and valine. The 1- and 2-hr plasma tryptophan concentrations were significantly greater in animals consuming both diets ($p < 0.001$) than in fasting controls. All brain tryptophan, serotonin, and 5-hydroxyindoleacetic acid concentrations were significantly greater in rats consuming the diet lacking the five large neutral amino acids (LNAAs) than in fasting controls ($p < 0.001$ for all but 1-hr serotonin, $p < 0.01$). Among animals eating the complete amino acid mixture, the 2-hr brain tryptophan concentration was significantly above that of the corresponding fasting controls ($p < 0.001$). [From Fernstrom and Wurtman (1972). Reprinted with permission from *Science* 178, 414–416. Copyright 1972 by the AAAS.]

19. Protein, CHO Intake and Brain Seratonin

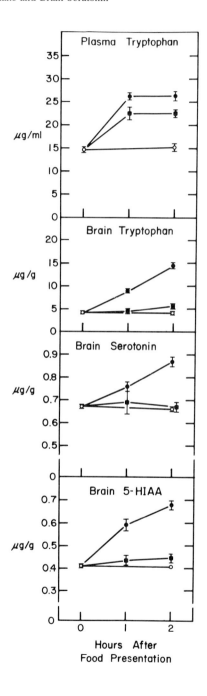

mimicked the normal casein-containing diet), but brain tryptophan level (or serotonin or 5-HIAA levels) did not rise appreciably. In rats consuming the diet lacking the competitors, however, remarkable increments occurred in brain tryptophan, brain serotonin, and brain 5-HIAA levels. In rats consuming the third diet, which lacked lysine and arginine (basic amino acids transported by their own carrier), blood tryptophan rose, but as for the complete diet, no appreciable increments occurred in brain tryptophan, serotonin, or 5-HIAA levels (Fernstrom and Wurtman, 1972).

These results, and those of many other experiments (e.g., Fernstrom and Wurtman, 1972; Fernstrom *et al.*, 1973, 1975a) led to the formulation of a model for predicting how the composition of a single meal would alter brain tryptophan uptake and serotonin synthesis (Fig. 3). In this model, the inges-

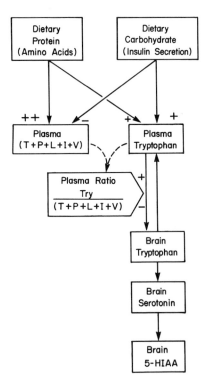

Fig. 3. Hypothesized model to describe diet-induced changes in brain serotonin synthesis. The ratio of tryptophan to the sum of the other large neutral amino acids [tyrosine (T) + phenylalanine (P) + leucine (L) + isoleucine (I) + valine (V)] in blood is thought to control tryptophan uptake into brain and brain tryptophan concentration. [From Fernstrom and Wurtman (1972). Reprinted with permission from *Science* **178**, 414–416. Copyright 1972 by the AAAS.]

tion of carbohydrates causes increased uptake of tryptophan by brain by changing competition for transport in tryptophan's favor. The level of tryptophan in blood rises, or at least does not fall. (Studies over the years have shown that insulin release does not always raise blood tryptophan in this paradigm, but it never lowers it.) The levels of tryptophan's competitors always fall, with leucine, isoleucine, and valine showing the greatest reductions. Hence, tryptophan gains a competitive advantage for brain uptake, and brain tryptophan level rises. With the ingestion of protein, blood tryptophan level rises, but so too do the blood levels of its competitors. Hence, for most levels of protein in the meal, the competitive standing of tryptophan does not change. In this case, brain tryptophan levels do not rise. The model shows as a central feature a serum ratio of tryptophan to the sum of its competitors. This ratio simply summarizes in a single number the changes in blood tryptophan relative to those in its competitors. Carbohydrate ingestion elevates the ratio (and brain tryptophan); protein ingestion does not alter the ratio (or brain tryptophan) (Fernstrom and Wurtman, 1972).

An experimental example that supports the model is shown in Fig. 4. In this study, again using animals fasted overnight, groups of rats ingested a single meal containing either carbohydrate alone, 18% protein, or 40% pro-

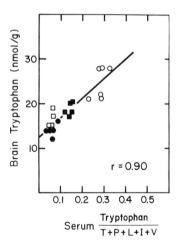

Fig. 4. Relationship between brain tryptophan level and the serum tryptophan ratio in individual animals consuming single meals of differing protein content. Groups of overnight-fasted rats received one of the three diets indicated below, and were killed 2 hr later. Closed circles, fasting controls; open circles, animals ingesting 0% protein (carbohydrate); closed squares, rats consuming 18% protein; open squares, rats consuming 40% protein. These data were analyzed by linear regression; the value of the correlation coefficient (r) is significantly different from 0, $p < 0.01$ (Student's t test). (From Fernstrom and Faller, 1978.)

tein. They were killed 2 hr later, and blood and brain samples analyzed for amino acids. A correlation was then made between brain tryptophan level and the serum tryptophan ratio. First, note that ingestion of the carbohydrate meal elevated the serum ratio and brain tryptophan. In addition, observe that the addition of protein to the meal antagonized the increment in brain tryptophan (and the serum ratio). The 40% meal was associated with no change in either the serum ratio or in brain tryptophan, when compared to fasting values. Finally, note that the correlation coefficient is excellent for this study. The high degree of correlation validates the serum ratio as a useful predictor of brain tryptophan changes *in this feeding paradigm* (Fernstrom and Faller, 1978).

In summary, single meals can lead rapidly to a change in brain serotonin formation via a mechanism that is well described. Ingesting a carbohydrate meal raises brain tryptophan levels and stimulates serotonin synthesis; consuming a meal containing moderate-to-high amounts of protein causes *no* changes in these parameters (compared to fasting values). The serotonin synthetic machinery is thus particularly sensitive to *physiologic*-sized changes in brain substrate levels. This is the *only* transmitter about which this can conclusively be said to be true.

III. DIET, BRAIN TRYPTOPHAN AND SEROTONIN, AND APPETITE

It is not sufficient to show that the formation of serotonin can be influenced by the composition of a meal. It is also important to demonstrate that the brain does something differently as a consequence. This is not an easy task, but in 1975, Anderson and associates began a line of investigation into the regulation of protein and energy intake using as a model rats that were allowed to select the amount of protein and energy they consumed each day. The specific interest they developed was in studying the relationship of self-selection of protein to the serum tryptophan ratio (Fig. 3) and brain serotonin. Several other laboratories have since pursued studies using this model. For simplicity, the results of this work can be considered in two parts, the first dealing with the regulation of protein (or energy) intake over a period of several weeks, and the second dealing with this regulation from meal to meal.

In the studies of regulation over several weeks, Anderson's group observed that when animals are given diet pairs, one high and one low in protein, they will elect to consume a level of protein characteristic of the particular protein source and its supplementations. The most interesting effect was seen in experiments of zein or partially corrected zein, to which tryptophan was added. As the amount of tryptophan added to the zein diets increased, the animals reduced the amount of protein, but not total energy, they consumed. In attempting to explain this apparent regulation, they invoked the model in Fig. 3. In particular, they noted that they could not account for the apparent regulation simply by looking at the blood levels of one or more of the essential amino acids. Instead, they found that if they correlated the amount of protein consumed with the serum tryptophan ratio, an excellent inverse correlation could be obtained (Ashley and Anderson, 1975a,b). They thus hypothesized that the ratio might be important in regulating protein intake, possibly by influencing serotonin (Anderson, 1977). In later studies (Woodger et al., 1979), diabetic and normal rats were compared for their ability to select for protein. In addition, half of the animals in each group selected from diets that included a supplement of tryptophan (1.45 or 4.5 g tryptophan/100 casein). The investigators reported that in both normal and diabetic rats consuming the diets supplemented with 4.5 g tryptophan/100 g casein, the intake of protein (but not total energy) was lower than in animals consuming less or no tryptophan, whereas the serum tryptophan ratios and brain tryptophan levels were higher. Only an equivocal rise in brain 5-hydroxyindoles (serotonin + 5-HIAA) was noted. (Since the measurements of serotonin and 5-HIAA were made at steady state, they indicate little about synthesis or turnover rates of serotonin, and can be ignored.) This result suggests that when tryptophan is consumed (and presumably serotonin synthesis is stimulated), animals consume less protein. The result is consistent with their original findings with tryptophan additions to zein diets.

In one related form or another, this hypothesis, that the serum tryptophan ratio (and brain tryptophan and serotonin) is inversely correlated with appetite for protein, and thus may influence it, has been advanced for almost 10 years. But this scheme is formulated in part on the model in Fig. 3, which derives from *single-meal* experiments. The protein-selection experiments have all been *chronic*. The animals consume the diets for 2–4 weeks, protein intake data are tabulated as total amount consumed during the entire feeding period (2–4 weeks), and biochemical data are reported for a *single* time, always 0900–1100 hr on the *final day* of the experiment. The assumption, therefore, is that the single meal model does apply in these chronic dietary conditions. But does it? The model in Fig. 3 posits connections between (1) the food consumed and the plasma tryptophan ratio, (2) this ratio and the level of tryptophan in brain, and (3) the brain tryptophan level and the rate of serotonin production. Though each of these connections has been amply affirmed in the *acute* model (Fernstrom, 1983), none of these connections has been established in the chronic model.

Therefore, we have recently begun to explore these biochemical relationships in a *chronic* dietary paradigm. The null hypothesis in these studies is derived from the above notions on the chronic control of protein selection: namely: (1) The regulatory loop for controlling protein intake includes a connection between the level of protein consumed and the serum tryptophan ratio. (2) This connection provides the key information transfer between a behavioral act (the election of how much protein to consume) and a chemical signal the brain can interpret (the ratio, which directly influences serotonin synthesis in brain). If these notions are true, then it should at least be relatively simple to demonstrate a dose/response curve between the level of protein an animal is allowed to consume in the diet and the serum tryptophan ratio. For example, if the acute model is used to form the hypothesis, one should find that the higher the level of protein in the diet, the lower the serum tryptophan ratio. Is this the case? To answer this question, we have studied the effects on the serum tryptophan ratio (and brain tryptophan and serotonin) of feeding rats for 10 days diets containing either 12, 24, or 40% casein. The results of one experiment are presented in Table I. In this study, animals had free access to the diets; to ensure that no effects would be missed, groups of rats in each diet group were killed at 4-hr intervals throughout the 24-hr period. Serum tryptophan levels were lowest in animals consuming the 12% protein diet, and highest in those ingesting the 40% protein diet (data not shown). However, neither the serum tryptophan ratio nor brain tryptophan levels showed a clear, systematic variation as a function of dietary protein content (Table I). Hence, the proposed chronic link between dietary protein intake and the serum tryptophan ratio, which does hold in the acute-feeding paradigm (Fig. 3), fails to obtain in the chronic dietary paradigm. The conclusion from this result seems inescapable: though animals in a free-feeding paradigm indeed appear chronically to regulate the amount of protein they consume, the chemical mechanism involved probably does not involve the serum tryptophan ratio and its effects on serotonin synthesis and release.

But what of the other connections posited in the acute model? Do they hold in the chronic experiments? The answer seems to be yes. Coincident with the serum tryptophan ratios being indistinguishable, regardless of dietary protein content, brain tryptophan levels (Table I), brain serotonin and 5-HIAA levels, and the *in vivo* rate of tryptophan hydroxylation (data not shown) were also invariant as a function of dietary protein content. (Even though one might point to the ratio values at 1AM and 5 AM for the 40% protein group as possibly being different from those for the other two groups, this difference is small in comparison to the variation in the ratio seen in Fig. 4. There is a six- or sevenfold variation in the serum tryptophan ratio in the acute experiment of Fig. 4; there is only a 25–50% difference in

TABLE I Chronic Effects of Dietary Protein (Casein) Content on Brain Tryptophan Levels and on the Serum Tryptophan Ratio in Rats[a,b]

Protein in diet (%)	Time of day					
	9 AM	1 PM	5 PM	9 PM	1 AM	5 AM
	Brain tryptophan (nmol/g)					
12	20 ± 2	19 ± 1	19 ± 1	17 ± 2	18 ± 1	20 ± 1
24	17 ± 1	21 ± 1	17 ± 1	20 ± 1	19 ± 1	20 ± 1
40	17 ± 1	20 ± 1	18 ± 1	17 ± 1	19 ± 1	18 ± 1
	Serum tryptophan ratio					
12	0.172 ± 0.005	0.173 ± 0.014	0.200 ± 0.015	0.203 ± 0.021	0.206 ± 0.014	0.250 ± 0.029
24	0.190 ± 0.012	0.220 ± 0.014	0.183 ± 0.006	0.168 ± 0.009	0.234 ± 0.038	0.242 ± 0.015
40	0.167 ± 0.014	0.183 ± 0.006	0.194 ± 0.016	0.198 ± 0.018	0.160 ± 0.015	0.164 ± 0.009

[a] From J. D. Fernstrom, M. H. Fernstrom, and P. E. Grubb (unpublished observations).

[b] Groups of male Sprague–Dawley rats ingested for 10 days either the 12%, the 24%, or the 40% casein diet *ad lib*. At the end of this time, they were killed in groups of 6, and blood and brain samples collected for amino-acid analysis (Beckman Model 6300). Tryptophan was analyzed fluorometrically. Lights were on between 7 AM and 7 PM daily. Data are the means ± SEM. The serum tryptophan ratio represents the serum values of tryptophan/tyrosine + phenylalanine + leucine + isoleucine + valine. In this study, no diet effect was noted on brain tryptophan analysis of variance (ANOVA); the diet did have an effect on the serum tryptophan ratio (ANOVA; $p < 0.01$), but there was no clear relationship to dietary protein content.

the ratio values at 1 AM and 5 AM in Table I. It is thus no surprise that the brain tryptophan levels are no different in this chronic experiment.)

It should be noted that these results do not detract from any of the data previously published. They simply raise questions about interpretation. The addition of tryptophan to a diet *may* chronically influence protein selection [though all do not think so (Peters et al., 1984)]. And the mechanism may involve serotonin. However, this type of "pharmacologic" finding does not seem to predict the mechanism of protein regulation when animals are consuming unadulterated protein. At least this is the message of the results in Table I.

More recently, the self-selection paradigm has been adapted for use in *short-term* experiments, both by Anderson's group and by others. Two general types of study have been pursued. One deals with the effects of drugs on the selection of protein (and carbohydrates). The second explores for effects of consuming small premeals on the animal's subsequent choice of nutrients.

The earliest studies involved determining the effects on protein and carbohydrate intake of administering anorectic agents (Wurtman and Wurtman, 1977, 1979). In these studies, animals were fasted 16 hr each day, and allowed to consume food during the remaining 8 hr. A drug was administered to rats at the onset of the daily feeding period, and the effects on subsequent food selection were followed. When fenfluramine was injected, a significant reduction in total calorie intake, but not protein intake, was noted over the first 4 hr of the feeding period. A similar effect was noted for fluoxetine and for MK-212 (a serotonin agonist). Because these drugs are all believed to facilitate transmission across serotonin synapses, the conclusion was drawn that the stimulation of serotonin receptors in brain selectively diminishes carbohydrate intake, while sparing protein consumption. A model was subsequently proposed based on these and other results, in which the serotonin neuron is reputed to regulate meal-to-meal selection of carbohydrates (Wurtman, 1983). The notion is that when an animal consumes carbohydrates, the serum tryptophan ratio rises, leading to an increase in brain tryptophan level and serotonin synthesis (these notions are based on the biochemical studies described in Section II). As a consequence of the increased serotonin release that follows, animals reduce their intake of carbohydrate at the next meal (this notion is drawn from the pharmacologic studies with fenfluramine, fluoxetine, and MK-212). This effect, which should lower the serum tryptophan ratio, would produce a reduction in brain tryptophan and a slowing of serotonin synthesis and release. The consequence of this would be to turn down the signal to reduce carbohydrate intake, such that more would be consumed at yet the next meal. By this mechanism, the animal would thus control his carbohydrate intake from meal to meal, and from day to day.

One logical prediction from this hypothesis has been that if an animal consumes a single carbohydrate meal, then in the next meal, in which he is allowed to choose how much protein and carbohydrate to eat, he should consume a larger-than-normal proportion of protein. Is this prediction borne out? Several experiments have tested the question.

1. In one experiment, Wurtman et al. (1979) took rats that were consuming a standard diet, and fasted them during the light period. At the onset of darkness, they gave half a small amount of carbohydrate to eat, and the other half a small amount of protein, carbohydrate, and fat. One hour later, they gave each group two diets that differed in carbohydrate, but not protein content. They noted that during the first 10 min after the foods were presented, the animals given the carbohydrate premeal consumed less carbohydrate than the group fed the mixed premeal.

2. In a study of similar design, Li and Anderson (1982) fasted rats during the light period. Just before the onset of darkness, they fed half a carbohydrate premeal, and half a 45% protein premeal. Soon thereafter, they allowed the two groups to select the level of protein they wished to consume, by giving them free access to two isocaloric diets differing in protein content. The effects they observed were as follows. During the first hour of feeding, animals that had consumed the premeal of 45% casein selected less protein, but also fewer total calories than those consuming the carbohydrate premeal. During the next 11 hr, animals consuming the 45% casein premeal ingested more total calories and proportionally more carbohydrate than the rats consuming the carbohydrate premeal. In a second, identical study, only one of these effects replicated: a reduction was noted in protein intake during the first hour of feeding in rats that had consumed the 45% casein premeal (and the effect was much smaller than seen in the first study). Otherwise, effects noted in the first study were opposite or did not occur in the second experiment. Finally, in a third study, in which the premeals were either carbohydrate or 70% casein, or 45 or 70% casein, in one study animals consumed less protein and calories if they had ingested the high-protein premeal, whereas in the other they ingested less protein, carbohydrate, and total calories.

3. Finally, Li and Anderson (1984) used a paradigm similar to the above, but fed rats premeals of either carbohydrate, or carbohydrate plus 15 mg tryptophan. The rats given tryptophan in their premeal consumed less carbohydrate in the first hour of free feeding than those that received no tryptophan. Otherwise, the two groups consumed similar total amounts of food and of protein during this period.

Do these results fit the prediction? In one sense, they do. In two of the experiments in which rats had consumed a carbohydrate premeal, they selected less carbohydrate for a short time after subsequent food presentation.

And, when tryptophan was added to a carbohydrate premeal, the result was a transient reduction in carbohydrate intake. But the results are not clear in all experiments presented. No doubt, part of the problem is in the size of the effects observed. They are generally short (10 min to 1 hr), and small (a gram or two difference). In fact, in Li and Anderson's 1982 studies, the data are reported as the composite of seven animals measured each day for 6 days, i.e., the group size used for statistics was apparently 42, and even then the effects were sometimes equivocal. Hence, it is sometimes difficult to accept that a clear effect is present.

More important in analyzing the model, however, are two assumptions underlying these experiments. The first is the often stated belief that the ingestion of a high-protein meal lowers brain tryptophan levels and serotonin synthesis (e.g., see Li and Anderson, 1983; Wurtman, 1983). The work cited in support of this is that from our own laboratory. However, as can be seen in Fig. 4, a high-protein meal (40% protein is quite high) does not lower brain tryptophan and serotonin; it *prevents* a carbohydrate-induced rise in these parameters in *fasting* animals. In other words, there are *no* data at present that support the belief that when a rat consumes a protein meal, his brain levels of tryptophan and serotonin *fall*. Such an effect is essential to the model proposed for regulation of carbohydrate and/or protein intake. The second assumption has to do with the belief that through continuing oscillation of protein and carbohydrate selection over the 24-hr period, a constant daily intake of protein or carbohydrate is achieved. That is, the data presented show that if a fasted animal consumes a carbohydrate premeal, his next meal will contain less carbohydrate (more protein) (Wurtman *et al.*, 1979, 1983; Li and Anderson, 1982, 1984). The mechanism advanced to explain this effect is that the carbohydrate meal causes insulin secretion, which leads indirectly to a rise in brain tryptophan and serotonin production, and then to increased serotonin release, which activates a mechanism for reducing appetite for carbohydrates. Let us suppose that this does take place. The animal then consumes a *second,* high protein meal, which should raise plasma levels of tryptophan's competitors, lower the serum ratio and thus brain tryptophan and serotonin (of course, such biochemical data do not exist, and some of the data discussed in Section II do not support this view; but ignore this for the argument's sake). This might conceivably happen, even though insulin continues to be secreted, because the input of branched-chain amino acids from the meal would more than overcome any continuing effect of insulin (it is presently unknown if this is true). But now in his *third* meal, the animal should desire a greater proportion of carbohydrate. If the model is to continue to work, this meal should lower serum competitor levels, drive up the tryptophan ratio, and thus increase the brain tryptophan level and serotonin synthesis. But will this in fact happen? It is doubtful because this effect requires insulin secretion sufficient to lower serum levels of the branched-chain amino acids. Since the animal has been consuming food for some time after his initial premeal, and insulin levels will probably be high due to continued secretion throughout his successive meals, there is likely to be no brisk rise in serum insulin levels after the third meal. Hence, serum branched-chain amino acid levels are unlikely to drop dramatically, nor is the tryptophan ratio likely to change. Thus, by the third meal, the animal can no longer modulate his intake of protein and carbohydrate intake by this model. Therefore, though the experiments presented above may conceivably explain an appetite effect that lasts 10 min to 1 hr after fasting rats ingest protein or carbohydrate premeals, they probably cannot be generalized to 24-hr regulation of protein and carbohydrate intake.

Of course, this analysis is hypothetical. But the importance of insulin secretion in generating the increase in brain serotonin is indisputable, And this effect is not likely to occur repeatedly if serum insulin levels are already high. There are some data showing that serum insulin levels are high at night in animals consuming food *ad libitum* (Fig. 5) (Fernstrom *et al.*, 1975b). That is, once the daily period of food intake has begun, insulin levels rise, and tend to stay elevated. Against a high serum insulin level, it seems unlikely that additional amounts of insulin secreted will so raise serum insulin levels

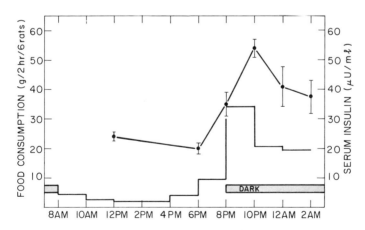

Fig. 5. Diurnal variations in serum insulin level and food intake in rats. Groups of male rats were given *ad libitum* access to a standard laboratory rat chow (Charles River) for 1 week prior to experimentation. The daily light period was 8 AM to 8 PM. On the day of the study, groups ($n=6$) were killed at noon, and then at 2-hr intervals beginning at 6 PM until 2 AM. Blood samples were obtained, and the sera harvested for radioimmunoassay of insulin. Food consumption/cage was also measured at each interval, and is reported as grams consumed/2hr/6 rats (each cage had 6 rats).

as to impact sizably on the serum ratio. Without this signal, the regulatory model does not work. The resolution of these important questions requires that a detailed analysis be undertaken of serum insulin and amino acid levels during succeeding meals in rats self-selecting for protein or carbohydrate. Perhaps the necessary oscillations occur on a small time scale and are not seen when only a handful of time points are studied over the 24-hr period. Only new experiments can provide the answers.

IV. SUMMARY AND CONCLUSIONS

This review has dealt with a putative biochemical link between carbohydrate/protein intake by rats, and their appetite for these specific nutrients. The analysis has focused in particular on the connection between carbohydrate and protein intake, insulin secretion, and the changes produced by these nutrients, when consumed, in the serum ratio of tryptophan to the sum of its competitors. This ratio predicts tryptophan uptake into brain (and serotonin synthesis) under certain specific conditions. The discussion suggests that, although protein and carbohydrate intake do appear to be regulated in rats, the connection between the act of eating and the brain's control of the intake of these nutrients may not involve changes in the serum tryptophan ratio, and thus in brain tryptophan uptake and serotonin synthesis. [This view is shared by others, e.g., Peters and Harper (1981).] The analysis may apply to both acute and chronic regulation of protein and carbohydrate intake.

However, the following points should be noted. First, the experiments offered here, showing no connection between chronic dietary protein intake and the serum ratio, represent one type of experimental paradigm. It is always possible that this is not the optimal model for testing this relationship. Hence, the data are offered only as one example indicating that the model may not be correct under these conditions. The hope is that the results will stimulate more careful investigation into each component of the proposed model for regulating protein/carbohydrate intake. Second, the analysis offered here of the model for meal-to-meal regulation of protein/carbohydrate should be correctly labeled at this point as speculation. It is hoped that this speculation will stimulate further, more careful biochemical work on this important question. And third, the focus of this review has been on *normal regulation* of protein and carbohydrate intake. It does not attempt to evaluate the importance of the pharmacologic work, such as that involving the administration of pharmacologic doses of tryptophan (Wurtman *et al.*, 1981), or of fenfluramine, fluoxetine, or MK-212. These anorectic agents may have the advertised effects [though all do not agree that they occur (Li and Anderson, 1984), or that they are mediated by the proposed serotonin mechanism (Carlton and Rowland, 1984; Davies *et al.*, 1983; Pinder *et al.*, 1975)]. But whether they do or not does not impact strongly on the question of normal, meal-to-meal regulation of nutrient intake. According to the analysis presented here, the arguments do not appear strong at this point that tie protein/carbohydrate selection to appetite for these nutrients via changes in the serum tryptophan ratio (and brain tryptophan and serotonin). Future work will hopefully provide a definitive answer.

ACKNOWLEDGMENTS

Some of the work described in this article was supported by a grant from the National Institute of Mental Health (MH38178). Dr. Fernstrom is the recipient of a Research Scientist Development Award from the National Institute of Mental Health (MH00254).

REFERENCES

Ashcroft, G. W., Eccleston, D., and Crawford, T. B. B. (1965). 5-Hydroxyindole metabolism in rat brain. A study of intermediate metabolism using the technique of tryptophan loading I. *J. Neurochem.* **12**, 483–492.

Anderson, G. H. (1977). Regulation of protein intake by plasma amino acids. In "Advances in Nutritional Research" (H. H. Draper, ed.), Vol. 1, pp. 145–166. Plenum Press, New York.

Ashley, D. V. M., and Anderson, G. H. (1975a). Food intake regulation in the weanling rat: Effects of the most limiting essential amino acids of gluten, casein, and zein on the self-selection of protein and energy. *J. Nutr.* **105**, 1405–1411.

Ashley, D. V. M., and Anderson, G. H. (1975b). Correlation between the plasma tryptophan to neutral amino acid ratio and protein intake in the self-selecting weanling rat. *J. Nutr.* **105**, 1412–1421.

Carlton, J., and Rowland, N. (1984). Anorexia and brain serotonin: Development of tolerance to the effects of fenfluramine and quipazine in rats with serotonin-depleting lesions. *Pharmacol. Biochem. Behav.* **20**, 739–745.

Davies, R. F., Rossi, J., Panksepp, J., Bean, N. J., and Zolovick, A. J. (1983). Fenfluramine anorexia: A peripheral locus of action. *Physiol. Behav.* **30**, 723–730.

Fernstrom, J. D. (1983). Role of precursor availability in the control of monoamine biosynthesis in the brain. *Physiol. Rev.* **63**, 484–546.

Fernstrom, J. D., and Faller, D. V. (1978). Neutral amino acids in the brain: Changes in response to food ingestion. *J. Neurochem.* **30**, 1531–1538.

Fernstrom, J. D., and Wurtman, R. J. (1971a). Brain serotonin content: Physiological dependence on plasma tryptophan levels. *Science* **173**, 149–152.

Fernstrom, J. D., and Wurtman, R. J. (1971b). Brain serotonin content: Increase following ingestion of carbohydrate diet. *Science* **174**, 1023–1025.

Fernstrom, J. D., and Wurtman, R. J. (1972). Brain serotonin content: Physiological regulation by plasma neutral amino acids. *Science* **178**, 414–416.

Fernstrom, J. D., Larin, F., Wurtman, R. J. (1973). Correlations between brain tryptophan and plasma neutral amino acid levels following food consumption in rats. *Life Sci.* **13,** 517–524.

Fernstrom, J. D., Faller, D. V., and Shabshelowitz, H. (1975a). Acute reduction of brain serotonin and 5-HIAA following food consumption: Correlation with the ratio of serum tryptophan to the sum of competing neutral amino acids. *J. Neural Transmission* **36,** 113–121.

Fernstrom, J. D., Hirsch, M. J., Faller, D. V. (1975b). Daily rhythm in serum immunoreactive insulin (IRI) levels in the rat. *Proc. Annual Meeting of the Endocrine Society, 57th,* p. 231.

Fernstrom, M. H. (1981). Lecithin, choline, and cholinergic transmission. *In* "Nutritional Pharmacology" (G. Spiller, ed.), pp. 5–29. Liss, New York.

Hess, S. M., and Doepfner, W. (1961). Behavioral effects and brain amine content in rats. *Arch. Int. Pharmacodyn.* **134,** 89–99.

Li, E. T. S., and Anderson, G. H. (1982). Meal composition influences subsequent food selection in the young rat. *Physiol. Behav.* **29,** 779–783.

Li, E. T. S., and Anderson, G. H. (1983). Amino acids in the regulation of food intake. *Nutr. Abst. Rev.: Rev. Clin. Nutr.* **53,** 171–181.

Li, E. T. S., and Anderson, G. H. (1984). 5-Hydroxytryptamine: A modulator of food composition but not quantity? *Life Sci.* **34,** 2453–2460.

Pardridge, W. M. (1977). Regulation of amino acid availability to the brain. *In* "Nutrition and the Brain" (R. J. Wurtman and J. J. Wurtman, eds.), Vol. 1, pp. 141–204. Raven Press, New York.

Peters, J. C., and Harper, A. E. (1981). Protein and energy consumption, plasma amino acid ratios, and brain neurotransmitter concentrations. *Physiol. Behav.* **27,** 287–298.

Peters, J. C., Bellissmo, D. B., and Harper, A. E. (1984). L-Tryptophan injection fails to alter nutrient selection by rats. *Physiol. Behav.* **32,** 253–259.

Pinder, R. M., Brogden, R. N., Sawyer, P. R., Speight, T. M., and Avery, G. S. (1975). Fenfluramine: A review of its pharmacologic properties and therapeutic efficacy in obesity. *Drugs* **10,** 241–323.

Woodger, T. L., Sirek, A., and Anderson, G. H. (1979). Diabetes, dietary tryptophan, and protein intake regulation in weanling rats. *Am. J. Physiol.* **236,** R307–R311.

Wurtman, R. J. (1983). Behavioural effects of nutrients. *Lancet* **i,** 1145–1147.

Wurtman, J. J., and Wurtman, R. J. (1977). Fenfluramine and fluoxetine spare protein consumption while suppressing caloric intake by rats. *Science* **198,** 1178–1180.

Wurtman, J. J., and Wurtman, R. J. (1979). Drugs that enhance serotoninergic transmission diminish elective carbohydrate consumption by rats. *Life Sci.* **24,** 895–904.

Wurtman, J. J., Wurtman, R. J., Growdon, J. H., Henry, P., Lipscomb, A., and Zeisel, S. H. (1981). Carbohydrate craving in obese people: Suppression by treatments affecting serotoninergic transmission. *Int. J. Eating Disorders* **1,** 2–15.

Wurtman, J. J., Moses, P. L., and Wurtman, R. J. (1983). Prior carbohydrate consumption affects the amount of carbohydrate that rats choose to eat. *J. Nutr.* **113,**70–78.

20

Time Course of Food Intake and Plasma and Brain Amino Acid Concentrations in Rats Fed Amino Acid-Imbalanced or -Deficient Diets

D. W. GIETZEN,[*,‡] P. M. B. LEUNG,[*,‡] T. W. CASTONGUAY,[†,‡] W. J. HARTMAN,[*,‡] AND Q. R. ROGERS[*,‡]

Departments of *Physiological Sciences and †Nutrition and ‡Food Intake Laboratory
University of California at Davis
Davis, California

I.	Introduction	415
II.	Time Course of the Food-Intake Response	420
III.	Feeding Patterns	423
IV.	Dietary Choice	425
V.	Amino Acid Concentrations in Plasma, Brain, and Cerebrospinal Fluid	433
VI.	Operant Response	437
VII.	Brain Areas Implicated	442
VIII.	Amino Acid and Neurotransmitter Concentrations in Brain	444
IX.	Monoamines in the Prepyriform Cortex	449
	References	452

I. INTRODUCTION

Dietary protein and amino acids are known to influence food intake. Most animals show reduced food intake and alterations in plasma and brain amino acid concentrations when they are fed diets imbalanced with respect to a particular essential amino acid, or diets devoid of one of the essential amino acids. The response is quite rapid and has been observed since the early 1900s. Also, the central nervous system is known to play a role in the behavioral response to these diets.

The food intake of animals fed diets containing various levels of protein or mixtures of amino acids has been well studied (reviewed in Harper et al., 1970; Rogers and Leung, 1973, 1977; Harper, 1976; Anderson, 1977; Harper and Peters, 1983). The responses to diets containing amino acid disproportionalities can be grouped into four general categories:

1. Rats will eat a diet within a certain range of protein levels (e.g., 20–30%) or the equivalent as a balanced amino acid mixture, without altering their food intake or without any apparent input from the food intake-controlling system.

2. Rats will reduce their intake of diets with a dietary protein level below or substantially above their requirement and will avoid such diets in favor of a diet of intermediate protein level. It should be noted, however, that animals resume eating normally as they adapt to the higher level of dietary protein after 1–2 weeks, but will not adapt to an inadequate level of dietary protein.

3. Rats will decrease their food intake if offered diets that are deficient in or devoid of one or more of the essential amino acids. Rats will also show reduced food intake with imbalanced amino acid diets, and if given a choice, will avoid diets that are deficient or imbalanced with regard to any essential amino acid, and choose, instead, a complete amino acid diet or even a protein-free diet. This choice for the protein-free diet has been termed an adverse choice, since the animals cannot grow or survive on the protein-free diet, but do grow slowly after adaptation to an imbalanced diet.

4. Food intake is also reduced in animals given diets containing excesses of single amino acids, and again animals will avoid such diets, selecting instead a control diet or even a protein-free diet.

In the absence of a choice, the animals do adapt (i.e., increase their previously low intake) to diets that are imbalanced, have excesses of single amino acids, or are very high in protein, but do not adapt to the protein-free or amino acid-devoid diets (Fig. 1). Animals have been shown to have a learned aversion to those diets that generated the food intake depression or that were avoided. It is clear, however, that the adaptation to the imbalanced diet is not the result of extinction of the learned aversion, since (1) animals adapted to the imbalanced diet for a long period of time will avoid the imbalanced diet when allowed a choice (Leung et al., 1968b), and (2) animals exposed to a cold environment, although they maintain their food

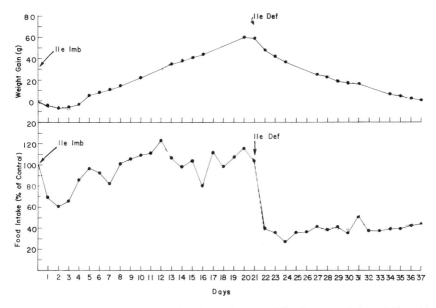

Fig. 1. Changes in food intake and body weight in rats fed isoleucine-imbalanced (Ile Imb) or -deficient, i.e., devoid, (Ile Def) diets. Arrows denote first day of feeding the indicated diet. [Adapted with permission from Leung and Rogers (1971b).]

intake on the imbalanced diet, will readily reject the imbalanced diet if given a choice (Harper and Rogers, 1966). However, a rapid extinction of the learned aversion is produced by ingestion of a small amount of a balanced amino acid diet. As little as 1 g of diet with a balanced amino acid pattern initiates consumption of an imbalanced diet that was previously rejected in favor of the protein-free diet (Zahler and Harper, 1972).

Of the four amino acid disproportionalities that have been studied in detail, we shall focus particularly on the early acute responses to diets that are imbalanced with regard to an essential amino acid or deficient in (devoid of) a particular essential amino acid.

Amino acid imbalance has been repeatedly demonstrated to be a general phenomenon with a variety of growth-limiting amino acids. Thus, it has been used as a model to study the influence of amino acid balance on the control of food intake (Rogers and Leung, 1977).

The mechanisms involved in the food intake responses to altered amino acid balance have not yet been determined. Several hypotheses have been proposed which postulate that the amino acids themselves, or some metabolite, may play a role in these responses. Individual amino acids, or alterations in the proportions of amino acids, could interact with a receptor system of some kind. The amino acids could act as neurotransmitters in the central nervous system, as some are known to do. It also has been suggested that ammonia, as a common metabolite of all the amino acids, may affect receptors either in the liver or the brain. Hormones, which have been stimulated either by amino acids or by their metabolites, could play a major role. In addition, several neurotransmitters, which are dependent on amino acid precursors—e.g., norepinephrine, dopamine, epinephrine, or serotonin— could be involved.

Another major issue that has received attention over the past several years is the location of this hypothetical receptor system. The olfactory system has been studied. Olfactory bulbectomy does not prevent the reductions of food intake in animals offered the amino acid-imbalanced or -deficient diets, even though it abolishes the animals' selection against a flavored control diet which they had avoided while intact (Leung et al., 1972). Thus, while the olfactory system may be used by intact animals to facilitate their discrimination against the imbalanced or deficient diet, olfaction has been shown not to be essential in the control of intake of the imbalanced amino acid diets.

In the rat, taste is not likely to be involved in the initial response to imbalanced amino acid diets. Food intake is rapidly depressed in animals that have been given an iv infusion of imbalanced amino acids, which bypasses oral input and therefore taste pathways (Peng and Harper, 1969). However, taste as an integral part of taste aversion may play a role in the adaptive phase of the control of food intake in animals ingesting amino acid-imbalanced or -deficient diets. Certain areas in the septum and hippocampus, the ablation of which affects the acquisition of taste avoidance (Krane et al., 1976), are not important in the initial depression but are involved in the behavioral adaptation of food intake in animals fed amino acid-imbalanced diets (Rogers and Leung, 1973; Leung and Rogers, 1979). Although rats with medial amygaloid lesions demonstrate an attenuated ability to acquire taste avoidance, such a deficit is not found in rats with lesions in the anterior prepyriform cortex (Meliza et al., 1981). Also, lesions placed in the thalamic taste relay do not prevent the initial depression of food intake, but facilitate the adaptive intake of the amino acid-imbalanced diet (Leung et al., 1982).

There have also been suggestions that the receptor system may be in the gastrointestinal tract or the liver. Perfusion of amino acids through the small intestine increases the rate of mesenteric nerve impulses (Sharma and Nasset, 1962). However, no direct behavioral effect or direct link of this increased electrical activity to food intake has been demonstrated. Humoral agents could be implicated: Certain amino acids, particularly phenylalanine and tryptophan, have been shown to release cholecystokinin when infused into the small intestine (Meyer, 1974). Also, releasers for bombesin-like peptide have not yet been determined, but if amino acids function to release this peptide from the small intestine, such a neurohormonal system could contribute to the reductions in food intake noted in these animals. To date, the effects of these substances on the food intake responses of animals fed disproportionate amounts of amino acids have not been determined. On the other hand, cortisol injection did abolish the reduced food intake response of rats fed an imbalanced diet. This was thought to be the result of a generalized release of amino acids into plasma such that the supply of the limiting amino acid is increased in the plasma of the animals (Leung et al., 1968c).

Recent evidence suggests that the liver, and particularly its vagal innervation, may play a role in the control of *ad libitum* eating behavior (Friedman and Sawchenko, 1984). Over a decade ago, Russek (1971) proposed that the control of protein intake may be mediated by receptors in the liver that are sensitive to the ammonia produced. However, Bellinger and co-workers (1977) reported that their dogs ate normally just 4 min after infusion of as much as 21.6% of their daily amino acid requirement via the hepatic portal vein. These results do not support the hepatoreceptor hypothesis of Russek. Using rats, Rogers and Leung (1977) infused ammonium acetate and did not find a depression in food intake of portal-cannulated rats until 4.67 μmol ammonium acetate per day was infused, and then there was a reduction of only 5% in food intake, whereas 0.001 μmol did cause a depression of food intake when infused into the carotid artery. Perhaps the acute high (millimolar) doses of ammonium chloride administered by Russek (1971) were sufficient to produce glycogenolysis in the liver via either a direct or indirect effect on glucagon secretion, causing sufficient glucose release to cause the satiety effect.

Peripheral infusion of an imbalanced mixture of amino acids can cause a reduction in food intake (Peng and Harper, 1969). Also, the food intake depression of rats (Leung and Rogers, 1969) and cockerels (Tobin and Boorman, 1979) fed amino acid-imbalanced diets was prevented if a small quantity of the limiting amino acid was infused into the carotid artery but not the jugular vein. These results suggest the importance of the brain in mediating the food intake response of the animals fed amino acid-imbalanced or -devoid diets, since oral ingestion and absorption by the gut may be bypassed and the reduced food intake will still be seen.

The most consistent biochemical changes that occur shortly (2–4 hr) after ingestion of amino acid-imbalanced diets are a fall in the concentration of the most limiting amino acids in the plasma and a rise in the concentrations of amino acids added to cause the imbalance in both plasma (Leung et al., 1968d) and brain (Peng et al., 1972). The present review will examine the time course of food intake responses as correlated with plasma and brain amino acid concentrations in rats fed amino acid-imbalanced or -deficient diets.

II. TIME COURSE OF THE FOOD-INTAKE RESPONSE

Diets deficient in one or more of the essential amino acids have been known for many decades to be associated with a marked depression of growth in animals (Willcock and Hopkins, 1906; Osborne and Mendel, 1914, 1916). Results from Willcock and Hopkins (1906) show (Table I) that the concept of essentiality of an amino acid began with observation of mortality as well as weight change in rodents fed diets containing different protein sources. These response parameters quickly gave way to the recognition of growth as the major criterion of amino acid adequacy. However, it eventually became clear that food intake was markedly affected. In 1916, Osborne and Mendel wrote: "We have frequently noted that when the protein concentration of the food becomes very low the animals do not eat satisfactorily." In the same year, feeding experiments by Mitchell (1916) demonstrated that daily food intake was reduced in animals fed a mixture of all the amino acids that were known at that time. The amino acids methionine and threonine were not yet discovered, and so, necessarily, were not included in the diet. In Fig. 2, taken from Mitchell's paper, daily food intake is depicted along with changes in body weight. The animals were fed the amino acid mixture every second day with a nitrogen-free ration on the intervening day. The food intake of the amino acid ration is indicated by the circled data points, and can be seen to be consistently less than the intake of the nitrogen-free diet. Rose (1931), in his attempts to get satisfactory growth using amino acid diets (prior to his discovery of threonine) suggested that amino acid deficiency was actually synonymous with a depression of food intake. Thus, Rose wrote, "Our experience with other types of deficiencies involving the nitrogenous portion of the ration has taught us to expect a marked

TABLE I Percentage Change in Weight and Survival in Days of Mice Fed Casein, Zein, or Zein + Tryptophan[a]

Casein (%)	Days	Zein (%)	Survival in days	Zein + tryptophan (%)	Survival in days
+68	14	−28	15	−18	16
+59	14	−32	15	−27	16
+27	14	−20	15	−26	16
+49	14	−23	12	−29	12
		−37	17	−24	16
		−33	17		
+51	14	−29	15	−25	15

[a] Results adapted from Willcock and Hopkins (1906).

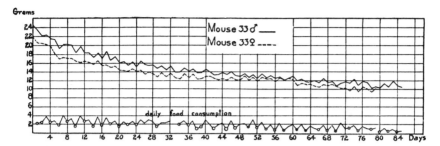

Fig. 2. Body weight and food intake of individual mice fed an amino acid diet on alternating days with a nitrogen-free diet. Days on which the amino acid diet was fed are denoted by circles around the food intake data points. [Taken with permission from Mitchell (1916).]

failure in appetite when the diet is completely devoid of an essential component."

A distinction should be made with regard to the differences between amino acid-imbalanced diets and those that are devoid of an essential amino acid. Harper (1976) has described an amino acid-imbalanced diet as follows: "In general, amino acid imbalances result from additions to a low protein diet of one or more amino acids, other than the one that is growth limiting, in amounts that individually are not toxic. They cause depressions in food intake and growth that are readily prevented by a supplement of the growth-limiting amino acid" (Harper, 1976). The devoid diet, on the other hand, is formulated by leaving one essential amino acid completely out of a diet that utilizes purified amino acids as the sole nitrogen source (Harper et al., 1970).

The reduced growth that occurs with the feeding of imbalanced amino acid diets has been shown to be the result of the depressed food intake in the rat (Leung et al., 1968a; Harper et al., 1970; Tews and Harper, 1982; Cieslak and Benevenga, 1984) as well as in the chick (Davis and Austic, 1982). If rats are force fed (Fig. 3, Leung et al., 1968a) an amino acid-imbalanced diet, they grow as well as rats fed the control diet. It therefore appears that the food intake depression and altered choice is a primary rather than a secondary effect of an amino acid imbalance or deficiency.

In addition to reducing food intake, if offered a choice, animals will avoid the amino acid-imbalanced or -deficient diet. Still, there are some factors that will abolish the response or reverse the choice. As indicated above, cortisol injections, which increase plasma levels of amino acids, will abolish the usual avoidance of the imbalanced diets (Leung et al., 1968c). Also, prefeeding a high-protein diet will eliminate the food intake depression (Leung et al., 1968b), since the animals have high levels of labile protein which provide a source of limiting amino acid during the initial period of

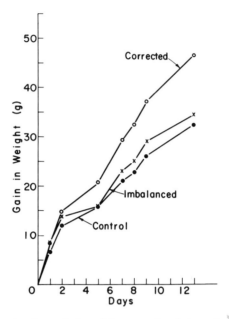

Fig. 3. Gain in weight of animals force-fed corrected, imbalanced, or control diets. Open circles, corrected; ×, imbalanced; closed circles, control diets. [Taken with permission from Leung et al. (1968a).]

feeding the imbalanced diet. For example, Ellison and King (1968) fed an amino acid-imbalanced diet to animals prefed the corrected diet, and did not see a clear reduction in food intake until after 8 hr. Thus, the prefeeding protocol is important in evaluating the time course of the food intake response. Animals prefed a low-protein diet always decrease their food intake rapidly when they are fed imbalanced amino acid diets. Kumta and Harper in 1962 showed (Fig. 4) that food intake was markedly reduced before 4 hr when animals were fed a histidine-imbalanced diet. When animals were force fed equal amounts of basal or histidine-imbalanced diets, the plasma level of histidine showed a marked decrease by 1 hr after the feeding period. Leung et al. (1968d) demonstrated that the decrease in intake of a threonine-imbalanced diet was seen as early as 2 hr (*ad libitum* feeding) and was marked by 3 hr (meal feeding) (Fig. 5, Leung et al., 1968d). Early food-intake depression was also reported by Zimmerman and Scott (1967) to occur by 2.5–3.0 hr in chicks fed a valine-deficient diet, while plasma valine levels were already reduced after 30 min.

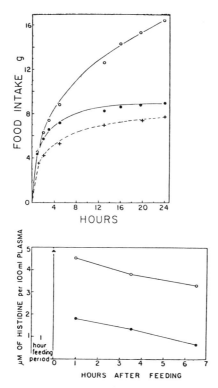

Fig. 4. Top: Food intake of protein-depleted rats, food deprived for 12 hr and then fed basal or imbalanced diets. Open circles, 6% fibrin; closed circles, 6% fibrin + 0.4% DL-methionine + 0.6% DL-phenylalanine; ×, 6% fibrin + amino acid mixture lacking histidine. Bottom: Plasma levels of histidine at intervals after feeding a single meal providing 16 mg histidine. Open circles, 6% fibrin; closed circles, 6% fibrin + amino acid mixture lacking histidine. [Taken with permission from Kumta and Harper (1962).]

III. FEEDING PATTERNS

A change in feeding patterns (reduction in both size and frequency of meals) has been demonstrated to occur within 3 hr after the ingestion of an amino acid-imbalanced diet (Rogers and Leung, 1973), showing a close association between the time of occurrence of the plasma amino acid changes and that of the food intake depression. Histograms from automatic electronic recordings made after each 100 mg of food was consumed (Fig. 6) show the responses of a typical rat offered a single diet: the isoleucine-basal diet (top), the first day of feeding the imbalanced diet (second line), the imbalanced

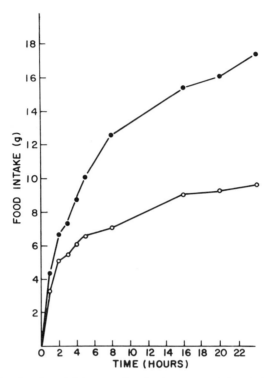

Fig. 5. Food intake of rats fed the control or the threonine-imbalanced diet. Closed circles, control; open circles, imbalanced diet. [Taken with permission from Leung *et al.* (1968c).]

diet after 14 days of feeding (third line), and the first day after feeding the isoleucine-corrected diet (bottom). This rat, previously adapted to the isoleucine-basal diet, ate 0.8 g of basal diet within 6 min after lights out. Total food intake for that day was 14 g. On the first day of presentation of the imbalanced diet (line 2), the animal ate 0.7 g during the first 6 min, an amount that was not different from intake of the basal diet, but its food intake was depressed by 2–3 hr after lights out. Total food intake for that day was just 6 g. By day 14, the rat consumed 13 g of the imbalanced diet in 24 hr, showing that food intake returned nearly to normal. This was the result of an increase in the size but not the frequency of the meals, with only one meal taken during the light period. The first day the rat was offered the corrected diet, he ate more frequent meals, with a larger first meal of 1.6 g and a 24-hr total of 22 g. By comparison, as seen in Fig. 7, line 2, a rat given a diet devoid of isoleucine reduced its 24-hr food intake from 17 to 6 g by decreasing meal size. Line 3 is representative of the lack of adaptation to the devoid diet, since after 14 days the food intake was still only 7 g, and the animal was

20. Time Course of Response to Amino Acid Imbalance 425

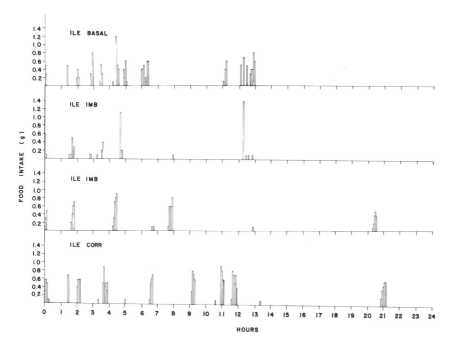

Fig. 6. Food intake record of an individual rat fed *ad libitum*. Consecutive lines indicate isoleucine-basal; -imbalanced, first day; isoleucine-imbalanced, fourteenth day; and isoleucine-corrected. Individual bars represent dietary intake in 3 min. Dark, 0–12 hr; light. 12–24 hr. [Taken with permission from Rogers and Leung (1973).]

eating several meals during the daylight hours. When isoleucine was added to "correct" the devoid diet, the rat ate about 2.6 g in the first 30 min, compared with less than 1 g of the devoid diet during the same period. The first day's intake of the isoleucine-corrected diet was 22 g. This is similar to the response of the rat that had been challenged with the isoleucine-imbalanced diet and then given the corrected diet, and demonstrates a very rapid response to the improved ration.

IV. DIETARY CHOICE

As noted above, amino acid imbalance and deficiency also affect dietary choices in animals. Rats, if offered a choice, will reject an amino acid-imbalanced or -devoid diet in favor of an alternative choice such as the low-protein basal diet or the corrected diet, which contains balanced amino acid patterns, or will even select a protein-free diet which does not support

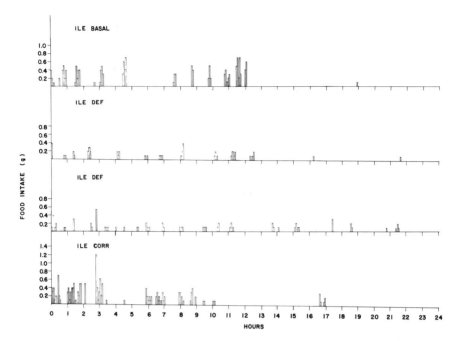

Fig. 7. Food intake for isoleucine-basal, -deficient, and -corrected diets. Conditions are the same as in Fig. 6. [Taken with permission from Rogers and Leung (1973).]

growth (Leung et al., 1968b). The degree of preference for the protein-free diet is associated with the severity and the nature of the imbalanced amino acid mixtures introduced (Leung et al., 1968b). Thus, choosing a protein-free diet over a diet with an imbalance of amino acids can be a sensitive measure of dietary amino acid imbalance.

In order to obtain information as to when and how the animals altered their feeding patterns of the amino acid-imbalanced or -devoid diet in favor of the alternative choices, and thus determine the time course of the food-intake control mechanism, detailed dietary choice patterns of rats were determined using a food-intake monitoring system which employed electronic balances (Mettler PL-300) interfaced with an on-line computer (PDP-11, Digital Equipment Corp.) as described previously (Castonguay et al., 1982). The animals were kept on a 12:12-hr light–dark cycle and offered a choice of 2 diets involving (1) the threonine-imbalanced diet and the threonine-basal diet, (2) the threonine-devoid diet and the threonine-basal diet, and (3) the threonine-imbalanced diet and the protein-free diet or, subsequently, the threonine-basal diet and the protein-free diet (Leung and Rogers, 1980b). All diets were offered at the beginning of the light cycle in an attempt to avoid inducing untoward interruptions in the dark feeding period in which the rats normally consume the majority of their daily food intake, as may be seen in Figs. 6 and 7 for diets first presented at the beginning of the dark cycle. All diets were rotated daily from side to side and among the animals tested so as to discourage side preference and individual markings, if any.

The feeding patterns were altered by offering the basal control diet in 2 cups rather than 1 as shown in Fig. 8. Simply offering the rats a choice of the same diet in 2 cups modified their feeding pattern such that the rats went back and forth eating small quantities of the diet from both food cups throughout the light period. The number of large meals did not change and the diet was eaten equally from both food cups.

The detailed first-day choice patterns of a representative individual rat allowed to choose between the threonine-imbalanced and threonine-basal diet following prefeeding of the threonine-basal diet are shown in Fig. 9. During the prefeeding period, the animals ate almost equally from the 2 food cups containing the same threonine-basal diet (Fig. 8). When the threonine-imbalanced diet was introduced for the first time together with the threonine–basal diet in the beginning of the light cycle, an intensive sampling process characterized by frequent, small sampling bouts was evident throughout the light period. The light-cycle food intakes of the threonine-imbalanced diet (20 sampling bouts) and the threonine-basal diet (8 sampling bouts) were 1.00 and 2.72 g, respectively, with a combined total of 3.72 g. The sampling process which occurred throughout the first light period would

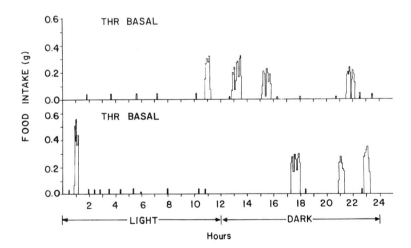

Fig. 8. Dietary choice patterns of a representative individual rat (prior to choice between basal and imbalanced diets) offered 2 identical cups of threonine-basal diet. Three minutes of nonfeeding was used as the minimal intermeal interval.

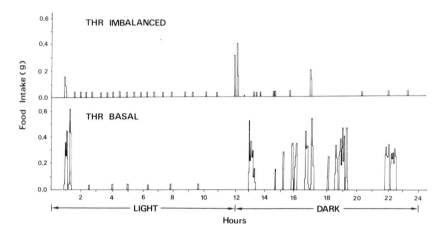

Fig. 9. Dietary choice patterns of a representative individual rat (day 1) offered the choice between threonine-imbalanced and threonine-basal diets. Upper and lower panels represent, respectively, the portions of intake from the diets indicated above. Three minutes of nonfeeding was used as the minimal intermeal interval.

indicate that 10–50 mg is not sufficient to initiate a food intake depression or avoidance of a threonine-imbalanced diet. An almost complete rejection of the threonine-imbalanced diet and the establishment of the choice for the threonine-basal diet occurred within 2 hr after the ingestion of about 1 g of the threonine-imbalanced diet in the beginning of the first dark feeding period (Fig. 9). The dark-cycle food intakes of the threonine-imbalanced and threonine-basal diets were 1.46 and 14.71 g, respectively, with a combined total of 16.17 g. The intake of the threonine-imbalanced diet in the first day of the choice regimen constituted 12% of the daily combined total. The almost exclusive preference for the threonine-basal diet was maintained throughout the experiment, with the portion of the threonine-imbalanced diet intake amounting to only 2% of the daily combined total at the end of the 10-day choice period.

The first-day choice patterns of a representative individual rat prefed the threonine-basal diet and offered a choice between the threonine-devoid diet and the threonine-basal diet are shown in Fig. 10. The animal showed no side-preference for the two identical food cups containing the same threonine-basal diet during the prefeeding period. The introduction of the threonine-devoid diet in the beginning of the light cycle evoked frequent, small sampling bouts, also showing the lack of avoidance of the threonine-devoid diet in the light period. The light-cycle food intakes of the threonine-devoid diet (28 sampling bouts) and the threonine-basal diet (4 sampling bouts) were 1.52 and 0.14 g, respectively, with a combined total of 1.66 g. The avoidance of the

20. Time Course of Response to Amino Acid Imbalance 429

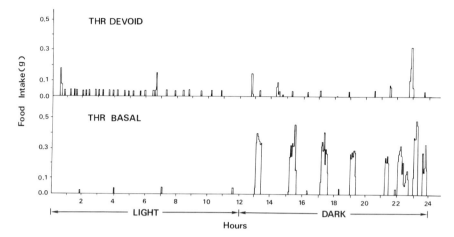

Fig. 10. Dietary choice patterns of a representative individual rat (day 1) offered the choice between threonine-devoid and threonine-basal diets. Conditions as in Fig. 9.

threonine-devoid diet and the selection of the threonine-basal diet at the beginning of the first dark cycle was rapid, occurring in less than 2 hr following the consumption of about 0.4 g of the threonine-devoid diet. The dark-cycle food intakes of the threonine-devoid and the threonine-basal diets were 1.53 and 15.01 g, respectively, with a combined total of 16.54 g. The animal maintained its favorable alternative choice of the threonine-basal diet throughout the 7-day choice period. The intakes of the threonine-devoid diet in the choice regimen accounted for 17 and 2% of the daily combined totals, respectively, on the first and the last days of the choice experiment.

The first-day choice patterns of a representative individual rat equilibrated on the threonine-basal diet and subjected to a choice between the threonine-imbalanced diet and the protein-free diet or, subsequently, between the threonine-basal and the protein-free diets, are shown in Figs. 11 and 12. The introduction of the threonine-imbalanced diet and the protein-free diet initiated the usual sampling activities of both diets throughout the first light period. The light-cycle food intakes of the threonine-imbalanced diet (11 sampling bouts) and the protein-free diet (8 sampling bouts) were 0.29 and 0.89 g, respectively, with a combined total of 1.18 g. The avoidance of the threonine-imbalanced diet and the choice for the protein-free diet was established within 2 hr following the ingestion of approximately 0.5 g of the threonine-imbalanced diet at the start of the first dark cycle. The food intakes of the threonine-imbalanced and the protein-free diets during the dark period were 2.24 and 14.58 g, respectively, with a combined total of 16.82 g

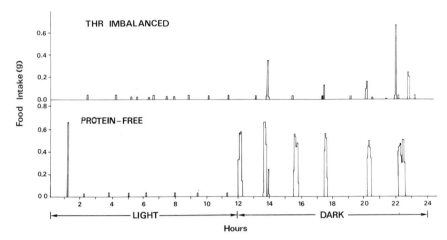

Fig. 11. Dietary choice patterns of a representative individual rat (day 1) offered the choice between threonine-imbalanced and protein-free diets. Conditions as in Fig. 9.

(Fig. 11). The portions of the threonine-imbalanced diet that were consumed constituted 14 and 18% of the combined daily totals, respectively, on the first and the eighth day of the choice experiment. However, the protein-free diet was avoided almost immediately during the first light cycle following the ingestion of one large meal of the threonine-basal diet at about 2 hr into the first light cycle (Fig. 12). The light-phase food intakes of the threonine-basal

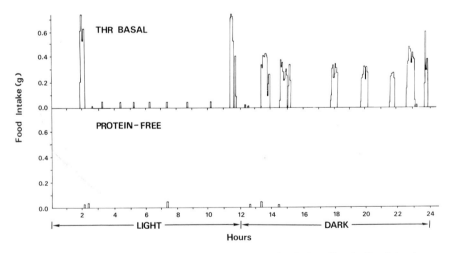

Fig. 12. Dietary choice patterns of a representative individual rat (day 1) offered the choice between threonine-basal and protein-free diets. Conditions as in Fig. 9.

and the protein-free diets were 7.84 and 0.07 g, respectively, with a combined total of 7.91 g. Apparently, the animal attempted protein repletion with the first introduction of the threonine-basal diet during the light period because of the prolonged choice for the protein-free diet during the previous choice regimen. The animal immediately selected the threonine-basal diet over the protein-free diet even though the latter was the favorable alternative choice when paired against the threonine-imbalanced diet. The dark-cycle food intakes of the threonine-basal and the protein-free diets were 17.46 and 0.05 g, respectively, with a combined total of 17.51 g. The food intakes of the protein-free diet represented 0.5 and 13% of the combined totals, respectively, on the first and sixth days of the choice experiment.

The sampling bouts of small quantities following the first introduction of the diets in the choice regimens may be an inherent investigative behavior whereby the physical or oropharyngeal properties of the diets are recognized, during the light period, when the animals are satiated. During this period, the animals have not eaten large enough meals to elicit the postingestive physiological responses that occur after eating the larger meals (Figs. 8–12). The rejection of the threonine-imbalanced diet or threonine-devoid diet during the choice regimens in favor of the threonine-basal diet or the protein-free diet was initiated within 2 hr after the ingestion of relatively large meals of the disproportionate amino acid diets during the beginning of the dark period. The establishment of the choices for the alternative diets in these choice situations thus provides additional information about the time course of the food intake control mechanisms in rats fed amino acid-imbalanced and -deficient diets.

To determine the shortest response time of rats to a threonine-devoid diet beginning in the dark phase, rats were adapted to a 12:12-hr light–dark cycle with the dark period from noon to midnight. They were trained over a 3-week period to eat threonine-basal diet for 6 hr beginning at noon. In Trial 1, rats received their initial exposure to threonine-corrected or threonine-devoid diet. They were given threonine-basal diet (6 hr daily) during the 4 days between Trials 1 and 2. In Trial 2, in a simple crossover design, the rats were given the other diet.

In Trial 1, the rats did not distinguish the threonine-devoid from the threonine-corrected diet during the first 15 min of exposure (Fig. 13). However, intake was significantly lower in the group fed the devoid diet during the second 15-min interval compared to either the first interval or the intake of the group fed the corrected diet. Thus, reduced intake of the devoid diet was noted by 30 min. Reduced food intake by the rats given the devoid diet was observed at every subsequent interval within 2 hr. Cumulative food intakes at 2 hr were 1.8 g in the devoid-fed group and 4.3 g in the group fed the corrected diet.

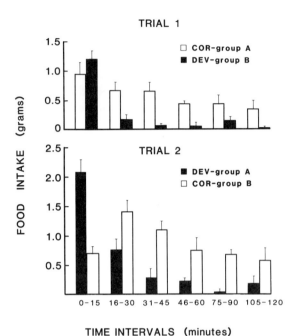

Fig. 13. Food intake during 15-min intervals of rats fed threonine-corrected (COR) or threonine-devoid (DEV) diet. Ten rats were divided into 2 groups with 5 rats per group. Group A received COR diet during Trial 1 and DEV diet during Trial 2, 5 days later. Group B received the diets in the opposite order. Rats were accustomed to eating threonine-basal diet 6 hr daily before the first trial, and received threonine-basal diet between trials as well. Only the initial 120 min of the feeding periods is shown, representing 1200 noon to 1400 hr. Fifteen-minute intakes were calculated by halving the 30-min intakes obtained at 90 and 120 min. Vertical lines represent SEM.

In Trial 2, the rats received their initial exposure to the other experimental diet. The group which has been given the threonine-corrected diet during the first trial showed enhanced intake of the threonine-devoid diet during the initial 15 min of exposure to the devoid diet. This indicates that the rats did not distinguish the devoid diet from the corrected diet for the first 15 min. However, intake of the threonine-devoid diet was less during the second 15-min interval of Trial 2 than during the first. Conversely, mean intake of the threonine-corrected diet during the second interval was double the intake during the first 15 min. Cumulative 2-hr food intakes in Trial 2 were 3.8 g for the devoid diet and 6.1 g for the corrected diet.

Therefore, control-fed, food-deprived rats can detect a lack of threonine in the diet between 15 and 30 min after initial exposure. Although previous learning may have affected the initial intake in Trial 2, it nevertheless appears that the rats begin to respond to a difference in the diet within 30 min.

V. AMINO ACID CONCENTRATIONS IN PLASMA, BRAIN, AND CEREBROSPINAL FLUID

Leung *et al.* (1968d) reported that alterations in plasma amino acids occurred very rapidly after animals were fed imbalanced amino acid diets. Plasma aminograms of rats trained to eat a single, 1-hr meal per day and fed control or threonine-imbalanced diets are shown in Fig. 14. Marked eleva-

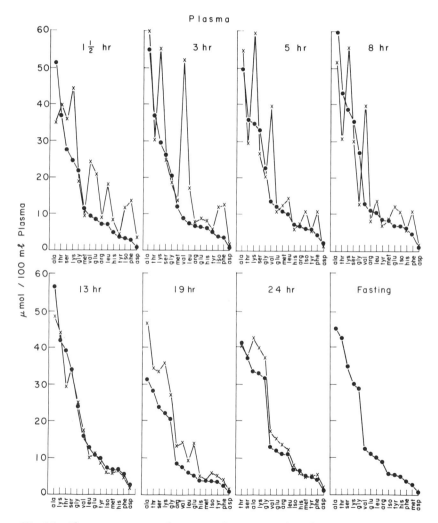

Fig. 14. Plasma aminograms of rats trained to eat a single, 1-hr meal daily and fed the control diet or the imbalanced diet. Closed circles, control; ×, imbalanced diet. [Taken with permission from Leung *et al.* (1968d).]

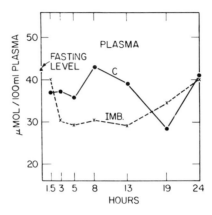

Fig. 15. Plasma threonine concentrations of rats fed the control diet and the imbalanced diet at different times after feeding. [Taken from Leung et al. (1968d).]

tions in the amino acids used to create the imbalance were noted at 1.5 hr, as indicated above, and a decreased concentration of threonine was noted by 3 hr after the meal (Fig. 15; Leung et al., 1968d).

Harper and co-workers found that the concentration of the most limiting amino acid decreased at least as much in the brain as in plasma, when rats were fed amino acid-imbalanced diets, as noted above and shown in Fig. 16 (Peng et al., 1972). They also found that the amino acids added to cause the imbalance increased markedly in plasma with some increase in brain, whereas a decrease occurred in threonine in both plasma and brain (Fig. 17). This reduction of the limiting amino acid in brain is thought to be, in part, a result of competition of the added amino acids with the limiting amino acid for the appropriate transport system at the blood–brain barrier (Tews et al., 1979). Figure 17 is representative of aminograms for 10 amino acids in the brains and plasma of rats at different time periods from 0 to 36 hr after they were force-fed a threonine-imbalanced diet (Peng et al., 1972). As can be seen from the aminograms, the concentrations of threonine were depressed in both plasma and brain at all time periods; the amino acids added to create the imbalance were markedly elevated in plasma, and increased, albeit less dramatically, in the brain.

Since many experiments had demonstrated the importance of the brain in mediating the food intake response of the animal to amino acid-imbalanced and -devoid diets, we questioned whether cerebrospinal fluid (CSF) threonine would also be decreased after feeding a diet devoid in threonine (Hartman et al., 1983). Changes in amino acids were determined after a single meal in rats which were trained to eat a threonine-limiting basal diet for 6 hr daily, beginning at noon as the lights went off. Three days before

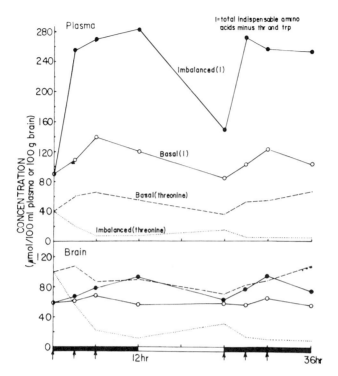

Fig. 16. Plasma and brain concentrations of threonine at different time periods after force-feeding threonine-basal (dashed line) or -imbalanced diets (dotted line). Open and closed circles also show the total indispensable amino acids minus threonine and tryptophan under basal and imbalanced conditions. [Taken with permission from Peng et al. (1972).]

killing, the rats were fed the threonine-corrected diet for their 6-hr feeding period, to accustom them to the taste of the experimental diets. Rats were assigned to one of three groups: food-deprived, threonine-devoid, or threonine-corrected diet. The corrected diet is identical to the threonine-devoid diet except that it contains 0.6% threonine. CSF, blood, and brain were sampled from the food-deprived group at noon, the beginning of the dark cycle, whereas the other rats were allowed to eat 2.0 g of corrected or devoid diet. All the rats ate the diet in 10–30 min, and CSF, blood, and brain were obtained 2 hr after the initiation of feeding. Plasma concentrations of threonine in rats 2 hr after feeding the devoid or corrected diets are compared with concentrations of threonine in rats that had been food deprived in Fig. 18. Plasma threonine increased 104% after feeding the corrected diet, whereas it decreased to 39% of that of the food-deprived group after feeding the threonine-devoid diet.

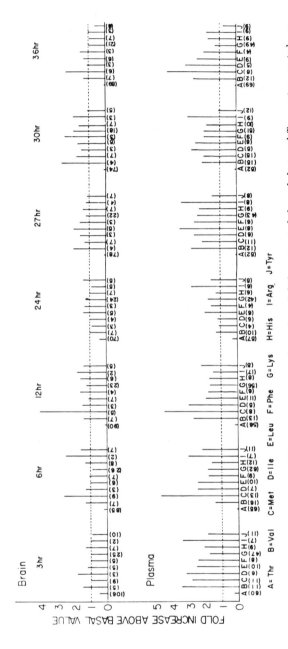

Fig. 17. Changes in brain and plasma aminograms of rats force-fed threonine-imbalanced diet at different time periods as indicated. [Taken with permission from Peng *et al.* (1972).]

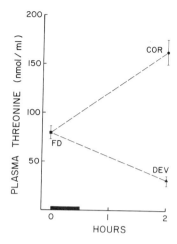

Fig. 18. Threonine concentrations in plasma of rats killed at the beginning of the feeding period without feeding (FD), at 2 hr after ingestion of 2.0 g threonine-corrected diet (COR), or 2 hr after ingestion of 2.0 g threonine-devoid diet (DEV). The solid black bar indicates the maximum time period allowed for consumption of 2.0 g food. Vertical bars indicate SEM. Number of rats was 7 or 8 per group. All values are significantly different ($p < 0.001$).

As can be seen in Fig. 19, brain threonine was also markedly reduced at 2 hr after the ingestion of 2 g of the threonine-devoid diet, to 66% of the food-deprived group, whereas the rats fed the threonine-corrected diet showed a 27% increase. Also, in Fig. 20, it can be seen that CSF values showed a similar response. The threonine-devoid group had a concentration of threonine in the CSF 48% of that of the food-deprived group, whereas the corrected group showed a 66% increase. Thus, the depressed concentrations of the missing amino acid are seen to be decreased in the CSF as well as in the plasma and brains of animals fed a threonine-devoid diet. These results demonstrate that not only plasma and brain, but also CSF amino acid concentrations can be altered by short-term dietary feeding. This is consistent with the rapidity with which animals can detect the amino acid-devoid diet. It is possible, therefore, that changes in CSF amino acid concentrations induced by diet may play a role in altering food intake in the rat.

VI. OPERANT RESPONSE

We have also used an operant approach to describe the time course of the onset of the reduction in food intake with a threonine-devoid diet. The onset of the behavioral change using this approach was again correlated with

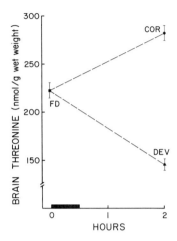

Fig. 19. Threonine concentrations in brain of rats killed at the beginning of the feeding period without feeding (FD), at 2 hr after ingestion of 2.0 g threonine-corrected diet (COR), or 2 hr after ingestion of 2.0 g threonine-devoid diet (DEV). The solid black bar indicates the maximum time period allowed for consumption of 2.0 g food. Vertical bars indicate SEM. Number of rats was 7 or 8 per group. All values are significantly different ($p < 0.001$).

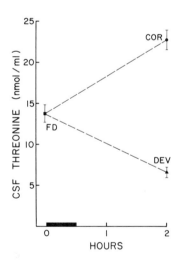

Fig. 20. Threonine concentrations in cerebrospinal fluid (CSF) of rats killed at the beginning of the feeding period without feeding (FD), at 2 hr after ingestion of 2.0 g threonine-corrected diet (COR), or 2 hr after ingestion of 2.0 g threonine-devoid diet (DEV). The solid black bar indicates the maximum time period allowed for consumption of 2.0 g food. Vertical bars indicate SEM. Number of rats was 7 or 8 per group. All values are significantly different ($p < 0.001$).

changes in plasma amino acid concentrations. Twelve rats were trained, over a period of 2–3 weeks, to earn over 80% of their total daily intake of food pellets (45 mg pellets of low-protein basal diet, pelleted by Bio-Serv, Inc., Frenchtown, NJ), according to a variable-interval, 40-sec schedule of reinforcement. The rats earned the food pellets, during a 2-hr session, by manipulating a small lever within a standard, sound-attenuated testing chamber. Pellets that were not eaten were collected and counted at the end of the session. Once stable patterns of performance for the daily 2-hr sessions were established, each rat was given one trial session in which threonine-devoid diet was substituted for the basal diet. Performance during this session was closely monitored, and 10 min after the onset of a disruption of the normal response, the animals were taken from the chamber, and 250 µl of blood was taken from the tail. The rat was returned to the testing chamber (usually within 3 min) and allowed to finish the 2-hr session. Some animals did not eat all the pellets that they earned; 25% of the animals left 10–12 pellets; the rest ate all they earned. The rats were tested the following day in the same chambers (after overnight *ad libitum* access to devoid diet) by allowing them to earn pellets of corrected diet. Following their single trial with the corrected diet, the rats were kept in the chambers for an additional 18 days, and readapted to the basal diet. Each was then bled at the same times it had shown a disruption on the devoid diet, returned to the test chamber for the remainder of the 2-hr session, and then bled a final time.

Eleven of the 12 rats tested with devoid diet displayed serious disruptions in lever pressing during the initial devoid-diet session. Five of the 12 rats stopped responding during the 2-hr session altogether. Of the remaining 7, 6 rats reduced their pellet acquisition rate to a small fraction of their normal (basal diet) rate. The cumulative records of 3 rats that were observed under the three dietary conditions are shown in Fig. 21. The baseline data used in the figure were collected on the day prior to the devoid diet test. Records from these 3 rats were representative of 3 of the 4 disruption styles that were observed. The fourth style (not presented) was for that of rat 112, that failed to show any signs of disruption. Rat 105 typically responded during the baseline session at an intermediate but steady rate, earning pellets of basal diet throughout the session. When given the devoid diet, the rat started the session at its characteristic rate, but then slowed and stopped responding. Upon being given access to the corrected diet, this same rat started to earn pellets at an initially higher rate than usual, but then stopped responding approximately 20 min before the end of the 2-hr session. Unlike rat 105, rat 102 responded to the devoid diet by gradually slowing down its rate of lever pressing during the devoid trial, stopping altogether almost 90 min into the 2-hr session. Finally, unlike either of the other 2 rats, rat 110, which had typically responded at very high rates under basal conditions, responded to

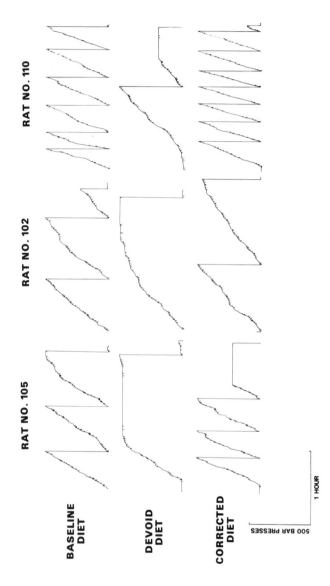

Fig. 21. Cumulative operant response patterns of 3 representative rats that earned food pellets containing baseline (basal), threonine-devoid, or threonine-corrected diet. Incidences of bar pressing and time elapsed are indicated on the scale in the figure.

the devoid diet by dramatically slowing down its rate of responding throughout the session. When given corrected diet on the following day, however, this rat resumed its characteristically high rate of responding.

During a typical basal diet session, the average rat ate 80 pellets over the 2-hr session, performing an average of 1007 responses per session. By contrast, during the devoid session, the average animal earned only 54 pellets, or 66% of the typical basal diet session. During the devoid trial, the average rat responded with only 512 responses on the lever (or 51% of the typical basal diet session). The average disruption time for the 12 rats was 59 min. Seven of the 12 rats displayed disruption times of less than 1 hr. Rat 107 stopped responding 28 min into the session.

Finally, the amino acid patterns of the plasma collected during disruption and again at the end of the 2-hr test sessions revealed that plasma amino acid concentrations of threonine were significantly lowered at the disruption time. Plasma threonine concentrations are shown in Table II for the 11 animals at the time of disruption during the devoid trial, and for all 12 animals at the end of the session. Note that the concentration of threonine was lower both at the disruption period and at the end of the 2-hr devoid-diet trial as compared to basal conditions.

Thus, it is clear from this and the *ad libitum* feeding trials that preceded it that the rat is capable of discriminating subtle differences in the quality of its diet very rapidly, i.e., within 30–90 min, even under restricted access conditions. The use of an operant method to study meal eating permitted an examination of the behavioral changes under a different set of circumstances. Because the rate of food delivery was only in part controlled by the rat, it was possible to monitor within-meal changes that may or may not be revealed in an analysis of moment-to-moment meal eating. It is interesting to note that the two approaches present results that agree closely, again demonstrating the very rapid response of the animals to amino acid deficiencies, and the correlation of the behavioral response with the plasma amino acid pattern.

TABLE II Plasma Threonine Concentrations (nmol/ml) of Rats Fed Devoid or Basal Diets under Operant Conditions

	Basal diet		Devoid diet	
	At disruption time ($N = 11$)	At end of session ($N = 12$)	At disruption time ($N = 11$)	At end of session ($N = 12$)
	130 ±14	86 ±10	54[a] ±8	41[a] ±5
% of basal	—	—	42	48

[a] Significantly lower than basal diet group, $p < 0.05$.

VII. BRAIN AREAS IMPLICATED

Control of food intake by the central nervous system has been studied in both hypothalamic and extrahypothalamic areas (Grossman, 1968; Hoebel, 1971). Lesions of the ventral medial hypothalamus (VMH) result in hyperphagic obesity (Hetherington and Ranson, 1940, 1942). However, ablation of the VMH nuclei, which does produce hyperphagia and obesity, does not prevent the depressive effects of amino acid imbalances and deficiencies on food intake during the dynamic (Leung and Rogers, 1970) or static phase (Scharrer et al., 1970) of hyperphagia. Overeating and increased gain in body weight which differ from those obtained by VMH lesions have also been reported to occur in rats following selective destruction of the ventral bundle of the ascending noradrenergic fibers (Ahlskog and Hoebel, 1973; Ahlskog et al., 1975). However, Leung and Rogers (1980a) have shown that hyperphagia induced by damage to the midbrain ventral tegmental nuclei, through which the ventral noradrenergic bundles pass (Ungerstedt, 1971; Lindvall and Bjorklung, 1974), does not alleviate the food intake depression of animals fed amino acid-imbalanced or -deficient diets. It remains to be determined whether lesions of the serotonergic food-intake inhibitory system, which can be placed in midbrain dorsal and medial raphe nuclei, would alter the control of amino acid intake in rats fed an amino acid-deficient diet.

Other extrahypothalamic neural areas such as the anterior prepyriform cortex and the medial amygdala have been demonstrated to be important in the control and/or adaptation of food intake in animals fed the amino acid-imbalanced (Leung and Rogers, 1971a, 1973) or -deficient diets (Leung and Rogers, 1971a). The food intake responses of animals with prepyriform cortical lesions fed an amino acid-imbalanced or -devoid diet (Leung and Rogers, 1971a) are shown in Fig. 22. The animals bearing such lesions did not reduce their food intake of an amino acid-imbalanced diet limiting in threonine or an amino acid diet completely devoid of threonine. The areas of the prepyriform cortex in a composite overlay of slides from 29 animals having effective lesions, i.e., animals that did not reduce their food intake when given the imbalanced diet (Rogers and Leung, 1973) are shown in Fig. 23. This specific brain area may be the site of the postulated receptor system responsible for the control of food intake in animals fed amino acid-imbalanced or -deficient diets. Long-term avoidance of the imbalanced diet in favor of the protein-free diet eventually leads to protein depletion and death (Leung et al., 1968b). However, animals with prepyriform lesions will select the imbalanced diet over the protein-free diet (Leung and Rogers, 1971a), even though there are no learned-aversion deficits found in the prepyriform lesioned rats, as shown in a study using a learned aversion to saccharin (Fig. 24; Meliza et al., 1981). Noda (1975) has also shown the lack of reduced

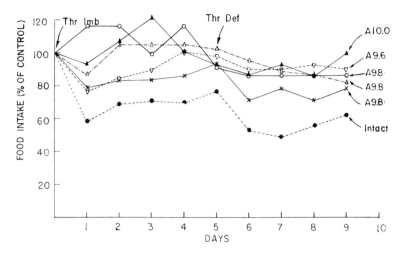

Fig. 22. Effect of prepyriform cortex lesions on daily food intake of rats fed threonine-imbalanced or devoid diet expressed as percentage of control. Numbers preceded by A represent stereotaxic coordinates for lesions of the prepyriform cortex. [Taken with permission from Leung and Rogers (1971a).]

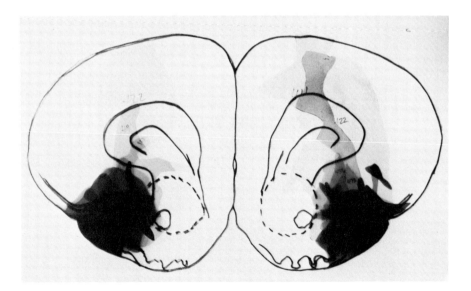

Fig. 23. Location of effective lesions of the prepyriform cortex which abolished the food intake depression of rats fed amino acid-imbalanced or -devoid diets. [Taken with permission from Rogers and Leung (1973).]

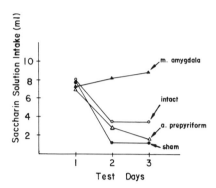

Fig. 24. Mean intake in milliliters of 0.2% saccharin solution by rats with various surgical lesions: m. amygdala, medial amygdala lesion; intact, unoperated; a. prepyriform, anterior prepyriform cortex-lesioned animals; sham, sham-operated animals. Rats received intraperitoneal injections of 0.12 M LiCl following ingestion of saccharin solution on days 1 and 2. [Reprinted with permission from *Physiology and Behavior*, Vol. 26, Meliza, L. L., Leung, P. M. B., and Rogers, Q. R. Effect of anterior prepyriform and medial amygdaloid lesions on acquisition of taste-avoidance and response to dietary amino acid imbalance. Copyright (1981), Pergamon Press.]

intake of an amino acid-imbalanced diet in rats with lesions of the prepyriform cortex.

Lesions of the medial amygdala have also been shown to abolish the depressed food intake response to the imbalanced diet, as shown in Fig. 25, days 1–4. However, it is interesting to note that the animals with lesions of the medial amygdala still reduce their food intake of an amino acid-devoid diet (see Fig. 25, days 6–10). Thus, the responses to the imbalanced diet may be abolished by either prepyriform cortical or amygdaloid lesions, but the responses to the devoid diet are not affected by lesions of the amygdala, but only by lesions of the prepyriform cortex (Leung and Rogers, 1971a).

VIII. AMINO ACID AND NEUROTRANSMITTER CONCENTRATIONS IN BRAIN

Whether protein intake is controlled by plasma amino acid ratios, brain amino acids, and/or monoamine neurotransmitter levels has been a matter of debate (Anderson, 1979; Chee *et al.*, 1981; Peters and Harper, 1981; Li and Anderson, 1983; R. J. Wurtman, 1983; Peters *et al.*, 1984). It is, however, clear that the brain is responsive to amino acids. For example, Panksepp and Booth (1971) observed that microinjections of a balanced amino acid solution into the dorsolateral perifornical hypothalamic area, where feeding behav-

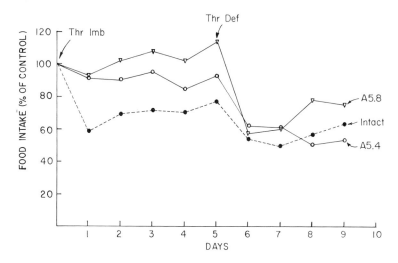

Fig. 25. Effect of medial amygdaloid lesions on daily food intake (percentage of control) of rats fed threonine-imbalanced or devoid diets. Intact, unoperated; numbers preceded by A, stereotaxic coordinates for amygdaloid lesions. [Taken with permission from Leung and Rogers (1973).]

iors are elicited by electrical stimulation, inhibit feeding in rats. As shown in Fig. 26, 1 µl of a balanced mixture of amino acids injected into the dorsolateral perifornical area depressed food intake, whereas saline or glucose did not. Also, microinjections of amino acids into the zona incerta and lateral hypothalamus increased the discharge frequency of firing of specific neurons (Wayner et al., 1975). Of the 11 amino acids tested, increases in discharge frequency were noted for all the amino acids except lysine and cystine. Cystine was unusual in that it was the only amino acid that decreased the spontaneous discharge frequency of the neurons. In addition, several amino acids are known to act as neurotransmitters in the central nervous system (McGeer et al., 1978), and may well be involved in feeding behavior, although their relationships to the control of protein and amino acid intake have not yet been elucidated. A number of other neurotransmitters associated with amino acids, such as γ-aminobutyric acid (GABA) and several peptides are currently thought to be involved in the control of food intake (Morley et al., 1983; Hoebel, 1984; Tews et al., 1984).

In order to determine whether amino acid concentrations were altered specifically in the prepyriform cortex area similarly to the alterations seen in plasma, CSF, and whole-brain homogenates, we prefed 3 groups of animals with the threonine-basal diet and adapted them to a 12-hr feeding schedule. At the beginning of the dark cycle on the day of the experiment, the rats

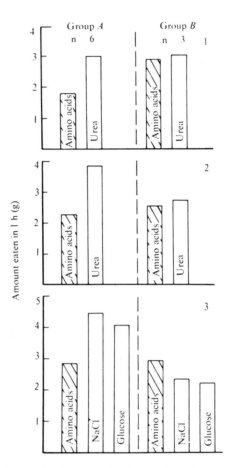

Fig. 26. Food intake in 1 hr of rats that had been subjected to microinjections of amino acids, urea, sodium chloride, or glucose into the dorsolateral perifornical area of the hypothalamus. [From Panksepp and Booth (1971). Reprinted by permission from *Nature (London)*, Vol. 233, pp. 341–342. Copyright © 1971 Macmillan Journals Limited.]

were given either the threonine-basal, -imbalanced, or -corrected diet. Thirty minutes after the food intakes were found to be significantly different among the 3 groups (i.e., at 2.5 hr), the brains were taken. Subsequently, the prepyriform cortex was carefully dissected out and the concentrations of the amino acids were determined in these small brain sections.

The results (Table III) show that threonine was markedly depressed in the prepyriform cortex of the animals fed the threonine-imbalanced diet. However, other amino acids were also reduced in that brain area: taurine, serine, glycine, alanine, ornithine, and GABA were significantly lower in the pre-

TABLE III Threonine Concentrations in the Prepyriform Cortex (nmol/g Wet Weight) of Animals Fed Basal, Imbalanced, or Corrected Diets for 2.5 hr

Basal	Imbalanced	% of basal	Corrected	% of basal
138 ± 20	79 ± 16[a]	57	250 ± 47	181
$N = 7$	$N = 6$		$N = 7$	

[a] Significantly lower than basal or corrected groups, $p < 0.05$.

pyriform cortex of the imbalanced group than in the other 2 groups. As seen in the aminogram for the prepyriform cortex (Fig. 27), several of the amino acids added to the imbalanced diet in amounts greater than those added to the basal diet were either unchanged or in slightly larger concentrations in animals fed the imbalanced diet. Therefore, in the prepyriform cortex, a brain area that has been shown to be involved in the initial response of animals to imbalanced amino acid diets, the limiting amino acid was decreased, with a pattern similar to the decreases shown in plasma and brain. In addition, several other amino acids were also depressed in prepyriform cortex showing alterations in the amino acid profile in these animals. It may be of interest that, although aspartate and glutamate—the predominant excitatory amino acid neurotransmitters in the prepyriform cortex (Harvey et al., 1975)—were not among the amino acids that were significantly altered in the animals fed the imbalanced diet, glycine and GABA, both of which have been shown to be neurotransmitters in the central nervous system (McGeer et al., 1978), were decreased in those animals. The role of amino acids as neurotransmitters in mediating the effects of imbalanced diets on the food intake of animals has not been determined.

Noda (1975) suggested that among the several metabolites of amino acids, ammonia might be involved in the response to imbalanced diets. Noda and Chikamori (1976) injected ammonium chloride into the prepyriform cortex and found that food intake was depressed as compared to rats injected with sodium chloride. They also reported that animals with lesions of the prepyriform cortex ate a diet containing ammonium chloride better than did controls. The use of ammonium chloride is not ideal since the chloride itself may alter electrolyte concentrations and thus affect food intake (Leprohon et al., 1979). However, ammonia concentrations were elevated to about the same extent, both in animals fed an imbalanced amino acid diet, with reduced food intake, and in animals fed the "corrected" diet, with normal food intake (Noda, 1975). It therefore seems unlikely that ammonia is the mediator signaling the effect of amino acid deficiency on food intake, although it may well be involved in the response to high levels of dietary protein. It has

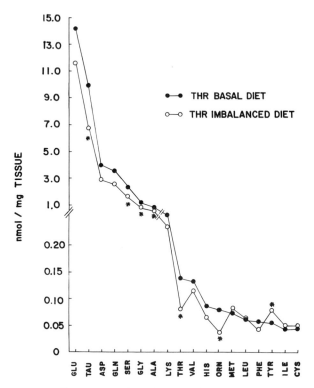

Fig. 27. Aminogram of free amino acids in the prepyriform cortex. Brains were taken 2.5 hr after presentation of threonine-basal or -imbalanced diets. Sections used for assay corresponded to the stereotaxic area shown as the shadow in Fig. 23.

yet to be determined whether ammonia concentrations in the blood vary to such an extent after a high-protein meal that the level of ammonia in the brain can trigger the neural regulation of food intake. Also, except for the suggestion of Noda's group, specific neural areas responsive to ammonia have yet to be identified.

The catecholamines and serotonin appear to participate in the control of feeding (Blundell, 1984; Hoebel, 1984). It has been proposed that shifts in the plasma tryptophan:neutral amino acid ratio, caused by either protein or carbohydrate and reflected in changes in the serotonin system, control aspects of the animal's selection for protein (Anderson, 1979) or carbohydrate (Wurtman and Wurtman, 1977), or the proportions of protein to carbohydrate (Li and Anderson, 1982; Blundell, 1983), as well as total food intake (Blundell, 1980). On the other hand, tyrosine, and thus the catecholamines, have been postulated to influence regulation of energy balance (Anderson,

1979) and food choice (Wurtman and Wurtman, 1979). The synthesis and turnover of serotonin and catecholamine neurotransmitters depend in part on the brain content of the amino acid precursors, and the plasma amino acid ratios appear to be the major determinants of precursor availability to the brain (Pardridge, 1979).

It is not known if or how the control of brain monoamine synthesis by diet and plasma amino acids relates to the control of food intake involving the ingestion of disproportionate amounts of dietary amino acids. However, it should be noted that tryptophan behaves as a typical limiting amino acid in the plasma when an amino acid imbalance is created with tryptophan as the most limiting amino acid (Pant et al., 1974).

IX. MONOAMINES IN THE PREPYRIFORM CORTEX

In an effort to determine whether the monoamines are altered in animals fed imbalanced or corrected diets, we measured the concentrations of norepinephrine, dopamine, and serotonin in prepyriform cortex samples by high-pressure liquid chromatography (HPLC). As reported above for the measurement of amino acid concentrations in the prepyriform cortex, the animals had been pretreated with a threonine-basal diet, and were adapted to a 12-hr feeding schedule. The food intake of the animals fed the threonine-imbalanced diet was significantly less than that of the animals fed either basal or corrected diets by 2 hr after access to the diets, and the brains were taken 30 min later. As seen in Table IV, the concentration of norepinephrine was reduced significantly, to 68% of control, in animals fed the imbalanced diet. No differences were found in the other catecholamines, although serotonin was reduced in the imbalanced diet group when com-

TABLE IV Monoamines (ng/g Wet Weight) in Prepyriform Cortex of Rats Fed Amino Acid Diets for 2.5 hr[a]

Monoamine	Basal	Imbalanced	Corrected
NE	430 ± 30 (7)	295 ± 26 (7)[b]	445 ± 38 (7)
DA	138 ± 35 (6)	123 ± 32 (5)	232 ± 76 (6)
5HT	884 ± 85 (7)[c]	578 ± 62 (7)[c]	710 ± 56 (7)

[a] Values are mean ± SE. NE, Norepinephrine; DA, dopamine; 5HT, serotonin; numbers in parentheses indicate number of animals per value.
[b] Significantly different from other values in row, $p < 0.01$.
[c] Significantly different from other values sharing the same letter, $p < 0.05$.

pared with the basal diet group. By contrast, in the anterior cingulate cortex, a brain area that has been shown to be involved in adaptive, but not immediate responses to imbalanced diets, there were no differences in the concentrations of any of the catecholamines. Serotonin was decreased in the anterior cingulate cortex of animals fed either the imbalanced or the corrected diet as it was in the prepyriform cortex, when compared with the concentration in animals fed the basal diet.

Pharmacological techniques have also been utilized to study the effects of neurotransmitter systems on protein ingestion. Depletion of serotonin by parachlorophenylalanine (PCPA), an inhibitor of serotonin synthesis, by 5,7-dihydroxytryptamine, a serotonin cytotoxin, or by lesion of the raphe nuclei, which contain serotonergic cell bodies, is associated with a reduction in protein and an increase in carbohydrate intake over several weeks in a choice situation (Ashley et al., 1979). Increased serotonergic activity after fenfluramine results in decreased carbohydrate but not protein intake (Blundell and McArthur, 1979; Wurtman and Wurtman, 1979). In humans, fenfluramine also results in preferential decreases in carbohydrate intake (Blundell and Rogers, 1980) and in frequency of carbohydrate snacks following a meal (Wurtman et al., 1981). Conversely, when fat intake is also monitored, fenfluramine decreases both protein and fat intake in rats, with no change in carbohydrate intake (Orthen-Gambill and Kanarek, 1982). However, that plasma amino acid ratios affect self-selection or control of protein intake has not been clearly established (Chee et al., 1981; Peters and Harper, 1981; Peters et al., 1984). Preliminary pharmacological studies in our laboratory have indicated that the serotonergic system may be involved in the response to amino acid-imbalanced diets. Depletion of serotonin by PCPA (300 mg/kg body weight, intraperitoneally) or 5,7-dihydroxytryptamine (200 μg/rat intracerebroventricularly) resulted in significantly exacerbated rapid depression of intake of an imbalanced amino acid diet by the treated rats. However, treatments designed to increase serotonin levels did not reverse this effect. Thus, other neural systems in addition to the serotonin system are also likely to play a role in the response to amino acid imbalance (Gietzen et al., 1984a). However, a residual effect of these serotonergic treatments was seen in facilitation of adaptation after tryptophan injection and long-lasting exacerbation of the decreased intake of an imbalanced diet several weeks after PCPA, when serotonin had returned to control levels (Gietzen et al., 1984b).

The noradrenergic system may participate in an inverse relationship with protein consumption. Thus, lowered noradrenergic activity may be associated with an increase in protein intake (Blundell, 1984). For example, reduction of noradrenergic tone by clonidine has been reported to result in an increase in protein as well as total food intake (Mauron *et al.*, 1980). On the other hand, amphetamine, which increases catecholaminergic activity levels, decreases total food intake, and has been reported to decrease protein intake either in proportion to total calories (Kanarek *et al.*, 1981) or more than total calories (Blundell and McArthur, 1979). It seems unlikely that the several neurotransmitter systems operate in isolation in their mediation of food intake. For example, increased dopamine and decreased norepinephrine, achieved by inhibition of dopamine β-hydroxylase (the enzyme that converts dopamine to norepinephrine), do not produce significant changes in feeding behavior, but when combined with a specific serotonin uptake inhibitor (L110-140), cause severe food intake depression (Rossi *et al.*, 1982).

Thus, the interactions between neurotransmitters and their precursors in the regulation of protein and amino acid intake are still not thoroughly understood. The specific neural areas that respond to deficiencies of amino acids remain to be fully elucidated, although ablations of the prepyriform cortex and the amygdala abolish the normal response to amino acid-imbalanced diets (Leung and Rogers, 1971, 1973). The concentrations and turnover rates of the neurotransmitters and of amino acids in the prepyriform cortex at the time associated with the food intake depression have been shown to be altered. The limiting amino acid was reduced, several other amino acid concentrations were reduced (including some with known neurotransmitter activity), and norepinephrine concentrations were decreased at the time food intake was depressed. Concentrations of amino acids in whole brain have been determined previously; increases in some of the amino acids used to generate the imbalance along with a decrease in the limiting amino acid have been shown (Rogers and Leung, 1977). There is much still to learn about the effects of the various neurotransmitter systems on the feeding response to amino acid imbalances and deficiencies. However, regimens are available in which the time course of the food intake response can be stated in minutes rather than hours, and alterations of amino acid concentrations in brain and CSF as well as plasma follow similar patterns. It would appear that some neuroactive metabolite(s)—if not the amino acids themselves—influence the food intake-controlling system.

ACKNOWLEDGMENTS

The previously unpublished work by the authors was supported in part by USPHS Research Grant AM-13252 and USPHS Training Grant T32 AM-07355 from the National Institute of Arthritis, Diabetes, and Digestive and Kidney Diseases. The authors appreciate the excellent secretarial assistance of Tracy Schuster.

REFERENCES

Ahlskog, J. E., and Hoebel, B. G. (1973). Overeating and obesity from damage to a noradrenergic system in the brain. *Science* **182**, 166–169.
Ahlskog, J. E., Randall, P. K., and Hoebel, B. G. (1975). Hypothalamic hyperphagia: Dissociation from hyperphagia following destruction of noradrenergic neurons. *Science* **190**, 399–401.
Anderson, G. H. (1977). Regulation of protein intake by plasma amino acids. *In* "Advances in Nutritional Research" (H. H. Draper, ed.), Vol. 1, pp. 145–166. Plenum Press, New York.
Anderson, G. H. (1979). Control of protein and energy intake: Role of plasma amino acids and brain neurotransmitters. *Can. J. Physiol. Pharmacol.* **57**, 1043–1057.
Ashley, D. V. M., Coscina, D. V., and Anderson, G. H. (1979). Selective decrease in protein intake following brain serotonin depletion. *Life Sci.* **24**, 973–984.
Bellinger, L. L., Birkhahn, R. H., Trietley, G. J., and Bernardis, L. L. (1977). Failure of hepatic infusion of amino acids and/or glucose to inhibit onset of feeding in the deprived dog. *J. Neurosci. Res.* **3**, 163–173.
Blundell, J. E. (1980). Pharmacological adjustments of the mechanisms underlying feeding and obesity. *In* "Obesity" (A. J. Stunkard, ed.), pp. 180–207. W. B. Saunders, Toronto.
Blundell, J. E. (1983). Problems and processes underlying the control of food selection and nutrient intake. *In* "Nutrition and the Brain" (R. J. Wurtman and J. J. Wurtman, eds.), Vol. 6, pp. 163–221. Raven Press, New York.
Blundell, J. E. (1984). Systems and interactions: An approach to the pharmacology of eating and hunger. *In* "Eating and Its Disorders" (A. J. Stunkard and E. Stellar, eds.), pp. 39–64. Raven Press, New York.
Blundell, J. E., and McArthur, R. A. (1979). Investigation of food consumption using a dietary self-selection procedure: Effects of pharmacological manipulation and feeding schedules. *Br. J. Pharmacol.* **67**, 436P–438P.
Blundell, J. E., and Rogers, P. J. (1980). Effects of anorexic drugs on food intake, food selection and preferences and hunger motivation and subjective experiences. *Appetite* **1**, 151–165.
Castonguay, T. W., Upton, D. E., Leung, P. M. B., and Stern, J. S. (1982). Meal patterns in the genetically obese Zucker rat: A reexamination. *Physiol. Behav.* **28**, 911–916.
Chee, K. M., Romsos, D. R., Bergen, W. G., and Leveille, G. A. (1981). Protein intake regulation and nitrogen retention in young obese and lean mice. *J. Nutr.* **111**, 58–67.
Cieslak, D. G., and Benevenga, N. J. (1984). The effect of amino acid excess on utilization by the rat of the limiting amino acid-threonine. *J. Nutr.* **114**, 1871–1877.
Davis, A. T., and Austic, R. E. (1982). Threonine metabolism of chicks fed threonine-imbalanced diets. *J. Nutr.* **112**, 2177–2186.
Ellison, J. S., and King, K. W. (1968). Mechanism of appetite control in rats consuming imbalanced amino acid mixture. *J. Nutr.* **94**, 543–554.
Friedman, M. I., and Sawchenko, P. E. (1984). Evidence for hepatic involvement in control of ad libitum food intake in rats. *Am. J. Physiol.* **247**, R106–R113.
Gietzen, D. W., Leung, P. M. B., Hartman, W. J., and Rogers, Q. R. (1984a). Treatment with either para-chlorophenylalanine or 5,7-dihydroxytryptamine exacerbates the decreased food intake of rats fed imbalanced amino acid diets. *Soc. Neurosci.* **10**, 304 (abstract No. 92.12).
Gietzen, D. W., Leung, P. M. B., Castonguay, T. W., and Rogers, Q. R. (1984b). Residual effects of p-chlorophenylalanine and L-tryptophan on food intake of rats fed an imbalanced amino acid diet. *In* The neural and metabolic bases of feeding. *Neurosci. Satellite Symp.* (abstract No. 38).

Grossman, S. P. (1968). Hypothalamic and limbic influences on food intake. *Fed. Proc.* **27**, 1349–1360.
Harper, A. E. (1976). Protein and amino acids in the regulation of food intake. In "Hunger: Basic Mechanisms and Clinical Implications" (D. Novin, W. Wyrwicka, and G. Bray, eds.), pp. 103–113. Raven Press, New York.
Harper, A. E., and Peters, J. C. (1983). Amino acid signals and food intake and preference: Relation to body protein metabolism. *Experientia* **44** (Suppl.), 107–134.
Harper, A. E., and Rogers, Q. R. (1966). Effect of amino acid imbalance on rats maintained in a cold environment. *Am. J. Physiol.* **210**, 1234–1238.
Harper, A. E., Benevenga, N. J., and Wohlhueter, R. M. (1970). Effects of ingestion of disproportionate amounts of amino acids. *Physiol. Rev.* **50**, 428–558.
Hartman, W. J., Calvert, C. C., and Rogers, Q. R. (1983). Decrease of threonine in CSF, plasma and brain of rats fed a threonine deficient diet. *Soc. Neurosci.* **9**, 202 (Abstract No. 62.9).
Harvey, J. A., Scholfield, C. N., Graham, L. T., Jr., and Aprison, M. H. (1975). Putative transmitters in denervated olfactory cortex. *J. Neurochem.* **24**, 445–449.
Hetherington, A. W., and Ranson, S. W. (1940). Hypothalamic lesions and adiposity in the rat. *Anat. Rec.* **78**, 149–172.
Hetherington, A. W., and Ranson, S. W. (1942). The spontaneous activity and food intake of rats with hypothalamic lesions. *Am. J. Physiol.* **136**, 609–617.
Hoebel, B. G. (1971). Feeding: Neural control of intake. *Ann. Rev. Physiol.* **33**, 533–568.
Hoebel, B. G. (1984). Neurotransmitters in the control of feeding and its rewards: Monoamines, opiates, and brain-gut peptides. In "Eating and Its Disorders" (A. J. Stunkard and E. Stellar, eds.), pp. 15–38. Raven Press, New York.
Kanarek, R. B., Feldman, P. G., and Hanes, C. (1981). Pattern of dietary self-selection in VMH-lesioned rats. *Physiol. Behav.* **27**, 337–343.
Krane, V. R., Sinnamon, H. M., and Thomas, G. J. (1976). Conditioned taste aversions and neophobia in rats with hippocampal lesions. *J. Comp. Physiol. Psychol.* **90**, 680–683.
Kumta, U. S., and Harper, A. E. (1962). Amino acid balance and imbalance. IX. Effect of amino acid imbalance on blood amino acid pattern. *Proc. Soc. Exp. Biol. Med.* **110**, 512–517.
Leprohon, C. E., Woodger, T. L., Ashley, D. V. M., and Anderson, G. H. (1979). Effect of mineral mixture in diet on protein intake regulation in the weanling rat. *J. Nutr.* **109**, 827–831.
Leung, P. M. B., and Rogers, Q. R. (1969). Food intake: Regulation by plasma amino acid pattern. *Life Sci.* **8**, 1–9.
Leung, P. M. B., and Rogers, Q. R. (1970). Effect of amino acid imbalance and deficiency on food intake of rats with hypothalamic lesions. *Nutr. Rep. Int.* **1**, 1–10.
Leung, P. M. B., and Rogers, Q. R. (1971a). Importance of prepyriform cortex in food-intake response of rats to amino acids. *Am. J. Physiol.* **221**, 929–935.
Leung, P. M. B., and Rogers, Q. R. (1971b). Effects of pituitary extract on food intake of intact and hypophysectomized rats fed imbalanced amino acid diets. *Nutr. Rep. Int.* **4**, 207–215.
Leung, P. M. B., and Rogers, Q. R. (1973). Effect of amygdaloid lesions on dietary intake of disproportionate amounts of amino acids. *Physiol. Behav.* **11**, 221–226.
Leung, P. M. B., and Rogers, Q. R. (1975). Disturbances in amino acid balance. In "Total Parenteral Nutrition" (H. Ghadimi, ed.), pp. 259–284. Wiley, New York.
Leung, P. M. B., and Rogers, Q. R. (1979). Effects of hippocampal lesions on adaptive intake of diets with disproportionate amounts of amino acids. *Physiol. Behav.* **23**, 129–136.
Leung, P. M. B., and Rogers, Q. R. (1980a). Hyperphagia after ventral tegmental lesions and food intake responses of rats fed disproportionate amounts of dietary amino acids. *Physiol. Behav.* **25**, 457–464.

Leung, P. M. B., and Rogers, Q. R. (1980b). Effect of amino acid imbalance and deficiency on feeding patterns of rats offered dietary choices. *Fed. Proc.* **39**, 794 (abstract No. 2783).

Leung, P. M. B., Rogers, Q. R., and Harper, A. E. (1968a). Effect of amino acid imbalance in rats fed ad libitum, interval fed, or force-fed. *J. Nutr.* **95**, 474–482.

Leung, P. M. B., Rogers, Q. R., and Harper, A. E. (1968b). Effect of amino acid imbalance on dietary choice in the rat. *J. Nutr.* **95**, 483–492.

Leung, P. M. B., Rogers, Q. R., and Harper, A. E. (1968c). Effect of cortisol on growth, food intake, dietary preference and plasma amino acid pattern of rats fed amino acid imbalanced diets. *J. Nutr.* **96**, 139–151.

Leung, P. M. B., Rogers, Q. R., and Harper, A. E. (1968d). Effect of amino acid imbalance on plasma and tissue free amino acids in the rat. *J. Nutr.* **96**, 303–318.

Leung, P. M. B., Larson, D. M., and Rogers, Q. R. (1972). Food intake and preference of olfactory bulbectomized rats fed amino acid imbalanced or deficient diets. *Physiol. Behav.* **9**, 553–557.

Leung, P. M. B., Hartman, W. J., and Rogers, Q. R. (1982). Effect of thalamic taste nuclei lesions on food intake responses and patterns of rats fed disproportionate amounts of dietary amino acids. *Fed. Proc.* **41**, 541 (abstract No. 1583).

Li, E. T. S., and Anderson, G. H. (1982). Self-selected meal composition, circadian rhythms and meal responses in plasma and brain tryptophan and 5-hydroxytryptamine in rats. *J. Nutr.* **112**, 2001–2010.

Li, E. T. S., and Anderson, G. H. (1983). Amino acids in the regulation of food intake. *Nutr. Abstr. Rev.* **53**, 169–181.

Lindvall, O., and Bjorklund, A. (1974). The organization of the ascending catecholamine neuron systems in the rat brain (as revealed by the glyoxylic acid fluorescence method). *Acta Physiol. Scand.* **412** (Suppl.), 1–48.

McGeer, P. L., Eccles, J. C., and McGeer, E. G. (1978). "Molecular Neurobiology of the Mammalian Brain," pp. 183–198. Plenum Press, New York.

Mauron, C., Wurtman, J. J., and Wurtman, R. J. (1980). Clonidine increases food and protein consumption in rats. *Life Sci.* **27**, 781–791.

Meliza, L. L., Leung, P. M. B., and Rogers, Q. R. (1981). Effect of anterior prepyriform and medial amygdaloid lesions on acquisition of taste-avoidance and response to dietary amino acid imbalance. *Physiol. Behav.* **26**, 1031–1035.

Meyer, J. H. (1974). Release of secretin and cholecystokinin. *In* "Endocrinology of the Gut" (W. Y. Chey and F. B. Brooks, eds.), pp. 241–252. C. B. Slack, Thorofare, New Jersey.

Mitchell, H. H. (1916). Feeding experiments on the substitution of protein by definite mixtures of isolated amino-acids. *J. Biol. Chem.* **26**, 231–261.

Morley, J. E., Levine, A. S., Yim, G. K., and Lowy, M. T. (1983). Opioid modulation of appetite. *Neurosci. Biobehav. Rev.* **7**, 281–305.

Noda, K. (1975). Possible effect of blood ammonia on food intake of rats fed amino acid imbalanced diets. *J. Nutr.* **105**, 508–516.

Noda, K., and Chikamori, K. (1976). Effect of ammonia via prepyriform cortex on regulation of food intake in the rat. *Am. J. Physiol.* **231**, 1263–1266.

Orthen-Gambill, N., and Kanarek, R. B. (1982). Differential effects of amphetamine and fenfluramine on dietary self-selection in rats. *Pharmacol. Biochem. Behav.* **16**, 303–309.

Osborne, T. B., and Mendel, L. B. (1914). Amino-acids in nutrition and growth. *J. Biol. Chem.* **17**, 325–349.

Osborne, T. B., and Mendel, L. B. (1916). The amino acid minimum for maintenance and growth, as exemplified by further experiments with lysine and tryptophane. *J. Biol. Chem.* **25**, 1–12.

Panksepp, J., and Booth, D. A. (1971). Decreased feeding after injections of amino-acids into the hypothalamus. *Nature (London)* **233**, 341–342.

Pant, K. C., Rogers, Q. R., and Harper, A. E. (1974). Plasma and tissue free amino acid concentrations in rats fed tryptophan-imbalanced diets with or without niacin. *J. Nutr.* **104**, 1584–1596.

Pardridge, W. M. (1979). The role of blood-brain barrier transport of tryptophan and other neutral amino acids in the regulation of substrate-limited pathways of brain amino acid metabolism. *J. Neural Transmis.* **15** (Suppl.), 43–54.

Peng, Y., and Harper, A. E. (1969). Amino acid balance and food intake: Effect of amino acid infusions on plasma amino acids. *Am. J. Physiol.* **217**, 1441–1445.

Peng, Y., Tews, J. K., and Harper, A. E. (1972). Amino acid imbalance, protein intake and changes in rat brain and plasma amino acids. *Am. J. Physiol.* **222**, 314–321.

Peters, J. C., and Harper, A. E. (1981). Protein and energy consumption, plasma amino acid ratios, and brain neurotransmitter concentrations. *Physiol. Behav.* **27**, 287–298.

Peters, J. C., Bellissimo, D. B., and Harper, A. E. (1984). L-Tryptophan injection fails to alter nutrient selection by rats. *Physiol. Behav.* **32**, 253–259.

Rogers, Q. R., and Leung, P. M. B. (1973). The influence of amino acids on the neuroregulation of food intake. *Fed. Proc.* **32**, 1709–1719.

Rogers, Q. R., and Leung, P. M. B. (1977). The control of food intake: When and how are amino acids involved? *In* "The Chemical Senses and Nutrition" (M. R. Kare and O. Maller, eds.), pp. 213–249. Academic Press, New York.

Rose, W. C. (1931). Feeding experiments with mixtures of highly purified amino acids: I. The inadequacy of diets containing nineteen amino acids. *J. Biol. Chem.* **94**, 155–165.

Rossi, J., III, Zolovick, A. J., Davies, R. F., and Panksepp, J. (1982). The role of norepinephrine in feeding behaviour. *Neurosci. Biobehav. Rev.* **6**, 195–204.

Russek, M. (1971). Hepatic receptors and the neurophysiological mechanisms controlling feeding behavior. *In* "Neurosciences Research" (S. Ehrenpreis and O. C. Solnitzky, eds.), Vol. 4, pp. 213–282. Academic Press, New York.

Scharrer, E., Baile, C. A., and Mayer, J. (1970). Effect of amino acids and protein on food intake of hyperphagic and recovered aphagic rats. *Am. J. Physiol.* **218**, 400–404.

Sharma, K. N., and Nasset, E. S. (1962). Electrical activity in mesenteric nerves after perfusion of gut lumen. *Am. J. Physiol.* **202**, 725.

Tews, J. K., and Harper, A. E. (1982). Food intake, growth and tissue amino acid concentrations in lean and obese (ob/ob) mice fed a threonine-imbalanced diet. *J. Nutr.* **112**, 1673–1681.

Tews, J. K., Kim, Y-W. L., and Harper, A. E. (1979). Induction of threonine imbalance by dispensable amino acids: Relation to competition for amino acid transport into brain. *J. Nutr.* **109**, 304–315.

Tews, J. K., Rogers, Q. R., Morris, J. G., and Harper, A. E. (1984). Effect of dietary protein and GABA on food intake, growth and tissue amino acids in cats. *Physiol. Behav.* **32**, 301–308.

Tobin, G., and Boorman, K. N. (1979). Carotid or jugular amino acid infusion and food intake in the cockerel. *Br. J. Nutr.* **41**, 157–162.

Ungerstedt, U. (1971). Stereotaxic mapping of the monoamine pathways in the rat brain. *Acta Physiol. Scand.* **367** (Suppl.), 1–48.

Wayner, M. J., Ono, T., DeYoung, A., and Barone, F. C. (1975). Effects of essential amino acids on central neurons. *Pharmacol. Biochem. Behav.* **3** (Suppl. 1), 85–90.

Willcock, E. G., and Hopkins, F. G. (1906). The importance of individual amino-acids in metabolism; observations on the effect of adding tryptophane to a dietary in which zein is the sole nitrogenous constituent. *J. Physiol. (London)* **35**, 88–102.

Wurtman, J. J., and Wurtman, R. J. (1977). Fenfluramine and fluoxetine spare protein consumption while suppressing calorie intake by rats. *Science* **198,** 1178–1180.

Wurtman, J. J., and Wurtman, R. J. (1979). Drugs that enhance central serotoninergic transmission diminish elective carbohydrate consumption by rats. *Life Sci.* **24,** 895–904.

Wurtman, J. J., Wurtman, R. J., Growdon, J. H., Henry, P., Lipscomb, A., and Zeisel, S. H. (1981). Carbohydrate craving in obese people: Suppression by treatments affecting serotoninergic transmission. *Int. J. Eating Disorders* **11,** 2–15.

Wurtman, R. J. (1983). Food consumption, neurotransmitter synthesis, and human behavior. *Experientia* **44** (Suppl.), 356–369.

Zahler, L. P., and Harper, A. E. (1972). Effects of dietary amino acid pattern on food preference behavior of rats. *J. Comp. Physiol. Psychol.* **81,** 155–162.

Zimmerman, R. A., and Scott, H. M. (1967). Plasma amino acid pattern of chicks in relation to length of feeding period. *J. Nutr.* **91,** 503–506.

PART IV

Discussion

Kissileff: Why did the level of serotonin go down in the animals fed the threonine-imbalanced diet?

Fernstrom: Branched-chain and aromatic amino acids are considered large neutral amino acids. Some include threonine in the list of large, neutral amino acids. If threonine is handled as such an amino acid, this could explain the drop in serotonin levels seen on a threonine-imbalanced diet.

Rogers: I did not state this, but one of the controls we looked at was cingulate cortex. In cingulate cortex, norepinephrine did not go down.

Kissileff: Have you looked at only one amino acid-imbalanced situation?

Rogers: That's right.

Kissileff: Would you expect similar results with other amino acid-imbalanced diets?

Rogers: From work we have done so far, it appears that there is one mechanism that handles amino acid deficiency in general. We're currently working on additional experiments particularly on altering the ratio of essential amino acids including in that paradigm, tryptophan. The ratios we use could lead to tryptophan decreases in brain.

Fernstrom: I have some questions for Dr. Rogers. First, of what relevance is a diet which is deficient in one amino acid only compared to what an animal like the rat may encounter in the wild, or, let's say, under normal circumstances? Can the results of your studies lead to some conclusions about what types of feeding changes the animal might undergo under normal conditions and normal vagaries in daily protein intake? Second, during the feeding of an imbalanced diet, rats will lose weight, but then begin growing in parallel to controls. Are they growing in the same way metabolically? Is their metabolic picture the same or is there more fat than protein for example? Third, why just focus on a plasma-brain amino acid? Maybe there is a metabolic change induced by the amino acid deficiency which would lead to a change in intake? Fourth, what is the overall model that you are postulating in terms of protein regulation? I can see that if protein is too low or too high, or there is an amino acid imbalance, there could be a change in amino acid metabolism and this may be part of the regulatory mechanism of food intake. And how do the acute studies fit into an overall regulatory scheme in terms of a general model?

Rogers: The altered choice that we see is, I believe, a survival technique. If you give an animal a limited choice, for example, between an amino acid-imbalanced diet and a protein-free diet, they choose the protein-free diet. We think of that as a improper choice. If you give them an imbalanced diet in a no-choice situation, they will grow. But on a protein-free diet they eventually die. We've kept animals for a couple of months on the point of death when right in front of them is a diet that if they eat it, they can grow. That's a very strong response. But all of this response to protein is secondary to a drive to eat for energy. If you do not give them a choice, they'll eat the imbalanced diet and grow. But then, let's say, 2 months later you again give them a choice, they'll eat the protein-free diet. In the natural situation, in a low-protein diet, there is always one amino acid that is limiting. If the protein gets low enough, an altered choice situation will arise. If the animal has a chance, he'll eat a higher protein diet. Only if you get into a situation like a famine, where the only other choices are lower protein diets, will the animal go to another source altogether, like a nonprotein diet. Under laboratory, strictly controlled conditions, this may seem to be an inappropriate choice, but now he's eating for energy, not protein. Metabolically, what does the imbalanced diet do? Under low-protein diets, where an amino acid must be limiting, or under an amino acid-deficient diet, there is not maximal protein synthesis in the liver or the muscle. If you give extra amino acids without the limiting one, you stimulate protein synthesis. And thus, the level of the limiting amino acid drops. But this is an acute effect. If you look after 2–3 days, there is less protein in the liver and muscle because they are not eating enough energy. In looking at all the metabolic parameters, the most consistent is a very rapid drop in the limiting amino acid in almost all compartments, especially peripherally in muscle, plasma, CSF. Because of our infusion studies into the carotid, we postulate that lack of it in the brain stimulates food intake. That is the primary effect.

Ramirez: There is a common cultural practice of combining two or more protein sources within a given meal in humans, such as potatoes with milk, corn with beans, rice with beans. Could it be, Dr. Rogers, that the mechanism you are studying in rats in which deficiency leads to an alternative choice behavior may have some parallel to the development of this practice in humans?

Rogers: To answer this, one needs to look for learned preferences. These are very difficult to document. What does occur more readily is learned aversions. Learned aversions are below the conscious level. If any of you have learned aversions, you know how very powerful they are. What may happen is that an animal will eat different diets and mix these by trial and error so as not to form a learned aversion.

Maller: Dr. Fernstrom, is there any information on the time course and levels of serotonin in the brain after a carbohydrate meal?

Fernstrom: In the rat, serotonin is elevated after feeding a fasted rat a carbohydrate meal for 2 to 3 hr after the meal. After 3 hr, levels are down.

Rogers: You showed in your diurinal study that tryptophan levels are constant. But valine concentration went up. Harper has shown that other branched-chain amino acids also increase. The other amino acids do not change. During this feeding of a high-protein diet, one might predict that tryptophan would go down. Did it in fact go down?

Fernstrom: In our hands, valine always changes the most. But the ratio did not change. So there was no change in brain tryptophan and no change in ratio as a function of dietary protein level.

Rogers: Harper has fed up to 60 to 80% protein, and he does get consistent increases in branched-chain amino acids. It would be interesting to determine what happens to brain tryptophan under these conditions of very high dietary protein.

Fernstrom: I think that 40% protein is about the highest level a rat might ever encounter in a food. Going higher will exaggerate the response, putting it in the pharmacologic range of response, rather than in the physiologic or normal range.

Grill: Dr. Torii had shown in his work that an animal given a lysine-deficient diet can pick out lysine from among 13 or so different amino acids. One would assume that to do this he is using an oral cue. Dr. Rogers, from your work can you say that an animal uses his oral chemoreceptors to detect different amino acids in food?

Rogers: If there is a clear difference in taste—be that related to the taste of the amino acid or not—the animal can make a clear choice, and this choice will be sustained. In most of our studies, we have just a very small difference in amino acid—maybe 0.2% threonine as opposed to 0.4%—in the diet. We don't think they can tell the difference by taste based on learned aversion studies, unless there is a taste cue in the diet. Odor of the diet is very important because if you bulbectomize the animal, they lose the aversion. The aversion returns but it takes 3–4 days. So, odor and taste are both very important in labeling these diets.

Grill: We still have to reconcile the way in which your animals are responding to a single amino acid deficiency and a general model of learning. The animal is not perceiving the amino acid per se but some general difference in flavor. This contrasts with the type of data Dr. Torii presented in which the animal can pick out that amino acid in which it was deficient from among many choices. The difference in the two experiments could be related to design wherein one of the animals is able to choose from among many amino acids and in the other there is only a choice of deficient and replete diet.

Rogers: If an amino acid is in solution, the animals do not make as clear-cut a choice. For example, with a histidine-deficient diet, if an animal is given a choice between a histidine-containing solution and water, they'll choose the histidine solution about 2:1 over water. In the dietary experiments, their choice is almost completely the replete diet.

Boyle: Dr. Fernstrom, you've shown your results using casein as a protein source. What about other protein sources such as Zein?

Fernstrom: We have not done this. In Dr. Anderson's work, using a Zein diet, graded levels of tryptophan were used. Harper has been unable to replicate that study. In particular he reported no correlation between dietary tryptophan level and subsequent ratio of tryptophan to neutral amino acids using the tryptophan-loaded diets.

Friedman: Dr. Norgren, what is known of the projection and distribution of the afferent visceral sympathetic fibers?

Norgren: Only within the last 5 years have investigators begun to look at this. Probable data from renal afferents are the most complete. It has been traced now into the spinal cord into the dorsal horn. It has not been looked at rigorously because it had been assumed, based on electrophysiology, that the afferents came to the dorsal horn and then distributed as did the somatosensory afferents. In fact, the afferents do go into the dorsal horn but then have restricted distributions within that. There is some information now that afferents from the kidney actually reach the caudal medulla and apparently terminate both in the somatosensory ganglion and in the solitary nucleus. Those sympathetic neurons coverged on the somatosensory neurons. There is a very large literature studying the convergence as a basis for referred pain.

Greeley: Dr. Fernstrom, do you believe there is a role for dietary phenylalanine and tyrosine and their neural metabolites in food and appetite regulation?

Fernstrom: As long as you are talking about normal dietary constituents, I would say probably no. There are situations in which you can administer phenylalanine and tyrosine and get the rate of catecholamines to increase. A very large dose of amino acid needs to be given. In fact, about 10 to 20 times as much needs to be given compared with the amount of tryptophan that needs to be given to get an increase in brain serotonin levels. Only activated neurons will be sensitive to tyrosine levels, but even with an activated system, the level of tyrosine needs to be high. There may be isolated areas of very high activity that would be sensitive to normal variations in tyrosine levels, but these have not been identified.

Greeley: What happens when dietary levels of tyrosine or phenylalanine become limiting?

Fernstrom: I don't know of any experiments in which you take a normal food constituent and generate a very low brain tyrosine level to which the ingestion of an adequate meal would lead to a big enough increase in brain tyrosine level to generate an increase in catecholamine production in a norepinephrine neuron. Someone may find such a situation, but to date I know of no model that is even remotely physiological that shows this.

PART V

Conclusion

21

Concluding Remarks

D. MARK HEGSTED
New England Regional Primate Center
Southborough, Massachusetts

It would not be useful for me to attempt to summarize the material which has been presented in this book. Rather, I will simply comment on some of the issues that occurred to me while I read it and other issues I have thought about in the past several years. My comments, of course, represent those of a nutritionist; I have no particular qualifications in the chemical senses, psychology, neurosciences, etc. These other fields will undoubtedly be important in clarifying the role of the chemical senses in nutrition. None of us can expect to be expert in everything germaine to the problems we are trying to solve. Solutions will come from collaborative efforts or by making others aware of the problems and the potentials that exist in all of these fields.

Modern nutrition began about 80 years ago with the discovery of the vitamins. The discovery of essential nutrients, which were required in only small amounts, appeared to offer the opportunity to finally define what we meant by a nutritious diet. It seemed clear to practically everyone in the field that we had a difficult but well-defined task ahead of us. We simply had to identify all the essential nutrients, determine how much of each was required to maintain health, and determine how much of each was present in the foods we eat. Most of us saw this as the ultimate goal of nutrition.

Progress has been substantial. We think we have now identified the essential nutrients—about 45 in all. Defining their requirements has not proceeded very well for many reasons, and I do not expect this situation to improve in the future. Our enthusiasm for this approach is dampened when we consider that the supermarket may have 10,000 foods which continually change. I believe that nutrition research is—or should be—in a transition stage, and the research agenda that has dominated nutrition for so long should be modified. This does not mean that we have answered all the questions we asked 50 years ago or that research in these areas should be abandoned. It is, however, clear that deficiency diseases in the population of the United States are not major public health problems, and I do not expect that we can attract funds for low priority areas when other major health issues that are nutrition related are obvious. Further, for practical purposes, we really know enough about deficiency diseases to prevent them.

I think everyone should be aware that the major health problems of our society are the so-called chronic diseases, not deficiency diseases. This is not a new area of research, of course, but solutions are far from being apparent. Since these diseases are related to overconsumption both of total food and of specific constituents and are problems of affluent societies in which there are minimal restraints on what is eaten, many areas of research which have been discussed here are relevant. Why do we eat what we eat? How can one modify food consumption without deprivation, real or imagined? There is no doubt that the food industry is succeeding in producing better foods—by trial and error I assume—but I think none of us doubts that if we knew some of the fundamentals, they could do a better job.

A general scientific approach to most problems is to continually simplify our test systems so that we can study each variable in isolation. The utility of this is clear to all of us; yet there also comes a time when the parts have to be put together again. The sum of the parts may be more or less than the whole. In nutrition we are now beginning to try to deal with the real world. Having identified some 45 essential nutrients, we now find that we cannot study them adequately in isolation. Total iron consumption does not appear to mean very much. One has to know about the form of iron in the food and consider the promotors and inhibitors of iron absorption to know much about iron nutrition. The fractionation of science into its specialties is a similar and necessary process. But I think part of our aim should be to try to put some of the parts together and see if we can deal with the whole.

When I hear of the various taste qualities—sweet, sour, bitter, and salty—I suspect that this field is about where we were when we talked about vitamins A, B, C, and D. We are now trying to deal with at least 50 nutrients and food constituents of nutritional importance, and the field will certainly become more complex. We speak often of dietary fiber, but this will certainly turn out to be a series of materials with different physiologic functions, and the general term may eventually not be meaningful. We believe that food factors which stimulate the mixed function oxidases that may be protective against various environmental carcinogens will probably need practical consideration. The relationships between diet and the neurotransmitters appears to be a field of promise, and all the new hormones that may be affected by diet or that may influence food consumption require more study.

21. Concluding Remarks

So I expect that researchers of taste and smell must "tighten their belts" and be ready for the explosion that must come. What we ultimately what to know is why we have certain food preferences and how they can be modified.

I am impressed with the kinds of studies mentioned by Drs. Beauchamp and Bertino, who show that the rat has no preference or aversion for salt in food but, of course, prefers salty water. A few years ago, my first reaction when I was told that a rat made hypertensive with a high salt diet still preferred salty solutions was that it was a pretty dumb animal. On further considering the fact that most of *us* prefer diets that cause coronary heart disease, hypertension, diabetes, etc., I think that "dumb" may not be the appropriate word. Nevertheless, it seems to me that their work raises important questions that had not occurred to me before. I though that when I was told that a rat preferred salt, he preferred salt in all vehicles, be they food or water. Since this is not the case, we must be careful how we define or even study preference. One issue is, of course, that we tend to develop systems that are convenient. It is easier to measure fluid consumption than food consumption, but this measure may not be appropriate for the problem in question. We need to consider other methodologies or at least how far we can generalize conclusions.

We also tend to get stuck with the few animal species in common use. The rat has contributed enormously to nutrition research, but that does not mean it is necessarily a good animal model for each specific problem. I speak from experience. I spent many years investigating the effects of various dietary fats and cholesterol on the serum cholesterol levels in rats. In retrospect, the main thing I learned was that for these types of studies, the rat is not a good model, at least not if I wanted to know something about how man responds to these dietary constituents. I wish others would learn that lesson. The best model for man is man, of course, but often he is not the most useful experimental animal. I concur, therefore, with one of the conclusions of the Discussion for Part I: that we need to pay more attention to the appropriateness of our animal models.

It is now generally accepted that in testing carcinogens we need data from at least two species before we generalize. Most of us do not doubt the wisdom of this, but it may be just as relevant in other fields. If we cannot test the proposition in man, as is often the case, perhaps the only approach is to try to determine if the finding is true in more than one species.

As I have already indicated, I think a major goal in nutrition is to find means to ameliorate the chronic diseases that are going to kill most of us. It seems to me that it would be worthwhile to search for species of varying susceptibility to such diseases and determine whether they have the capacity of protecting themselves from self-destructive dietary practices. The sand rat, for example, develops diabetes when fed an ordinary chow-type rat diet. Does it have the capacity of protecting itself when offered a choice of foods?

High fat diets are now implicated in coronary heart disease, cancer, diabetes, obesity, and hypertension, yet seem to be universally preferred by human populations. I expect this reflects our evolutionary development when availability of food or the efficient utilization of food determined survival. People with the "thrifty gene" were more likely to survive. High fat foods were generally available only at times of a "feast," but now we can literally "feast" three times a day. In any event, it seems urgent to initiate studies on why we like fat, whether we can develop flavors that make foods lower in fat equally acceptable, whether other species have a similar preference, etc. Recommendations to lower fat consumption are not likely to be very successful until we know more about why we prefer high fat diets.

Experimental nutritionists have paid almost no attention to taste or flavor. I suppose the reason for this is that the common laboratory species appear to eat almost anything. I think I recall Morley Kare saying years ago that rats hate casein. If this is true it has little effect on how we feed rats and does not appear to have much effect on the food consumption of rats. One might ask if the rat is too nondiscriminating to be useful and how we bridge the gap between preference testing and overall food consumption?

It seems to me that the cafeteria diet studies have potential here. The studies reported by Naim and colleagues appear to be the first that have tried to specifically link the increased food consumption found in the cafeteria arrangement to flavor. It will be of obvious interest to know whether such findings can be generalized. Are the preferred flavors the same with different basal diets? Will greater variation in flavors maximize such effects? Since very small increases in food consumption and food efficiency are of great importance in animal production, one wonders whether such studies have commerical possibilities.

Nutrition research developed largely as an outgrowth of biochemistry. The traditional curriculum for training nutrition researchers has been largely built around the essential nutrients and their biochemical functions. Some of the newer areas of nutrition, such as the role of the chemical senses, now appear to offer more opportunities than the traditional field, yet many of these areas are still foreign territory to nutritionists, and shifts in emphasis are often resisted by traditional nutritionists. I do not suppose that nutritionists are unique in displaying this conservative attitude. Each area of research develops its own traditions, publishes its own journals, and develops its own unwritten rules about what is or is not appropriate. Each field expands so that you have to be a specialist to succeed. Yet there are important problems that cannot be solved—or do not appear to be solvable—by

21. Concluding Remarks

this approach. The merit of this book is that it may help us develop enough appreciation of the problems in nutrition to stimulate experts in other fields to think about them. We will also, I trust, develop a cadre of young people who have enough expertise in more than one field to bridge some of these gaps directly.

Index

A

Acetylcholine, gastric secretion and, 183
Acids
   composition of saliva and, 9
   salivary flow and, 7, 8
   recognition thresholds and, 17
Acid taste, effect of ethynyl estradiol on detection threshold after ovariectomy
   discussion, 83–84
   methods, 75
   results, 79–80
Adaptation, to salivary components, 12–15
Adrenalin, food intake and, 378, 379, 382
Age
   as predictor of chemosensory preference, 98, 99
   protein and MSG intake and, 51
Aldosterone, salivary composition and, 9, 11
Alkaline phosphatase, zinc deficiency and, 112
Alliesthesia, satiation and, 384
Alloxan, feeding behavior and, 352
Amino acid(s)
   competition with tryptophan for transport, 400–402
   concentrations in brain, amino acid-imbalanced diets and, 444–449
   concentrations in plasma, brain and cerebrospinal fluid, amino acid-imbalanced diets and, 433–437
   taste preferences for, 47–48
      dietary protein content and, 48–49
   tastes of, 46
   utilization, fasting and, 377
γ-Aminobutyric acid, food intake and, 445, 447
Ammonia, injection into prepyriform cortex, food intake and, 447
Amphetamine, food intake and, 451

Amygdala, intake of amino acid-imbalanced diets and, 444
Animals, effects of sham feeding in, 184–186
Anosmia, gastric secretion and, 189–190
Appetite
   diet, brain tryptophan and serotonin and, 404–412
   intragastric feeding and, 155–156
Apple juice, palatability of, 305
Arginine, intake, diet and, 56
Atropine
   gastric secretion and, 183
   prandial drinking and, 226
   salivary composition and, 17
Autonomic nervous system, taste and, 323–325
   afferent limb, 325–329
   central projections, 334–339
   efferent limb, 329–334

B

Banana colada yogurt drink, palatability of, 306–307
Bethanechol, prandial drinking and, 226
Bicarbonate, gastric secretion and, 186–187
Bitter taste, effect of ethynyl estradiol on detection threshold after ovariectomy
   discussion, 83–84
   methods, 75
   results, 79–80
Blood glucose, food intake schedule and, 218, 228
Blood pressure, feeding and, 234
Blood urea nitrogen, as predictor of chemosensory preference, 95, 98–99
Bombesin
   activity of DVN neurons and, 204
   satiation and, 386

Brain
  areas implicated in responses to amino acid-imbalanced diets, 442–444
  intake of imbalanced amino acid diets and, 419, 434–437
Buffers, salivary, stimulus modification and, 16, 17–20

C

Cafeteria feeding
  experiments with nutritionally controlled diets, 281–289
  as model for dietary obesity, 271–273
Caffeine, in saliva, 11
Caloric intake, intragastric feeding and, 155–156, 167
Carbohydrate ingestion, tryptophan and serotonin levels and, 397–398
Catecholamines, control of feeding and, 448–449
Caudal brainstem
  input from oral exteroceptors, 348–349
  integration of input from food deprivation and insulin-induced hypoglycemia with oral afferent information, 365–368
  production of discriminative responses to taste and, 358–364
  reflex connections in
    cephalic insulin response, 355
    mastication, 357–358
    salivation, 354
    sympathoadrenal hyperglycemic reflex, 355–357
  as site of metabolic interoceptors, 349–354
Cephalic insulin response, CBS reflex connections and, 355
Cephalic phase, of gastric secretion, 181–184
  contribution of different senses to, 189–191
  sham feeding and, 184–189
Chemosensory loss, in elderly, review of literature, 89–93
Chemosensory preference, and biochemical indexes in elderly
  experiment 1, 93–96
  experiment 2, 96–101
Children, responses to sugar and salt in, 29–34

p-Chlorophenylalanine, food intake and, 450
Cholecystokinin
  activity of DVN neurons and, 204
  food intake and, 386
  in decerebrate rats and, 368
  gastric distention and, 176–178
  MMC activity and, 197
Chow intake, stomach and, 175–176
Chromaffin cells, hepatic, 382
Cimetidine
  gastric secretion and, 183–184
  periprandial drinking and, 226
Clonidine, food intake and, 450–451
Conditioning, intragastric feeding and, 152–153
Copper
  deficiency
    D-penicillamine and, 114
    stimulus intake and, 115–117
  toxicosis, tastant intake and, 113
Corn zein, amino acid composition of, 53
Corticosterone
  thiamine deficiency and, 110
  zinc deficiency and, 112–113
Cortisol, food intake on imbalanced diets and, 419, 421
Cumin, food palatability and, 307, 309–310

D

Decerebration, autonomic nervous system and, 336
Dehydration, salivary flow rate and, 8–9
2-Deoxy-D-glucose
  food intake and, 379
  responses to, 350, 353
Depression, gustation in NIDDM and, 138, 139
Diabetes, see also Noninsulin-dependent diabetes
  taste preference and, 61
Dichloroisoproterenol, food intake and, 382
Diet(s)
  amino acid-imbalanced, food intake and, 416–418
  brain tryptophan and serotonin and appetite and, 404–412
  brain tryptophan uptake and serotonin synthesis and, 396–404
  caloric content, satiation and, 385

Index   471

influences on saliva
  composition, 9–11
  flow rate, 6–9
Dietary choice, amino acid-imbalanced diets and, 425–432
Dietary obesity, cafeteria feeding and, 271–273
Digestion, intragastric feeding and, 153–154
Digestive tract, historical view of, 193–194
5.7-Dihydroxytryptamine, food intake and, 450
Dopamine, food intake and, 451
Dorsal vagal nucleus, gut–brain axis and, 201–202
Drinking
  patterns in horses, 215
  patterns in pigs, 225–226
  periprandial, mechanism underlying, 226

E

Egg protein, amino acid composition, 53
Elderly
  chemosensory loss in, review of literature, 89–93
  chemosensory preference and biochemical indexes in
    experiment 1, 93–96
    experiment 2, 96–101
  nutritional status in, 87–88
Electrolytes, in saliva, 5–6
Enovid, see Oral contraceptive
Ensure diet, palatability of, 305
Enteric nervous system, see Gut brain
Ethynyl estradiol, effect on detection threshold for bitter and acid modalities of taste after ovariectomy
  discussion, 83–84
  methods, 75
  results, 79–80
Exteroceptors, oral, input to CBS, 348–349

F

Fasting, digestive tract activity during, 195–196
Fat
  dietary preferences and, 278–279

emulsified, intragastric feeding of, 157–158
intragastric feeding and, 156–162
Feeding
  hepatic hypothesis and, 379–381
  patterns in horses, 212–215
  patterns of pigs, 224–225
Feeding history, of infants
  response to sugar and, 30–32
  response to salt and, 40–41
Feeding patterns, amino acid-imbalanced diets and, 423–425
Fenfluramine, food intake and, 408, 450
Flavors, food preference tests for, 273–276
Fluoxetine, food intake and, 408
Food(s)
  acceptance of salt in, developmental shifts in, 38–40
  changes in pleasantness
    influence of low-calorie foods on, 258–259
    time course of, 255–258
  discussion of, gastric secretion and, 190–191
  gastric secretion and, 187
  intestinal motor activity and, 199
  palatability, effect of eating on, 249–250
  palatability and intake in man
    definition, 295–296
    objectives, 293–295
    previous studies, 296–297
  pleasantness changes in a varied, four-course meal, 250–253
  preference tests for flavors and textures, 273–280
  stimulation of salivary flow and, 8
  uneaten, changes in palatability of, 259–263
  variation in sensory properties, effects on intake, 263–266
Food intake
  control of, hepatic receptors and, 377–379
  in decerebrate rats, 365–367
  effect of addition of maltose to NaCl solution on
    discussion, 84–85
    methods, 75–76
    results, 80–82
  factors affecting, 129–130
  hypoglycemia in decerebrate rats and, 366
  physiological responses in horses, 216–224

Food intake (cont.)
  relationship to intrinsic palatability
    methods, 308–309
    objectives, 307
    results, 309–313
  social factors as stimulants, 212
Food-intake response, time course of, 420–423
Forebrain, autonomic projections to, 336–339
Free fatty acids, feeding and, 232, 234
Fructose
  food intake and, 378–379
  secretion of salivary amylase and, 9–10

G

Gastric distention
  cholecystokinin and, 176–178
  secretion and, 183
Gastric distention receptors, satiation and, 384–385
Gastric pressure, feeding and, 234–235
Gastric secretion, cephalic phase of, 181–184
  contribution of different senses to, 189–191
  sham feeding and, 184–189
Gastrin
  MMC activity and, 197
  prandial drinking and, 226
  secretion of, 183, 187–188
Gastrointestinal chemoreceptors, satiation and, 385–386
Genetic predisposition, relation to protein intake and taste preference, 61–63
Glucagon
  food intake and, 378, 379, 382, 385
  MMC activity and, 197
Glucoreceptors, intestinal, response of brain and, 202–203
Glucose
  central neural control of level, 349–350
  concentration, gastric emptying time and, 169–170
  consumption, stomach and, 173–175
  food intake and, 378, 379
Glycemia, regulation of, 375–377
Glycerol, free in plasma, intragastric fat feeding and, 159

Glycine, intake, diet and, 49, 50, 54, 55
Growth, intragastric feeding and, 155
Guanethidine, food intake and, 382
Gustatory assessment, study of NIDDM and, 132
  discussion, 138–140
  magnitude judgments, 136–137
  results, 133–134
Gustatory hedonics, in NIDDM, 135
Gut brain
  functions of, 195–200
  interdependence with peptides, 205–206
Gut–brain axis, function of, 200–205

H

Heart rate, anticipation of feeding and, 221
Hedonic reactions, to salt by newborns, 28
Hepatic receptors, control of food intake and, 377–379
Histamine
  gastric secretion and, 183–184
  periprandial drinking and, 226
Homeostasis, of energy, neural control of, 348
Hormones, gastrointestinal, satiation and, 386
Horse, as model herbivore
  feeding and drinking patterns in, 212–215
  physiological responses to feeding, 216–224
  social factors in stimulation of intake, 212
Horseradish peroxidase
  afferent neurons to nucleus of solitary tract and, 327–328
  vagus nerve and, 330
Humans, effect of sham feeding in, 186–189
Humoral factors, intake of imbalanced amino acid diets and, 418–419
Humoral responses, to feeding in pigs, 226–234
Hydrochloric acid, secretion of, 183
Hyperglycemia, 5-thioglucose and, 352
Hyperparakeratosis, of oral mucosa, zinc deficiency and, 112
Hyperphagia, food variety and, 272
Hypertension, taste preference and, 61–63
Hypoglycemia, insulin-induced, food intake in decerebrate rats and, 366

Index   473

Hypothalamus
    activity, gastric distention and, 205
    glucoresponsive neurons in, 350
    importance of, 338–339
Hypovolemia, food intake and, 218

I

Infants, older, responses to sugar and salt in, 29–34
Insulin
    feeding and, 228, 232, 297
    regulation of glycemia and, 375–376
    tryptophan and serotonin levels and, 397, 411–412
Insulin response, CBS reflex connections and, 355
Interoceptors, metabolic, CBS as site of, 349–354
Intestine
    control of gastric emptying and, 170–171
    Thiry–Vella segments, MMC activity and, 197–199
Intragastric feeding, review of,
    appetite, 155–156
    conditioning and learning, 152–153
    digestion and growth, 153–155
    fat, 156–162
Iron, deficiency, stimulus preferences and, 113
Irritants, stimulation of salivary flow and, 6–7
Isoproterenol, food intake and, 378

L

Learning, intragastric feeding and, 152–153
Leu-enkephalin, activity of DVN neurons and, 204
Lincomycin, MMC activity and, 206
Lingual lipase, intragastric fat feeding and, 162, 163
Lipostasis, satiation and, 387
Liquids
    as diet, salivary flow and, 8
    gastric emptying of, 168–170
Liver, receptors, control of protein intake and, 419

L-Lysine
    deficiency, changes in taste preference during, 52–61
    taste preference for, 59, 60

M

Maltose
    determination of preference threshold for
        discussion, 83
        methods, 73–74
        results, 77–78
    effect of addition to NaCl solution on intakes of NaCl, water and food
        discussion, 84–85
        methods, 75–76
        results, 80–82
    intake, effect of chronic treatment with Enovid on
        discussion, 82–83
        methods, 72–73
        results, 76–77
    preference threshold concentration during treatment with Enovid
        discussion, 83
        methods, 74
        results, 78–79
Mastication, CBS reflex connections and, 357–358
Medulla oblongata, gastric secretion and, 182
Micronutrients
    deficiencies, causes and effects, 108–109
    effects of deficiency on taste preferences
        B vitamins, 110–111
        iron and copper, 113
        D-penicillamine and, 114
        zinc, 111–113
Migrating myoelectric complex, feeding and, 196–197
MK-212, food intake and, 408
Monoamines, in prepyriform cortex, 449–451
Monosodium glutamate
    intake
        age and, 60
        diet and, 49–50, 51, 54–55
        zinc and pyridoxine deficiency and, 120
Motilin, MMC activity and, 197, 206

N

Naloxone, gastric secretion and, 184
Neophobia, acceptance of salt and, 41
Neurotensin, MMC activity and, 197, 206
Neurotransmitters
  concentrations in brain, amino acid-imbalanced diets and, 444–449
  zinc deficiency and, 113
Newborns, responses to salt and sugar in, 27–29
Non-insulin-dependent diabetes mellitus (NIDDM)
  dietary compliance and, 130
  methods of study of gustation and olfaction in
    gustatory assessment, 132–133
    olfactory assessment, 133
    subjects, 131–132
  results of study of gustation and olfaction in
    gustatory assessment, 133
    gustatory hedonics, 135
    olfactory assessment, 135
    olfactory hedonics, 135
    relative magnitude judgments, 136–138
Norepinephrine
  amino acid-imbalanced diet and, 449–450, 451
  salt intake and, 113
Nucleic acid(s), metabolism, zinc deficiency and, 112
5'-Nucleotides, taste and, 46
Nucleus ambiguus, efferent autonomic neurons and, 330–332
Nucleus tractus solitarius
  afferent autonomic axons and, 326–329
  gut–brain axis and, 201–202
Nutritional status, in elderly, 87–88

O

Olfaction, control of intake of imbalanced amino acid diets, 418
  decline with age, 91
  gastric secretion and, 189–190
Olfactory assessment, study of NIDDM and, 133
  discussion, 140
  magnitude judgments, 137, 138
  results, 135
Olfactory hedonics, in NIDDM, 135
Operant response, amino acid-imbalanced diets and, 437–441
Opiate peptides, gastric secretion and, 184
Oral contraceptive
  effect of chronic treatment on spontaneous intake of NaCl and maltose,
    discussion, 82–83
    methods, 72–73
    results, 76–77
  preference threshold for maltose solution during treatment
    discussion, 83–84
    methods, 74
    results, 78–79
Oropharyngeal receptors, satiation and, 383–384

P

Palatability
  brief-exposure taste test for measuring intrinsic palatability ratings, 299–307
    procedure, 298–299
    subjects, 299
  food intake in man and
    definition, 295–296
    objectives, 293–295
    previous studies, 296–297
  of foods, effects of eating it, 249–250
  intrinsic, 295
  relationship to food consumption
    methods, 308–309
    objective, 307
    results, 309–313
D-Penicillamine, nutrient deficiencies and, 114
Pentagastrin, gastric acid secretion and, 186–187
Peptides, interdependence with gut brain, 205–206
Phlorizin, feeding behavior and, 349, 351–352, 354
Pig, as model omnivore
  eating and drinking patterns, 224–226
  gastric pressure due to feeding, 234–235
  humoral response to feeding, 226–234

Index 475

mechanisms underlying preprandial drinking, 226
Plasma osmotic pressure, food intake schedule and, 218
Plasma protein, food intake schedule and, 218
Potassium chloride intake, severe pyridoxine deficiency and, 121–123
Preference threshold, definition of, 72
Prepyriform cortex
 intake of amino acid-imbalanced diets and, 442–444, 446–447
 monoamines in, 449–451
Propranolol, gastric secretion and, 183
Protein
 dietary, tryptophan levels and, 399
 intake
  relation to taste preference and genetic pre-disposition, 61–63
  taste preference in rats during growth and, 47–52
Pyridoxine, deficiency
 copper deficiency and, 115–117
 D-penicillamine and, 114
 tastant intake and, 110, 120–124
 zinc deficiency and, 117–121
Pyruvate, hepatic control of feeding and, 379–381

Q

Quinine, response of decerebrate rats to, 359
Quinine sulfate intake, copper and pyridoxine deficiency and, 116–117

R

Ratings, of intrinsic palatability
 real food and drinks, 304–307
 repeated testing before rinse, 302–304
 water and solutions, 299–302
Receptors, for cholecystokinin, 177–178

S

Saliva
 dietary influences on
  composition, 9–11
  flow rate, 6–9

 influences on taste perception, 11–12
 modification of taste stimuli, 15–20
 sensory adaptation to salivary components, 12–15
 proteins in, zinc deficiency and, 112
Salivary glands
 composition of secretion, 5–6
 flow rate and, 5–6
 flow rate
  parotid glands, 4, 5
  sublingual glands, 4
  submandibular glands, 4
 innervation, composition of saliva and, 6, 10
Salivation, CBS reflex connections and, 354
Salt, see also Sodium
 responses to in early development, 26–27
  newborns, 27–29
  older infants and children, 29–34
  summary, 34–35
 taste preference for, 61–63
Salt acceptance, developmental shifts in
 discussion, 40–41
 in foods, 38–40
 in water, 35–37
Salt intake
 deficiency in copper and pyridoxine and, 116
 deficiency in zinc and pyridoxine and, 118–119
 effect of addition of maltose to NaCl solution
  discussion, 84–85
  methods, 75–76
  results, 80–82
 effect of chronic treatment with Enovid on
  discussion, 82–83
  methods, 72–73
  results, 76–77
 D-penicillamine and, 114
 severe pyridoxine deficiency and, 120–123
 thiamine deficiency and, 110
Satiation
 gastric distention receptors and, 384–385
 gastrointestinal chemoreceptors and, 385–386
 gastrointestinal hormones and, 386
 lipostasis and, 387
 oropharyngeal receptors and, 383–384
 preabsorptive, hepatic pyruvate and, 382–383

Satiety
  nutrient-specific, 253–255
  role of sensory properties of foods in
    influence of low-calorie foods on pleasantness changes, 258–259
    time course of changes in pleasantness, 255–258
  sensory-specific
    effects of eating food on palatability, 249–250
    methods, 248–249
    pleasantness changes in a varied, four-course meal, 250–253
Senses, various,
  contribution to cephalic phase of gastric secretion, 189–191
Sensory properties
  role of in satiety
  influence of low-calorie foods on pleasantness changes, 258–259
  time course of changes in pleasantness, 255–258
  variation in, effects on food intake, 263–266
Serotonin
  amino acid-imbalanced diets and, 449–450
  brain, diet, brain tryptophan and diet and, 404–412
  carbohydrate selection and, 408
  control of feeding and, 448–449, 451
  response to tryptophan administration, 395–396
  synthesis, diet and brain tryptophan uptake and, 396–404
Serum albumin level, as predictor of chemosensory preference, 95–96, 98
Sham feeding
  investigation of cephalic phase of gastric acid secretion and
    in animals, 184–186
    in humans, 186–189
    of ponies, 223–224
Sight, gastric secretion and, 190
Smell, gastric secretion and, 189–190
Social factors, as stimulants of feeding, 212
Sodium, *see also* Salt
  dietary, salivary composition and, 10–11
  detection level, saliva and, 16
  intake, dietary amino acid and protein and, 49–50
  as stimulus, salivary adaptation and, 12–15

Stomach
  chow intake and, 175–176
  glucose consumption and, 173–175
  stretch receptors in, gut–brain axis and, 200–201, 203
Stomach emptying
  intestinal control of, 170–171
  intragastric fat feeding and, 161–162
  of liquids, 168–170
  two phases of, 171–173
Sucrose
  response of decerebrate rats to, 359
  palatability and, 297
  palatability ratings, 300–302
Sucrose intake
  copper and pyridoxine deficiency and, 116
  zinc and pyridoxine deficiency and, 119
Sucrose polyester, dietary preferences and, 280
Sugar(s)
  elevation of insulin levels and, 355
  responses to in early development, 26–27
    newborns, 27–29
    older infants and children, 29–34
    summary, 34–35
Sympathoadrenal hyperglycemic reflex, CBS reflex connections and, 355, 357

T

Taste
  autonomic nervous system and, 323–325
    afferent limb, 325–329
    central projections, 334–339
    efferent limb, 329–334
  discriminative responses to, CBS connections and, 358–364
  intake of imbalanced amino acid diets and, 418
  intragastric feeding and, 154
    fat and, 159–161, 162
  modification by saliva, 15–20
  parasympathetic innervation and, 326
  vitamin A and, 109
Taste perception, influences of saliva on, 11–12
  modification of stimulus properties, 15–20
  sensory adaptation to salivary components, 12–15

Taste preference
  changes during lysine deficiency, 52–61
  effects of micronutrient deficiencies on
    B vitamins, 110–111
    iron and copper, 113
    D-penicillamine and, 114
    zinc, 111–113
  protein intake in rats during growth and, 47–52
  relation to protein intake and genetic predisposition, 61–63
Textures, food preference tests for, 276–277
Thiamine, deficiency, stimulus intake and, 110
5-Thioglucose, feeding response to, 350, 351–352
Thought, of food, gastric secretion and, 190–191
Threonine, nitrogen sparing by, 59–60
Triglyceride tolerance curve, intragastric fat feeding and, 157–158
Triiodothyronine
  in cafeteria fed rats, 285–286
  hepatic pyruvate levels and, 387
  level, food intake schedule and, 218, 234
Tryptophan
  brain, diet, brain serotonin and appetite and, 404–412
  brain/serum ratio, dietary regulation and, 405–408
  brain uptake, diet and serotonin synthesis and, 396–404
  transport, competitive nature of, 399–403
Tryptophan hydroxylase, serotonin levels and, 396
Tuna fish salad, palatability of, 306

U

Urea, in saliva, 11

V

Vagus nerve, 326
  behavior patterns and, 201–202
  gastric secretion and, 182–183, 185–186
Variety, hyperphagia and, 272
Vitamin A, taste and, 109

W

Water
  acceptance of salt in, developmental shifts in, 35–38
  and solutions, rating of intrinsic palatability, 299–302
Water intake, effect of addition of maltose to NaCl solution on
  discussion, 84–85
  methods, 75–76
  results, 80–82
Wheat gluten, amino acid composition of, 53

Y

Yogurt, palatability of, 304–305

Z

Zein, tryptophan-supplemented, regulation of dietary intake and, 404–405
Zinc, deficiency
  D-penicillamine and, 114
  stimulus intake and, 111–113, 117–120